W9-BAT-804

The STARGAZER'S Handbook

The STARGAZER'S Handbook

Giles Sparrow

Quercus

Contents

Introduction

Step outside on a clear night, and as you gaze up at the stars, you are also looking out into the depths of the Universe. Early astronomers thought the Earth was at the centre of everything, surrounded by a huge spherical shell of stars, with the planets, Sun and Moon moving on crystalline spheres in the space between Earth and the great outer shell. Some said that the stars were pinpricks in the sphere, allowing in light from the mysterious realms beyond the Universe. Others believed that the planets, moving at their own pace among the stars, were manifestations of the gods themselves, watching over and influencing events on Earth.

Today, we know that the truth is stranger still – our planet is a tiny world orbiting an average star, on the outskirts of a middling galaxy in a small galaxy cluster, itself in thrall to an enormous supercluster that is merely a speck in the true scale of the Universe. And yet on this insignificant world, a miracle has happened – life has evolved, with enough intelligence to stare across the cosmos and dimly comprehend its true extent.

Almost everything we know about the cosmos comes from direct observation – stargazing. Over the centuries, successive discoveries with ever-improving instruments, methods and techniques have informed our understanding of our place in the Universe, the workings of other planets and the life cycles of stars and galaxies. Even the most obscure theories of the origin, shape and fate of the cosmos stand or fall on how well they describe the observed Universe around us.

And yet, despite the incredible sophistication of modern observatories, computers and satellites, astronomy has remained the most democratic science. Everyone can appreciate the wonder of a starlit night, the beauty of the Moon's cratered landscape or Saturn's distant rings. And understanding these phenomena does not diminish their beauty – it only adds to the sense of awe at the workings of the cosmos.

But the sky can seem intimidating at first – especially to town and city-dwellers confronted with the scintillating majesty of a truly dark sky for the first time. How can one hope to begin?

This guide, we hope, offers one solution. In its pages you will find a simple introduction to the basics of astronomy, guides to locating and observing the planets, and charts of the entire sky, showing the stars visible from your location at various times of year. The centrepiece, however, are the constellation charts. These combine traditional charts of the sky, with labelling and constellation lines joining up the various patterns of the sky, with a more realistic rendering of the stars devoid of such references. Comparing the two and learning to recognize the constellations devoid of lines and labels should stand you in good stead when confronted with the confusion of the 'perfect sky' for the first time.

On the brightness of things

Anyone consulting an astronomy book for the first time needs to know a little astronomical jargon, and perhaps the most important term of all is 'magnitude'. This is a measurement system that can be used to record the brightness of any object in the sky. Simple though that may seem, measuring perceived brightness is not as easy as it might appear, since the human eye has an unusual response to varying levels of light. Just 1/1000 of the light of Sirius, the brightest star, is still enough to register with the average observer, and it takes a difference in actual light levels of roughly 2.5 times for an object to be perceived as twice as bright.

The magnitude system used today measures stars according to this 'logarithmic' relationship. The brighter an object is, the lower its magnitude, until the very brightest objects in the sky have magnitudes below zero. A difference in magnitude of 5.0 is equivalent to a difference in 'true' brightness of 100 times, so the faintest naked-eye stars have magnitudes of roughly 6.0, while brilliant Sirius has a magnitude of -1.4, and the blazing Sun has a magnitude of -26.7.

Of course, the magnitude of an object only indicates the brightness of its light reaching Earth. Increasing distance from a star, and the presence of light-absorbing material such as dust clouds, all decrease its light. As a result, the magnitude measured from Earth is often termed 'apparent magnitude'. This is sometimes contrasted with 'absolute magnitude', which records the brightness of a star as seen from a standard distance of 32.6 light years. However, in the pages that follow, we shall use more direct comparisons with the overall energy output or 'luminosity' of a familiar star – our own Sun.

While the brightness of a star is one of its most important characteristics, there are others – most significantly its colour, which can reveal the temperature of its surface. Just as an iron bar heated in a furnace glows first red hot, then yellow hot, then white and finally blue hot, so the colours of stars go from red to blue as they get hotter. When the true luminosity of a star is compared to its surface temperature, it can reveal other characteristics, such as its size (since the temperature of a star depends on the amount of energy passing through each bit of its surface – and so a luminous but cool star must have a much larger surface area to spread out the energy).

Astronomical instruments

The heavens offer much that can be appreciated with the unaided eye alone – from the movements of the Sun, Moon and bright planets to the patterns of the constellations, the variations of stars, the band of the Milky Way (the plane of our own galaxy) and the light from remote galaxies such as the Andromeda Spiral and the Magellanic Clouds. But sooner or later, most people will want to look at the sky in more detail.

Binoculars and telescopes both work on the same principle – they use a lens or mirror with a larger diameter than the human eye to collect light and form an image that can then be magnified by another lens – the 'eyepiece'. In many ways, the magnification offered by an instrument is secondary to its light-collecting power or 'light grasp', since there are many large objects in the sky, such as faintly glowing gas clouds, that would be easily visible with no magnification if they were not so faint.

Many people already have a pair of binoculars, and these are an ideal next step in your exploration of the cosmos. A good pair will reveal stars down to about magnitude 9 – about 16 times fainter than the human eye alone can manage. They can also usefully magnify an image by about 10 times – much more than that, and their limited light grasp, coupled with the inevitable shake from the hands that hold them, makes them unusable.

Small telescopes come in two forms – refractors, which use a large lens to collect their light, and reflectors, which use a mirror. Refractors are generally more portable, and due to various factors, produce a sharper image than a reflector of the same diameter and light grasp. However, a fixed budget will usually buy a larger reflector than it will a refractor, bringing with it benefits such as increased light grasp. In the terms of this book, a 'small' telescope is generally a 5–7-cm refractor or a 10-cm reflector, while a medium-sized instrument is a 10-cm refractor or a 15-cm reflector.

Seasoned amateur astronomers know that any instrument is only as good as its mount. A stable tripod support is a must for a fully functional telescope (although some modern computer-controlled devices can be set up on a flat table top). An 'equatorial' mount, aligned with the axis of the sky's rotation, used to be a must to avoid complicated adjustments in two axes at the same time (see chapter 1), but computerized instruments can continuously adjust a telescope's orientation to compensate for the rotation of the sky. Purchasing a telescope can be the gateway to a greater enjoyment of the night sky, but it is a serious investment – fortunately, there are many other books and magazines that offer more detailed advice.

Whether you are stargazing with the naked eye, scanning the heavens with binoculars or focusing on distant objects with a telescope, the night sky is an almost infinite source of wonder and delight. We hope that this book may be a gateway to a lifetime's starry nights.

Key to constellation maps

Greek alphabet

α	alpha	ν	nu
β	beta	ξ	xi
γ	gamma	ο	omicron
δ	delta	π	pi
ε	epsilon	ρ	rho
ζ	zeta	σ	sigma
η	eta	τ	tau
θ	theta	υ	upsilon
ι	iota	φ	phi
κ	kappa	χ	chi
λ	lambda	ψ	psi
μ	mu	ϖ	omega

Stars

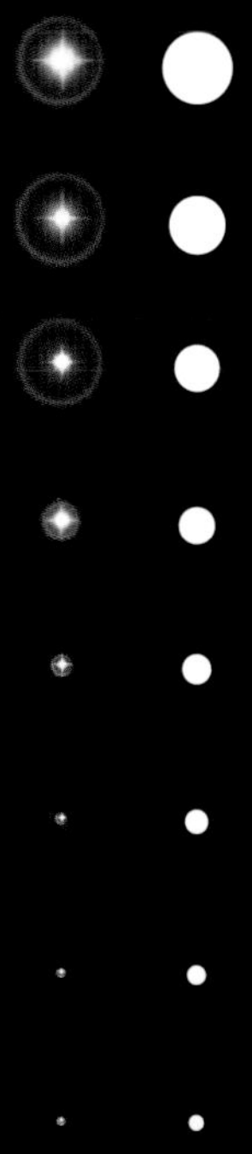

Magnitude
<0
<1
<2
<3
<4
<5
<6
<7

Deep sky objects

- Galaxy
- Globular cluster
- Open cluster
- Planetary nebula
- Diffuse nebula
- Variable star
- Black hole

Fuhrmann
Ziege
(Capella)
Kl. Ziegen
Messier
Algenib
Perseus
Algol
Cassiopeja
Aldebaran
Hyaden
Plejaden
Stier
Fliege
Triangel
Andromeda
Widder
Wendekreis
Walfisch
Ekliptik
Fische
Gerade Aufsteigung
Frühlings
ellatrix
eteigeuze
15
30
45

1 The changing skies

The shifting motions of objects in the sky are largely a reflection of our planet's own movement through space. With a basic understanding of them, the heavens become a far less intimidating challenge, and we can begin to find our way around their wonders.

The Earth in space

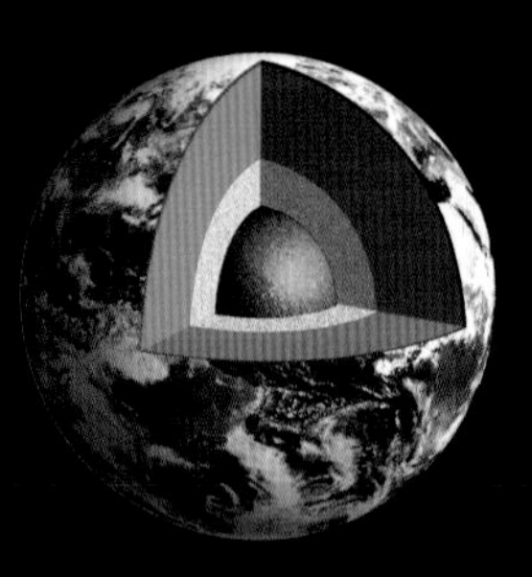

Earth is the largest of the rocky planets, with a substantial core of molten iron, and separated inner and outer mantles of mobile rock. Uniquely in the solar system, its crust of silicate rocks is fractured into plates that drift around on top of the mantle, constantly reshaping the planet's surface.

The moving heavens

Thanks to our planet's motion in space, Earth's skies are a scene of constant change. Even those who never look at the night sky are still ruled over by these celestial changes – most notably the seasons that are the driving force behind our planet's changing climate.

The moving Sun

Because the Earth spins on its axis every 23 hours 56 minutes, but also moves more than 2.5 million kilometres along its orbit in that time, each point on our planet's surface must spin a little further each day to return to the same point in relation to the Sun – and as a result, a full 'day' in solar terms takes 24 hours. So because our daily point of reference is the Sun, the rest of the sky appears to rotate slowly in relation to it, returning to the same point once a year.

Throughout each year, as Earth moves around the Sun, the Sun appears to change its position against the background stars. Over the course of 12 months, it passes through 13 constellations. Twelve of these are the 'signs of the zodiac', the thirteenth is an interloper called Ophiuchus (of which more below). The blazing orb of the Sun, and the daylight it brings with it as its light scatters through the atmosphere, ensures that the constellations behind close to the Sun are lost to view, while those on the opposite side of the sky are visible all night – they rise as the Sun sets and set as the Sun rises. Through the course of a year, most of the sky will be visible – constellations drift westwards across the sky, appearing in the pre-dawn twilight, passing through a phase of 'opposition' when they are visible all night, and finally drawing closer to the Sun and sinking into the west shortly after sunset.

Below: As Earth orbits the Sun (moving from left to right in this illustration), the Sun appears to move eastwards across the sky, appearing against different zodiac constellations and returning to the same position after one year.

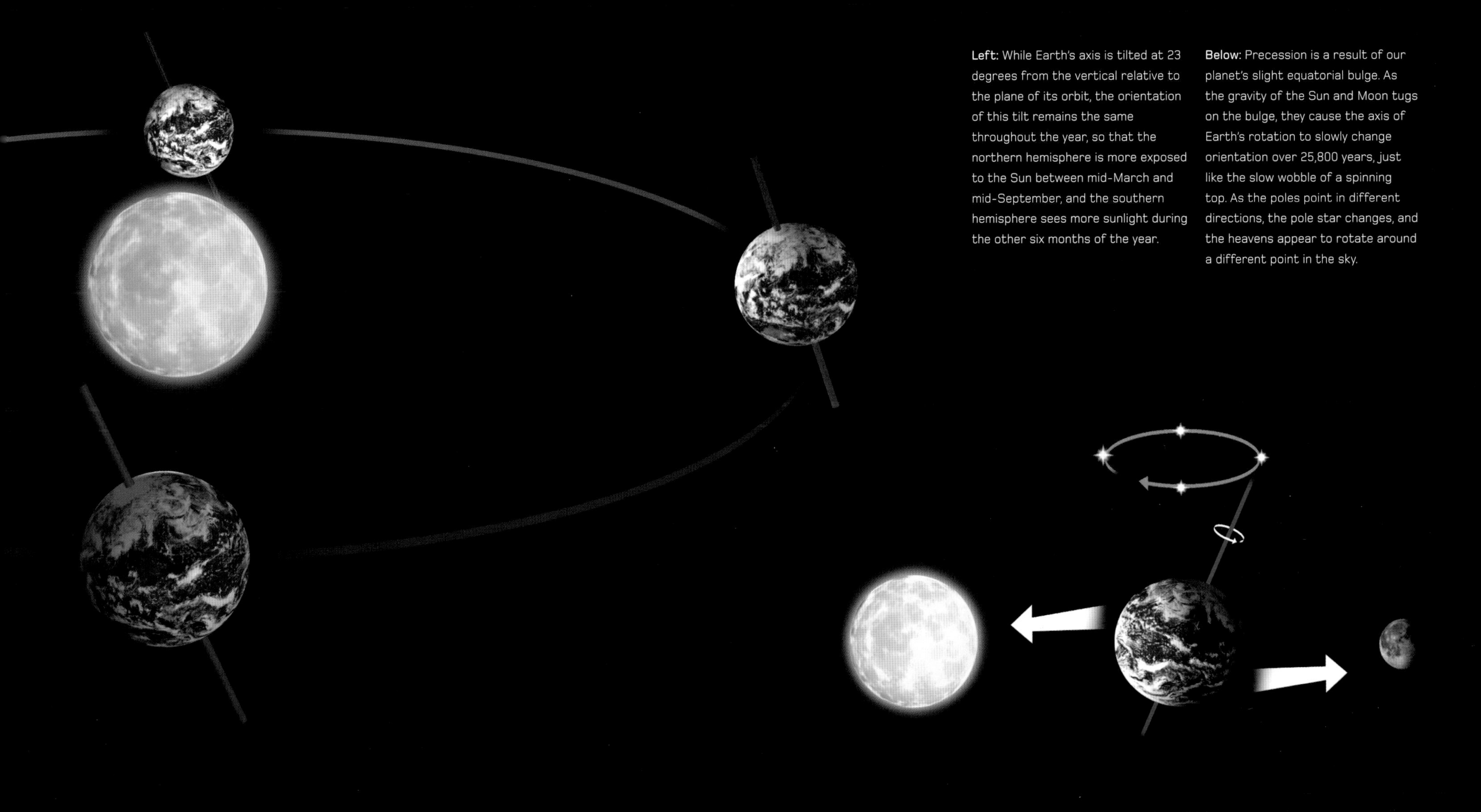

Left: While Earth's axis is tilted at 23 degrees from the vertical relative to the plane of its orbit, the orientation of this tilt remains the same throughout the year, so that the northern hemisphere is more exposed to the Sun between mid-March and mid-September, and the southern hemisphere sees more sunlight during the other six months of the year.

Below: Precession is a result of our planet's slight equatorial bulge. As the gravity of the Sun and Moon tugs on the bulge, they cause the axis of Earth's rotation to slowly change orientation over 25,800 years, just like the slow wobble of a spinning top. As the poles point in different directions, the pole star changes, and the heavens appear to rotate around a different point in the sky.

The Seasons

This might be the only aspect of the Sun's motion in the sky, were it not for the fact that our planet is tilted on its axis: the Earth does not sit 'bolt upright' in its orbit, and the Sun does not, in general, lie directly above the equator. If it did, our planet would be a far less hospitable place, its poles even colder and its tropics completely roasted.

As it is, Earth's poles are tilted over at 23.5 degrees from the vertical. They remain oriented towards the same point in distant space, even as they orbit the Sun, and as a result the northern and southern hemispheres are angled towards the Sun for half a year in turn. As the Sun comes into more prominent view from each hemisphere, it is above the horizon for more than half the time, and is higher in the sky at midday. When the axis of rotation is pointed directly towards the Sun, it reaches its peak in the sky, at that hemisphere's midsummer, before gradually slipping southward again, crossing back to the other hemisphere at the equinox, a point in the year when day and night are equal in both hemispheres. As the Sun sinks lower in the winter skies of one hemisphere, it rises to its summer height in the other. The wave-like line that the Sun traces through the background stars is known as the ecliptic – it is a projection of the Earth's own orbital plane onto the sky.

Precession of the equinoxes

While the seasons affect the position of the Sun, they do not affect the orientation of the stars, since these are determined by the 'celestial sphere', an extension of the Earth's geography into space (see over). However, the Earth is also subject to a long, slow 'wobble' of its axis – the direction of its tilt remains almost static in relation to the Sun over the course of a year, but it does move very slightly. If the direction at right angles to Earth's orbit is considered as 'straight up', then the poles would describe a circle around it every 25,800 years. This wobble is called 'the precession of the equinoxes', and it causes the orientation of the stars to change while, ironically, the fast-moving Sun maintains more or less the same path through the sky. As a result, the path of the Sun's motion through the sky changes against the background constellations over very long periods. One effect of this is that the First Point of Aries, the reference point for all our celestial co-ordinates (see over), slowly moves around the sky along the path of the Sun. Another, more immediate consequence is the intrusion of Ophiuchus into the zodiac – four thousand or more years ago, when the first astronomers were compiling their lists of constellations, the Sun's path took it straight from Scorpius into the southern part of Sagittarius. But as the constellations have swung south, the Sun now has to cross a gap between the northern reaches of the constellations, occupied by Ophiuchus. In a few more millennia, the ecliptic will pass out of some zodiac constellations altogether, returning to its original state in perhaps 20,000 years.

The celestial sphere

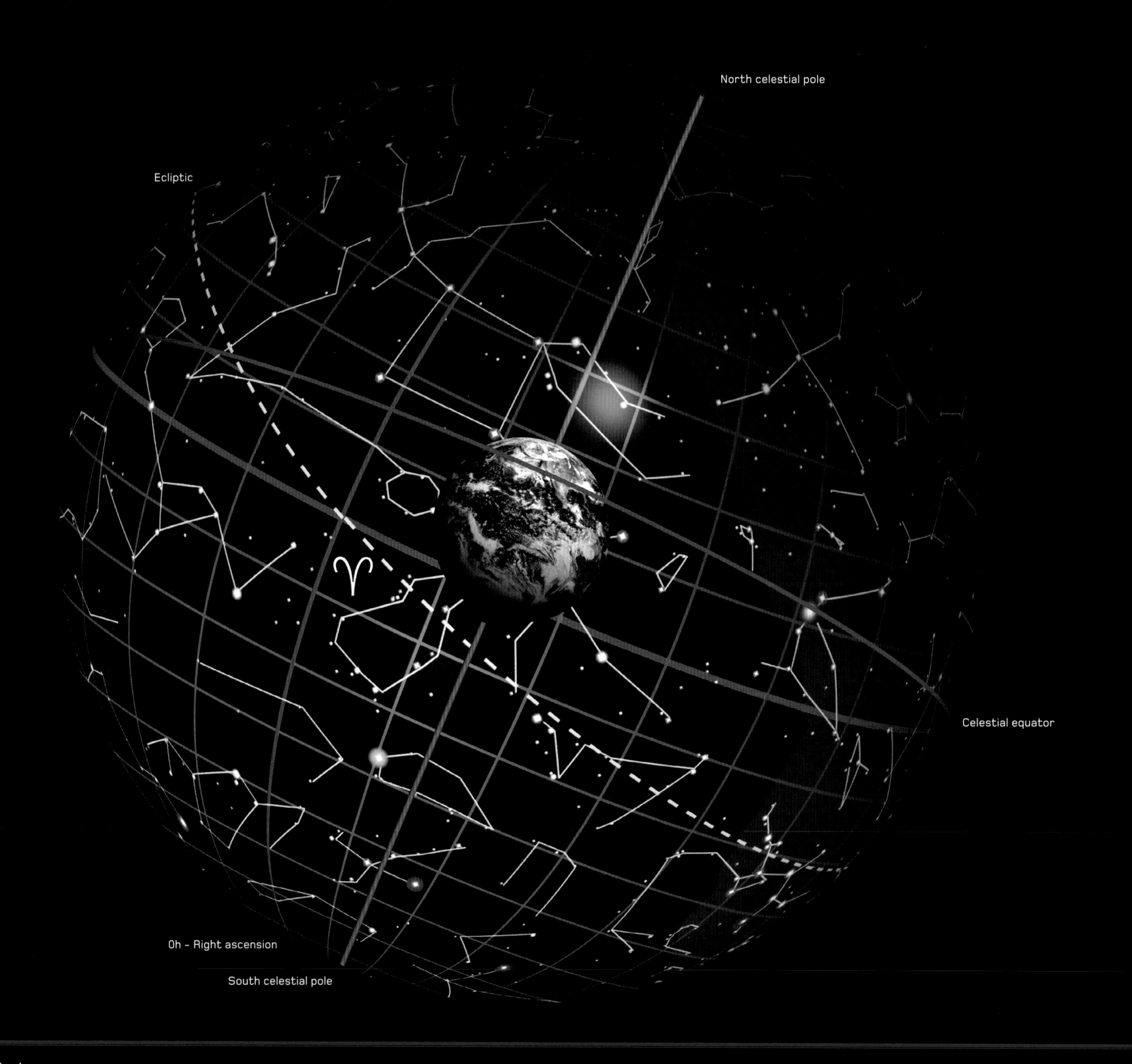

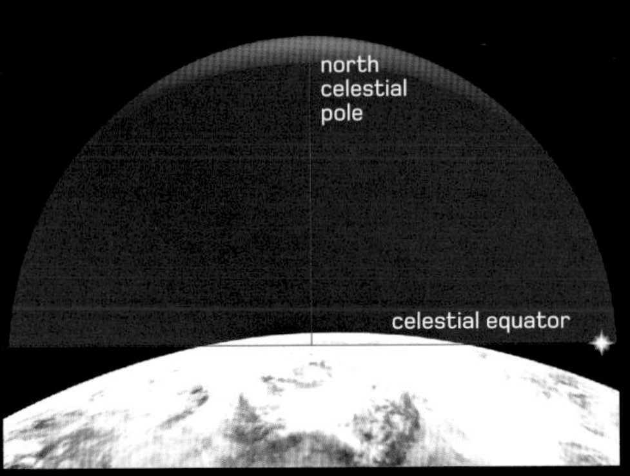

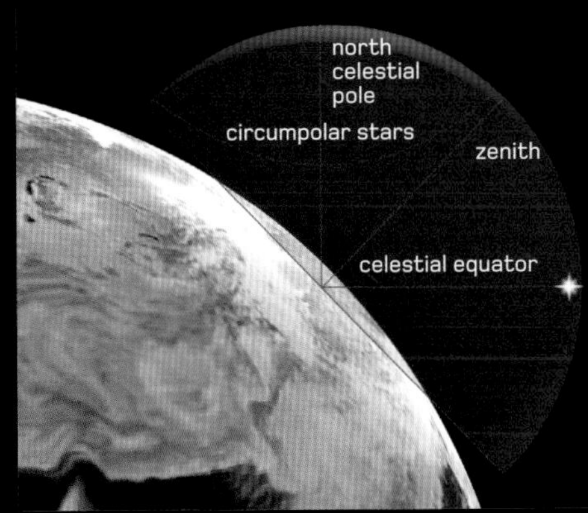

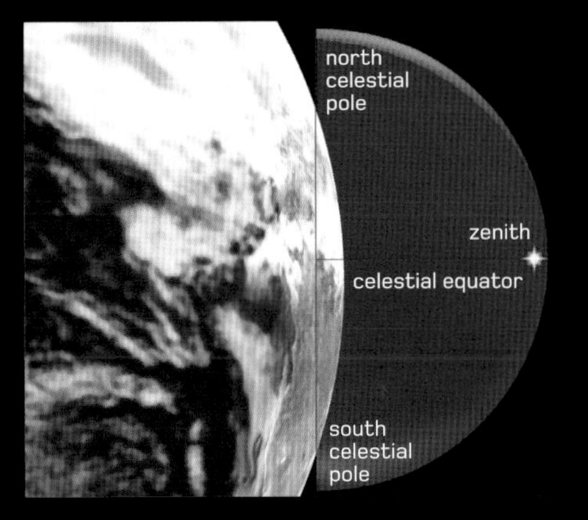

Left: The visible hemisphere of the sky changes with an observer's position on the Earth's surface, because local reference points (the zenith and horizon) shift while the directions of the celestial poles and equator remain the same.

Below: As a star spins across the sky in the course of a night, different alt-azi co-ordinates are needed to describe each position (top). In contrast, its right ascension and declination remain the same relative to the celestial equator and poles, and the First Point of Aries (bottom).

In order to look for objects in the sky, we need a co-ordinate system – a map reference to a star, cluster or nebula's location. Although they are now fully aware of Earth's true place in the Universe, when it comes to mapmaking astronomers still find it useful to revert to the old idea that our planet is at the centre of everything, surrounded by a spherical shell on which stars, Sun, Moon and planets move. This is the celestial sphere – the extension of Earth's geography into space.

Systems of measurement

The simplest way of locating any object in the sky is to measure its angle up from the horizon, and its direction in relation to a fixed point (usually due North). This simple system is known as altitude-azimuth or 'alt-azi', but it has a serious flaw. Earth's rotation means that objects are continuously changing their altitude and azimuth, so alt-azi co-ordinates are only useful for the precise moment of observation.

A better system, universally adopted, is to ignore the rotation of the Earth, 'freezing' the positions of the stars and measuring the location of objects by a form of celestial latitude and longitude. Since the positions of the background stars relative to Earth only change on the very long timescales of precession, they remain in the same positions on the celestial sphere for a long time.

The celestial equivalents of Earth's latitude and longitude are called 'declination' (Dec) and 'right ascension' (RA). Declination is easily understood – it is simply a measurement of an object's inclination relative to the 'celestial equator' (the imaginary ring around the sky directly above Earth's equator).

Right ascension, meanwhile, is a measure of celestial 'longitude', and just as longitude on Earth is measured relative to the Royal Observatory in Greenwich, London, right ascension in the sky is measured relative to a point called the 'First Point of Aries'. No bright star marks this location, and confusingly (thanks to the endless cycle of precession), the First Point actually lies in Pisces, but it is not a completely random choice of reference point – it marks the place where the Sun passes from the southern to the northern half of the sky at the northern spring equinox – the intersection of the ecliptic and the celestial equator.

Declination is measured in degrees, minutes and seconds 'of arc' north or south of the celestial equator (a minute of arc is one sixtieth of a degree, and a second of arc one sixtieth of that). Right ascension, meanwhile, is measured with a clock-like system of hours, minutes and seconds. An object's right ascension indicates how far it lags behind the First Point of Aries in its journey across the sky – defined by the length of time between the First Point crossing an observer's meridian (the north-south line across the sky), and the object following it. Because the sky rotates from east to west, the lines of right ascension go 'anticlockwise' around the sky, so an object a little to the east of the First Point may have a right ascension of 1 hour, while one just to its west may have a right ascension of 23h, taking almost an entire day to follow the First Point across the meridian.

Because there are 360 degrees 'of arc' in a complete circle, one hour of right ascension is equivalent to 15 degrees of declination, so (confusingly) one minute of RA is equivalent to 15 minutes of arc, and one second of RA equivalent to 15 seconds of arc. RA co-ordinates aside, all other angles in the sky (including those mentioned in this book) are measured in degrees, minutes and seconds of arc.

The visible sky

Unfortunately, we can only ever see one hemisphere of the sky at a time – the bulk of the planet beneath our feet blocks out the other side of the sky. The exact area of sky visible at any time depends on which direction our part of the planet is facing – the north pole, for instance, only ever points in one direction in the sky, and only ever sees the stars of the northern hemisphere. Likewise the south pole only ever sees half the sky. Observers on the equator get the best of both worlds – their view of the sky consists of half of each celestial hemisphere, so at one time or another, they can view every star in the sky.

For skywatchers at mid-latitudes, however, the sky is a hybrid. Our hemisphere's 'celestial pole' occupies a fixed point in the sky, with the stars wheeling around it. They rise in the east, pass high across the 'meridian' (the north-south line across the sky) in the south (for northern hemisphere observers) or north (for those in the southern hemisphere), and set in the west. Stars and constellations close enough to the celestial pole never set at all, and are termed circumpolar, but the changing stars include many from the opposite hemisphere of the sky.

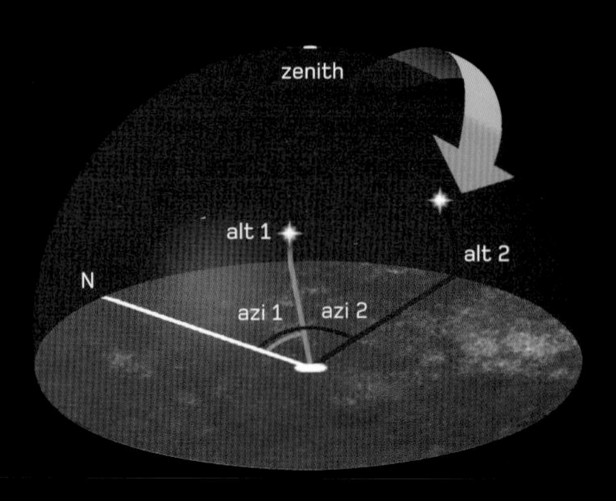

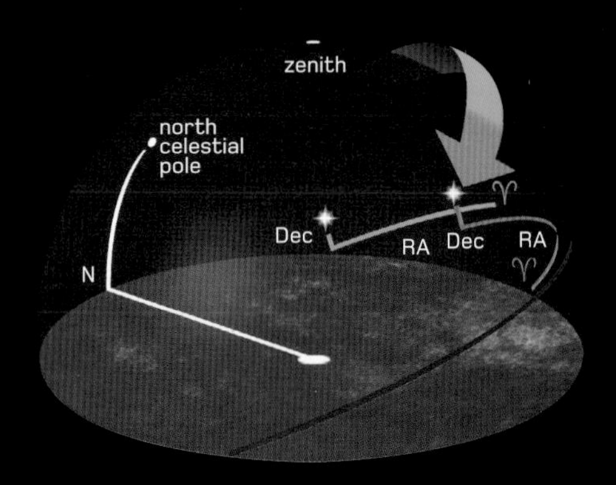

Circumpolar stars

Northern hemisphere

The charts on these pages show the stars around each celestial pole, which are generally circumpolar (never setting) for observers in the mid-latitudes of each hemisphere, and remain visible throughout the year.

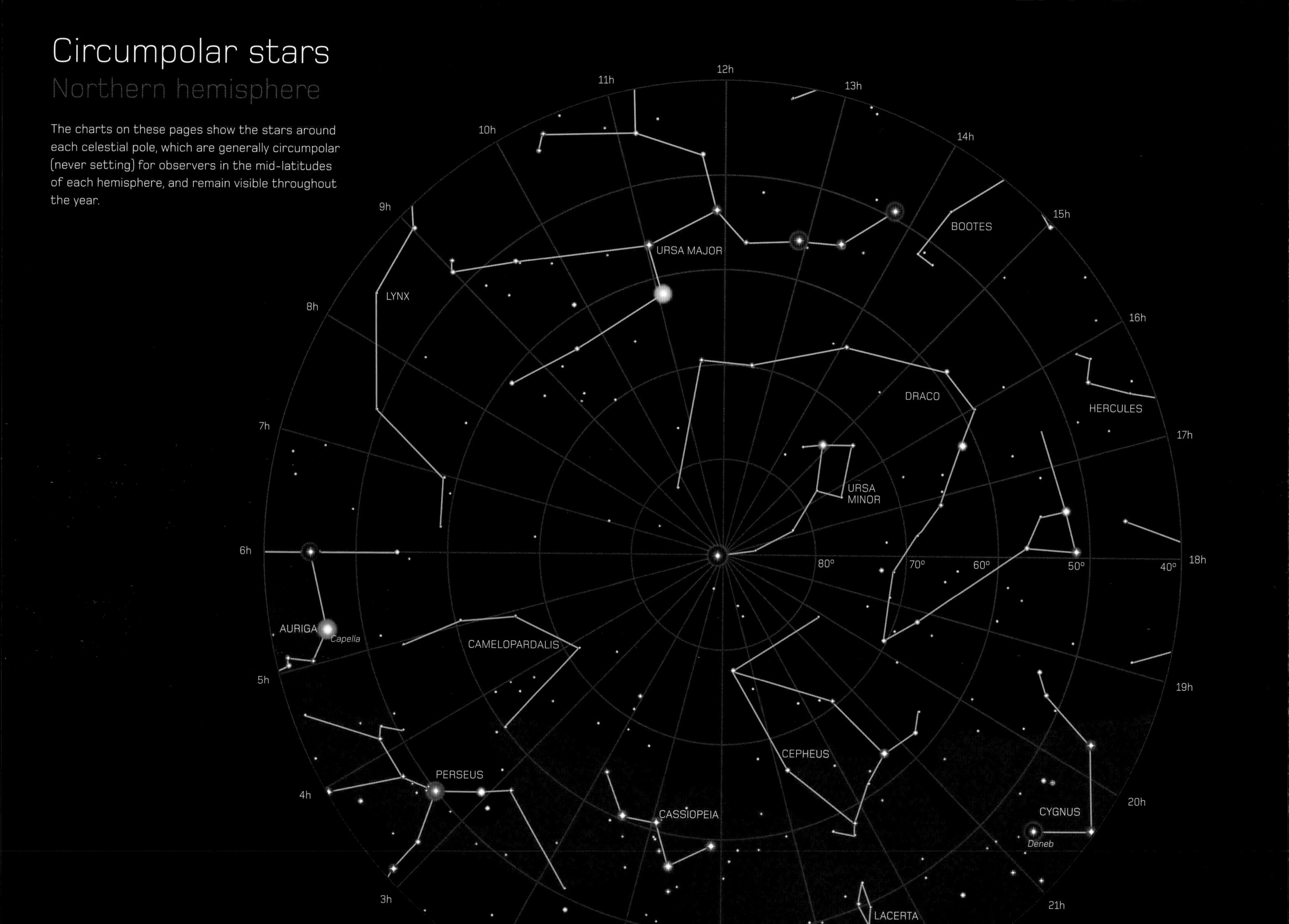

Southern hemisphere

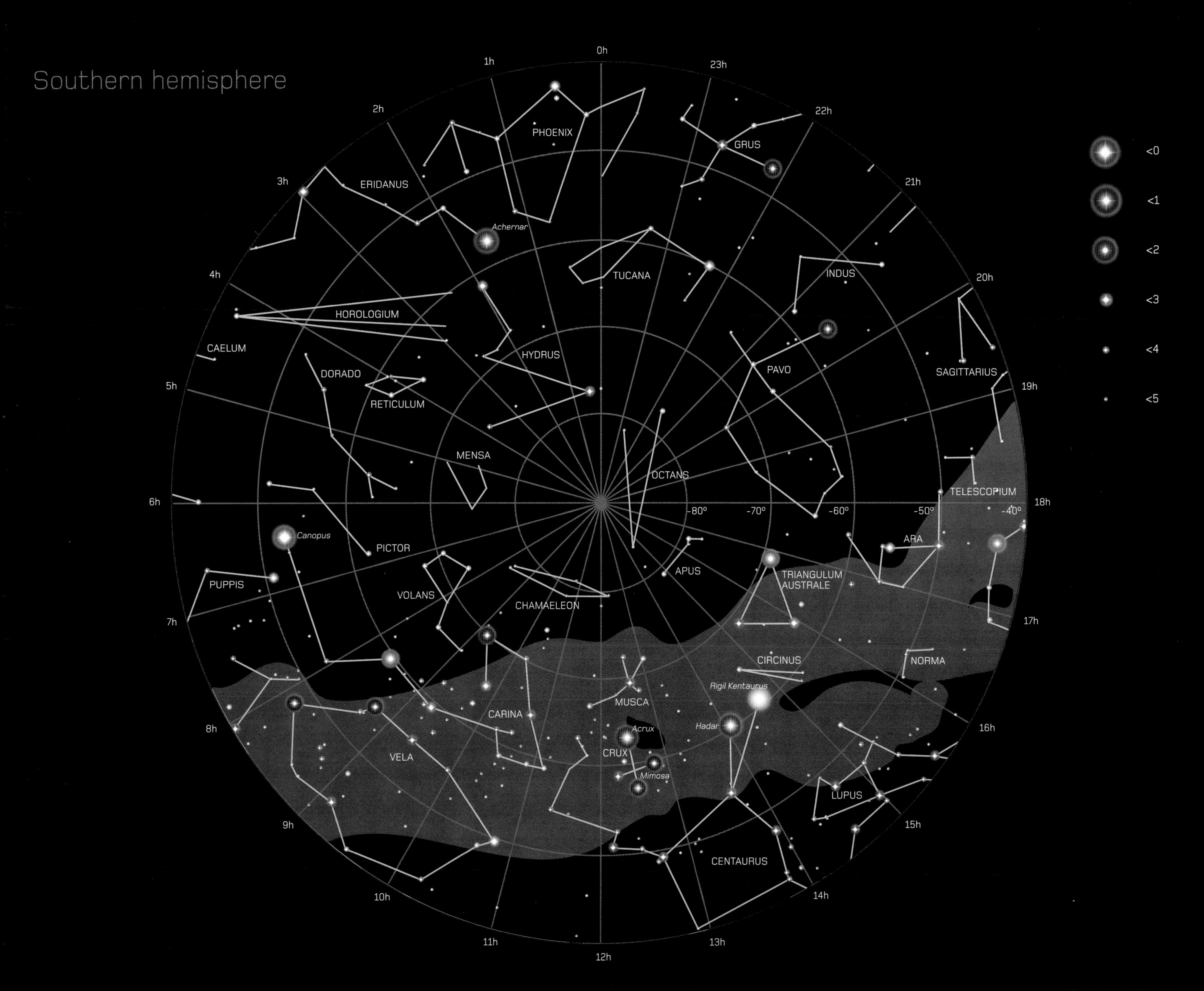

Equatorial stars

6h right ascension

The charts on these pages show the regions around the celestial equator centred on right ascensions of 0h and 6h. These are the stars that can be seen from mid-latitudes of each hemisphere at different times of night depending on the time of year. Stars around 0h lie opposite the Sun and are visible through the night around September, while those around 6h are in 'opposition' around December.

0h right ascension

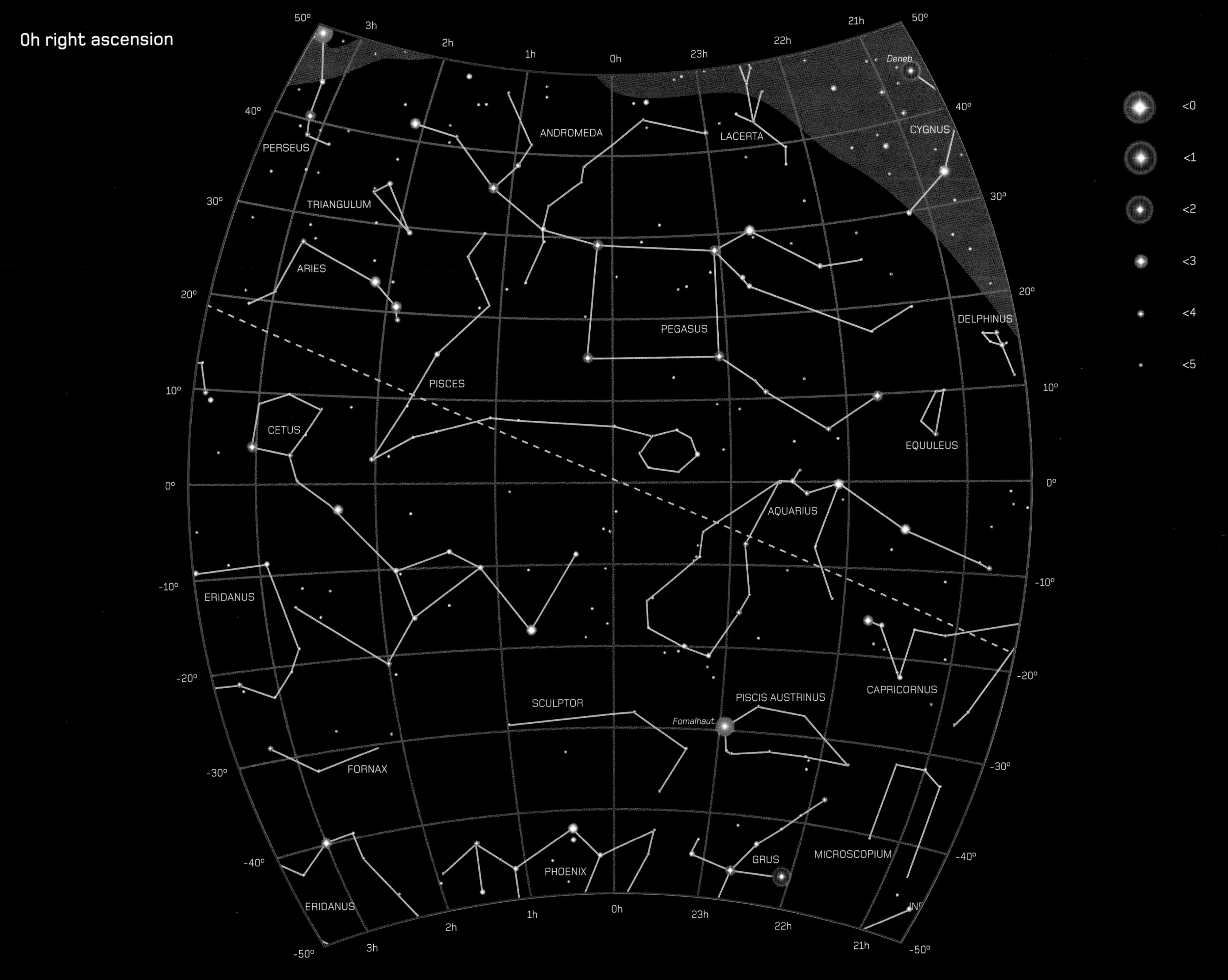

Equatorial stars

18h right ascension

The charts on these pages show the regions around the celestial equator centred on right ascensions of 12h and 18h. These are the stars that can be seen from mid-latitudes of each hemisphere at different times of night depending on the time of year. Stars around 12h lie opposite the Sun and are visible through the night around March, while those around 18h are in 'opposition' around June.

12h right ascension

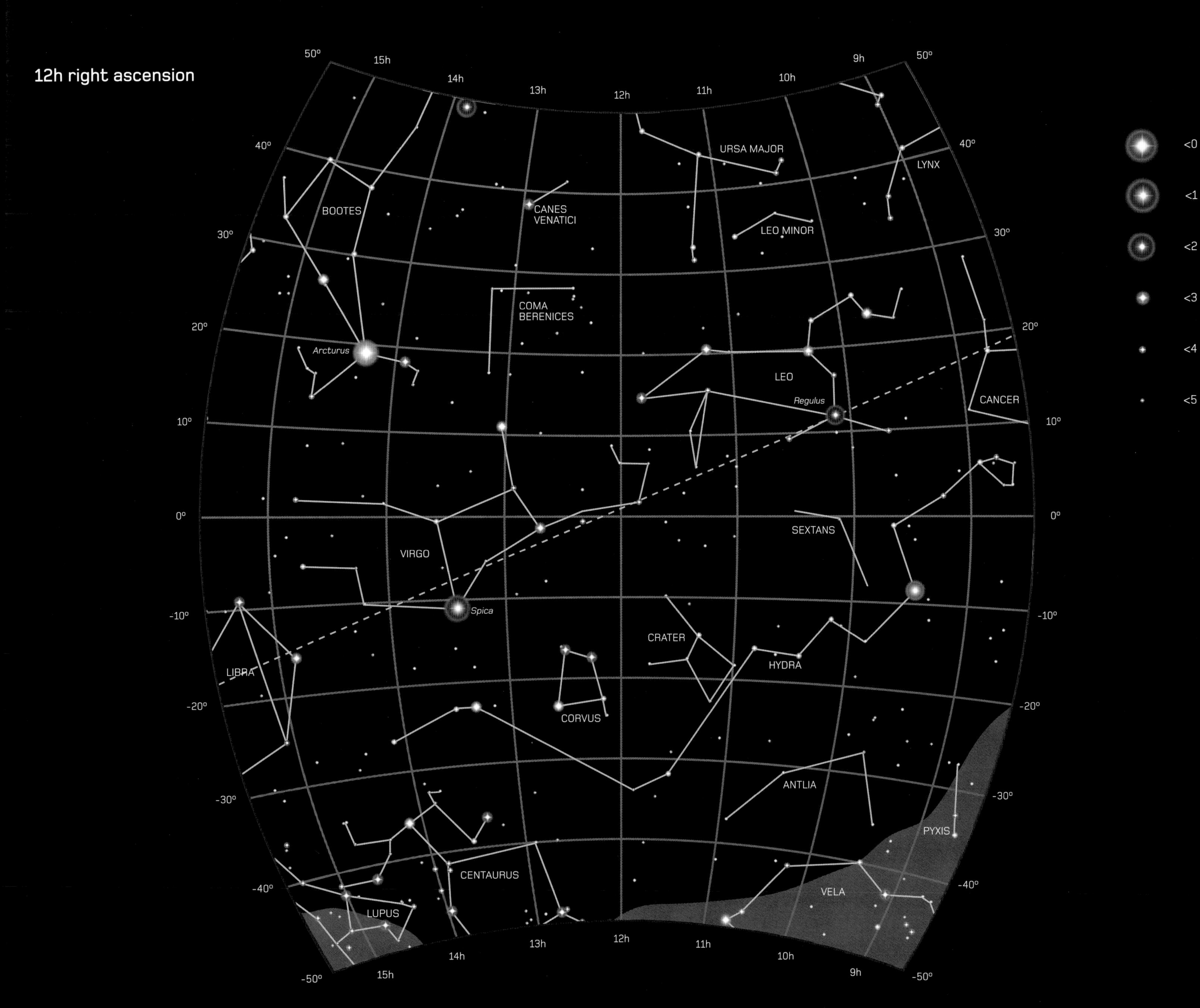

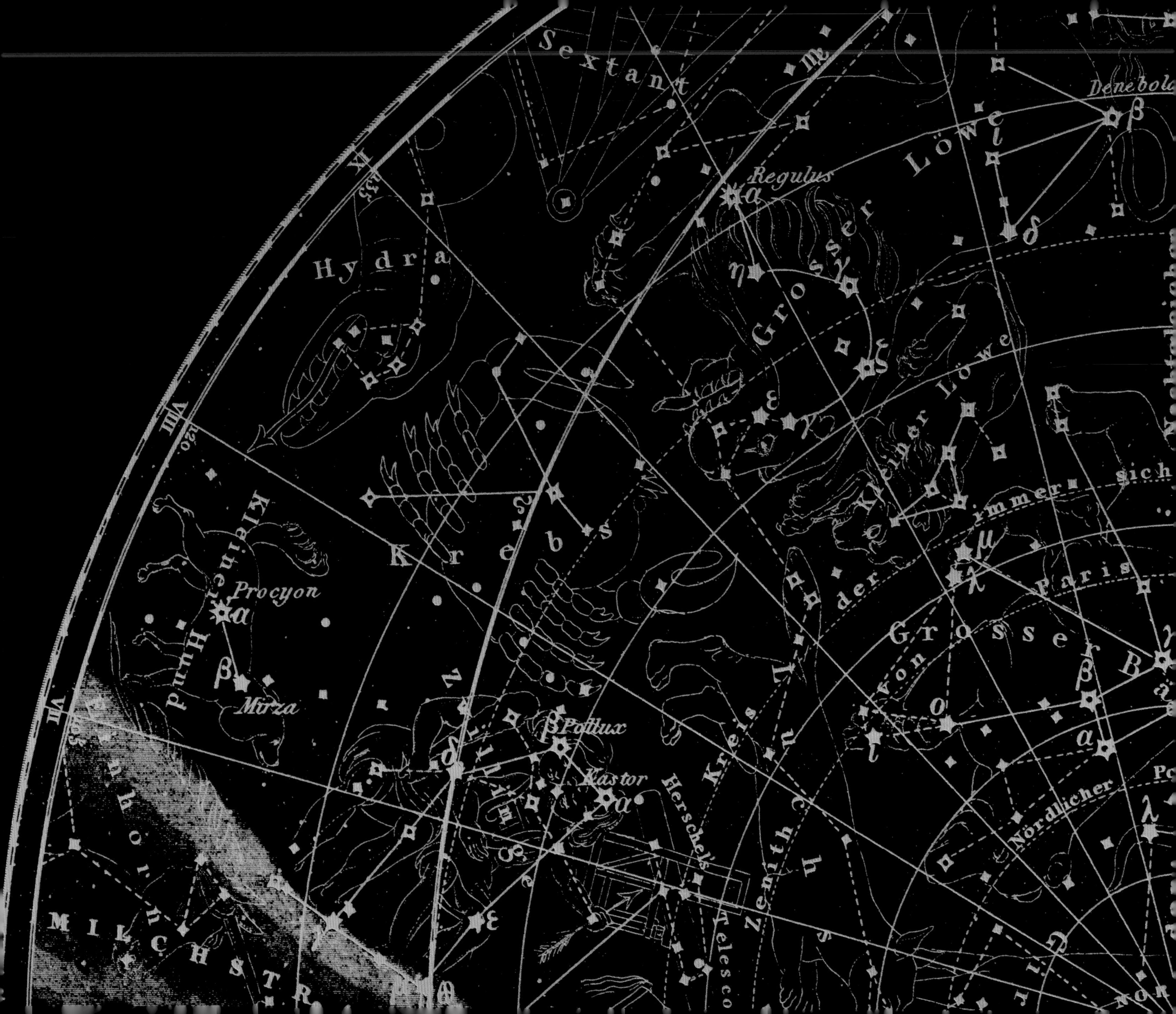

Sextant
Denebola
Löwe
Regulus
Hydra
Grosser
Kleiner Löwe
Krebs
Kleiner Hund
Procyon
Mirza
Pollux
Kastor
Herschels Telesco
Zenith
Nördlicher
Grosser B
Paris
MILCHSTR

2 The solar system

The solar system is our cosmic neighbourhood, and its residents some of the most fascinating objects in the sky – always rewarding however familiar they become. The Sun is our local star, and the only one of all the billions in our galaxy that we can see close up. The planets range from the rocky to the tenuous, the tiny to the gargantuan, with a range of moons almost as diverse. And the space between and beyond these major bodies is filled with smaller bodies, ranging in size from dwarf planets to dust particles. The pages that follow offer some advice on getting the most out of our solar system.

Introducing the solar system

Our solar system is the kingdom of the Sun – a roughly spherical area of space about 19 million million km across in which our local star's gravity is the governing influence. It contains eight major planets and billions of smaller objects, and is so vast that the Sun's light takes a year to reach its outer edge.

Just a handful of these worlds are visible to the naked eye, and perhaps a few dozen lie within the reach of amateur telescopes. The most obvious are the major planets, which huddle close to the centre (so that light takes only four hours to reach even Neptune). Earth is the third of these planets, and from our position 150 million km or eight 'light minutes' from the Sun, we can see the other planets as they shine by reflected sunlight.

Planets come in two types – small, rocky worlds like Earth close to the Sun, and much larger but more tenuous giants, composed of gas and liquid, further out. All three of our rocky neighbours – Mercury, Venus and Mars – are visible to the naked eye, as are the two closest giants, Jupiter and Saturn. The outer giants, icy Uranus and Neptune, are visible only with binoculars or a telescope.

The fact that the planets move against the background stars is the origin of their name – 'planet' comes from the ancient Greek word for 'wanderer'. Most ancient civilizations saw the moving planets as manifestations of their gods – the names of the planets today come from ancient Graeco-Roman deities.

The planets all move around their orbits in roughly the same plane – at most just slightly tilted to the plane of Earth's own orbit. As a result, they always stay close to the ecliptic, the path on which the Sun appears to move around the sky. The following pages contain charts that show the positions of the planets in years to come, and offer advice on how best to observe them.

All the planets except Mercury and Venus have natural satellites or moons in orbit around them, the most familiar of which is undoubtedly Earth's Moon (with a capital 'M'). While Mars's moons are a pair of tiny rocks only visible through the largest telescopes, the outer gas giants have huge families of satellites in orbit around them, including some (similar in size to our Moon, if not larger) that can be seen with modest equipment – binoculars or a small telescope. Jupiter has four such satellites – the so-called Galilean moons – and Saturn has several, though they are harder to see because of

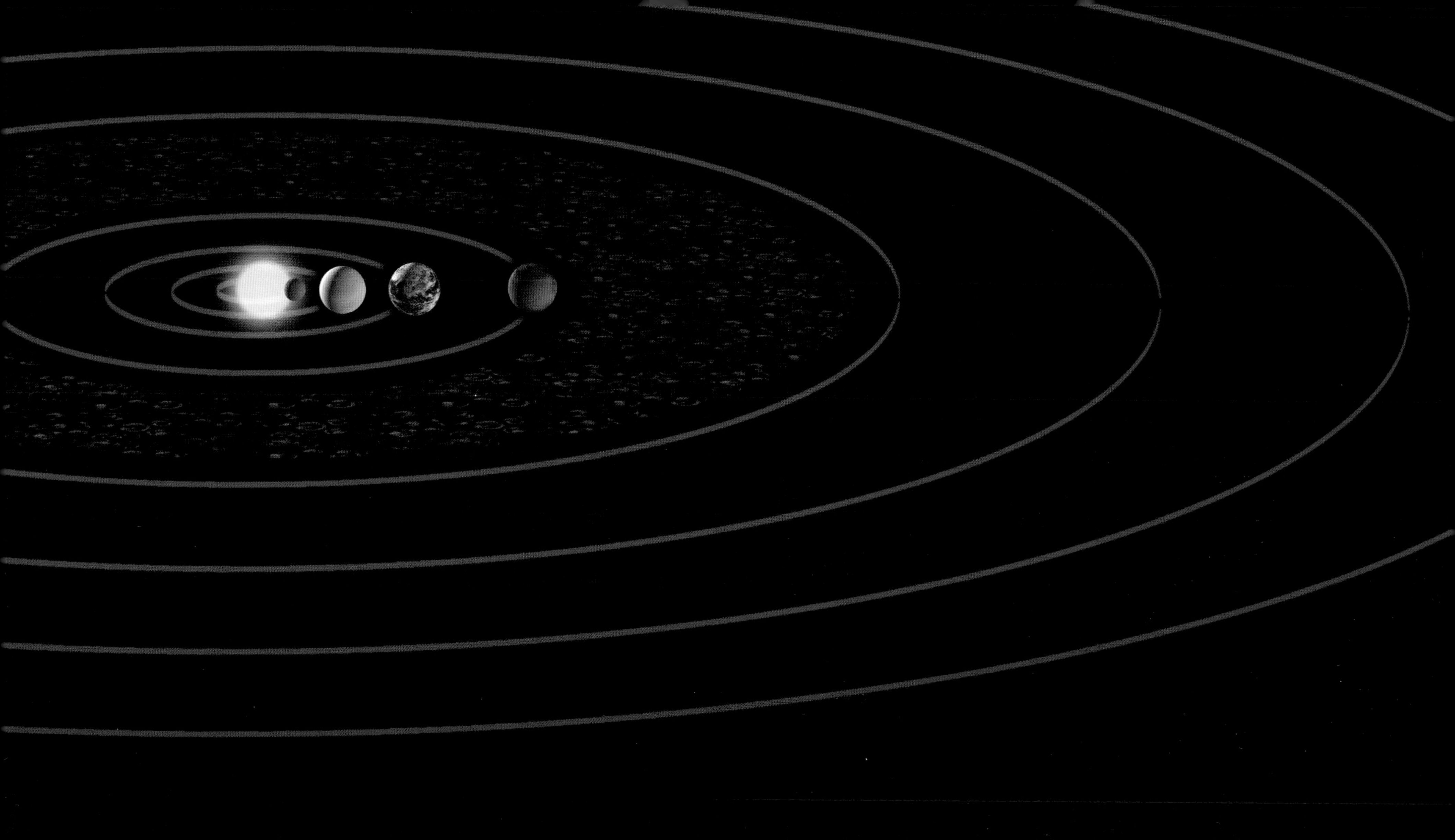

their greater distance. The same goes for the mid-sized satellites of Uranus and Neptune.

In between and beyond the orbits of the planets orbit huge numbers of other objects – rocky asteroids close to the Sun, icy comets further out (though sometimes swinging closer in), and particles of dust and ice everywhere. This is the debris of the solar system – remnants left over when the major planets came together, and shrapnel created by countless interplanetary collisions in the 4.6 billion years since then.

Most asteroids are confined in a belt that separates the inner planets from the outer giants, although some stray closer to the Sun. Nearly all these worlds are just outsized boulders, perhaps a few tens of kilometres across, but a handful are more substantial, and can be seen with basic equipment. The easiest are Ceres (the largest asteroid and technically a 'dwarf planet') with a diameter of 950 km, and Vesta, which is smaller at 525 km, but brighter – at times it can even reach naked-eye visibility. The best way to locate these objects is by consulting an online ephemeris or star guide (see p.269).

Comets are the icy equivalent of asteroids. They originate further from the Sun, mostly in a huge and undetectable spherical halo called the Oort Cloud that surrounds the solar system at the limit of the Sun's gravitational influence. Sent spinning inwards by collisions or the gravity of other stars, they enter the inner solar system at high speeds on highly elliptical orbits. As they pass through the realm of the rocky planets, heat from the Sun melts their ice, blasting dust and vapour off their surfaces to form an extended atmosphere called a coma, and often a tail that is driven away from the Sun on the solar wind. Comets are brief and mostly unpredictable visitors, but occasionally a close encounter with one of the giant planets can swing a comet into a new orbit that sees it return to the inner solar system every few decades. Halley's Comet, which returns every 76 years, is the brightest and best known of these 'short-period' comets.

As comets pass through the inner solar system and asteroids occasionally collide and fragment, they leave trails of small particles behind them, which are swept up by the planets. The smallest of these particles burn up in Earth's atmosphere as shooting stars, but more substantial rocks make it to the surface as meteorites, and the largest meteorites of all can create major impact craters, similar to those which pepper the surfaces of smaller and less well-protected worlds.

At the outer edge of their orbits, comets such as Halley's linger in the Kuiper Belt – an icy equivalent of the asteroid belt where countless dwarf worlds orbit beyond Neptune. The brightest and best-known member of the Kuiper Belt is Pluto, demoted in 2006 from a true planet to a 'dwarf', but even this is well beyond the reach of most amateur telescopes.

Observing the Sun

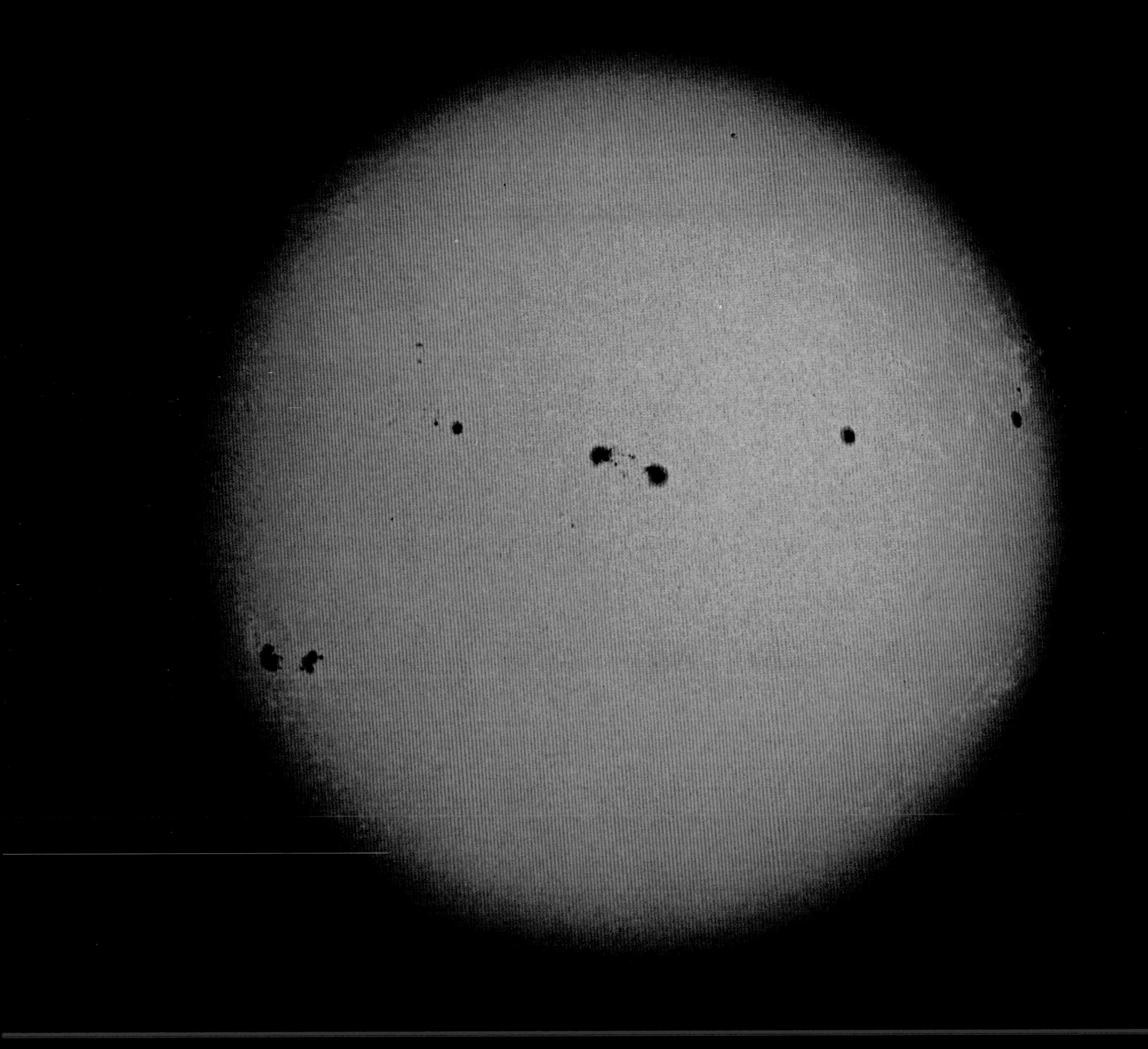

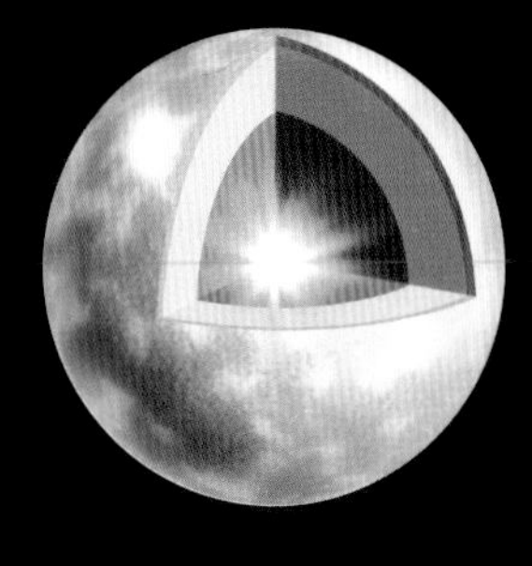

The interior of the Sun consists of layers of gas at high temperatures and pressures. Nuclear reactions in the core release energy that flows out until it escapes at the Sun's visible surface, the photosphere.

Top: Eclipses occur when the Earth, Sun and Moon line up precisely. Because the Moon's orbit is tilted slightly to the ecliptic, this does not happen at every new and full Moon. When Earth's shadow is cast onto the Moon, it darkens its appearance, often turning it blood red.

Above: For anyone directly in the path of the Moon's shadow on Earth, a total solar eclipse is a far more impressive experience.

The Sun dominates Earth's skies and affects our day to day lives like no other celestial body. A blazing ball of exploding gases, it spreads heat and light across the solar system – from our location on Earth, it appears as a blazing orb, half a degree across and too bright to look at without risking permanent damage to your eyesight.

Projecting the Sun

The only safe way to look at the Sun is by projecting its image onto a screen. NEVER look directly at the Sun, as this risks permanent blindness. In order to project the Sun's image, place a screen such as a piece of card behind the eyepiece of a small telescope (or binoculars with one eyepiece covered). Align the instrument so it is pointing at the Sun, not by looking through it, but by referring to the shadow that it casts back onto the screen. When the alignment is perfect, the bright disc of the Sun should appear, and the size and focus of the image can be adjusted by moving the screen back and forth. Once the image is in focus, any dark sunspots will be obvious, and these can be tracked from day to day as they drift with the Sun's slow rotation. Sometimes brighter streaks on the Sun's surface may also be visible, though these have far less contrast with the light background.

Some telescopes come supplied with solar filters that screw into the eyepiece, but these are not to be trusted, since a telescope's optics can act like a burning glass and focus the Sun's heat to such intensity that the filter may suddenly crack. Even heavy-duty filters that block most of the Sun's light before it enters the telescope should be treated with caution, and for the same reason it can be dangerous to leave even the projected image of the Sun unattended.

Nevertheless, with proper care and attention, the projection method can also create clear images of the Sun for observing phenomena such as transits of Venus and Mercury.

Eclipses

The only way to see more detailed features of the Sun from Earth is to wait for a total eclipse of the Sun, when the Moon passes between Earth and Sun and briefly blocks out its brilliant disc, allowing fainter features of its upper atmosphere to be seen, including flame-like loops of cool gas called prominences, and the tenuous but extremely hot corona that extends for millions of kilometres beyond the Sun.

Unfortunately eclipses are rare events – they rely on a precise alignment of Sun and Moon, and the coincidence that the two bodies are almost exactly the same size in Earth's skies. The lunar shadow cast on Earth covers only a few tens of kilometres and moves rapidly, so at best a total eclipse may last for a few minutes (partial eclipses, when the Moon blocks out just part of the Sun, can be seen over a much wider area, but they do not produce such spectacular results). Frustratingly, even when the alignment is perfect, an eclipse may not be total – if it happens at the wrong time of month, the Moon may be at the outer limit of its orbit, appearing too small to block the Sun completely, and creating a ring-like 'annular' eclipse.

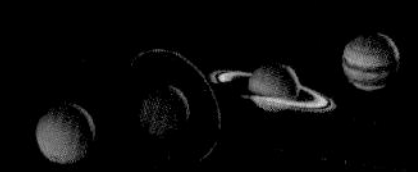

The Sun dominates a huge area of space. In gravitational terms, the solar system extends to the Oort Cloud, up to one light year away – more than 2,000 times more distant than the outer planet Neptune.

Above left: The Sun's behaviour goes from quiescent to violent and back in an 11-year cycle. At its peak of activity, enormous eruptions such as this coronal mass ejection (CME) are created by the Sun's tangled magnetic field. When the energetic particles in CMEs and flares reach Earth, they cause aurorae and radio interference in our planet's atmosphere.

Above right: Filaments are plumes of hot gas up to 100,000 km long that rise out of the Sun's turbulent surface and help tranfer heat to the sparse but hot outer corona.

Opposite: Sunspots range in size from 'small' pairs about the size of Earth, to huge groups that dwarf Jupiter. They are about 2,000°C cooler than their surroundings, and so appear dark by comparison. Like flares and CMEs, sunspot numbers also wax and wane with the 11-year solar cycle.

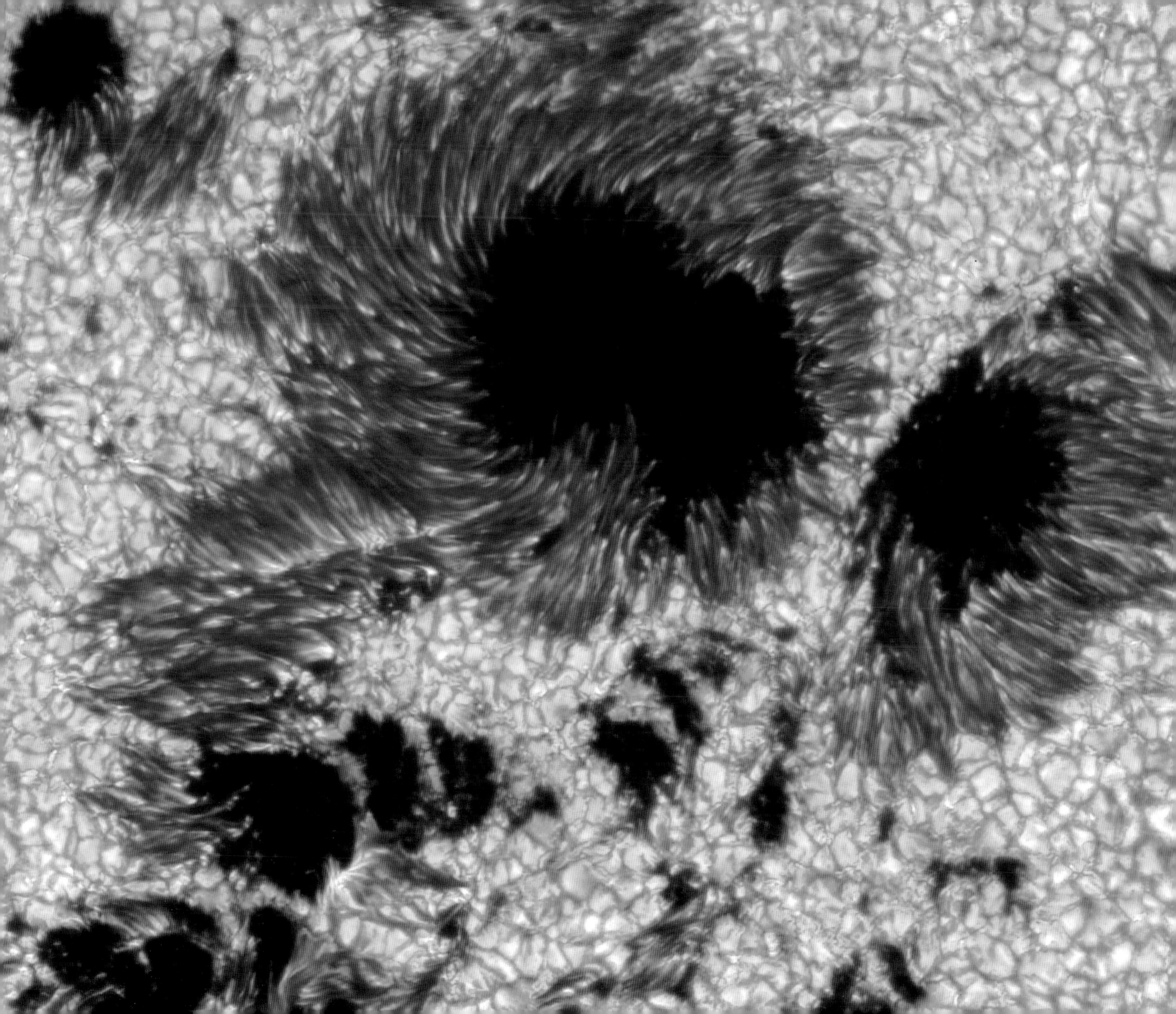

Observing the Moon

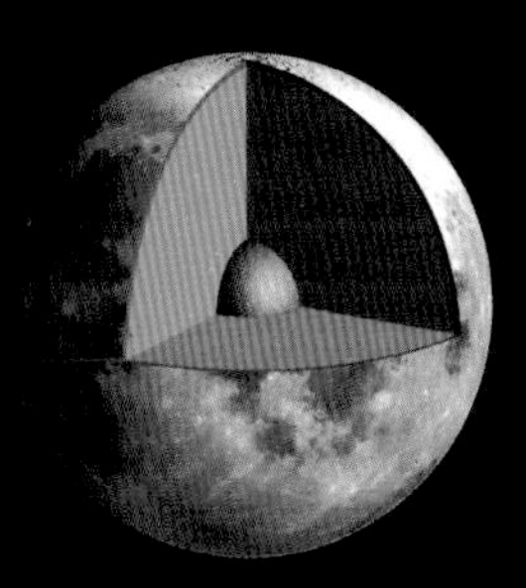

The Moon's internal structure is fairly simple, consisting of a rocky crust and mantle with a denser core at the centre. One oddity is that Earth's gravity has 'pulled' the core slightly closer to the surface of the near side, explaining the greater signs of volcanic activity on the Earth-facing half of the Moon.

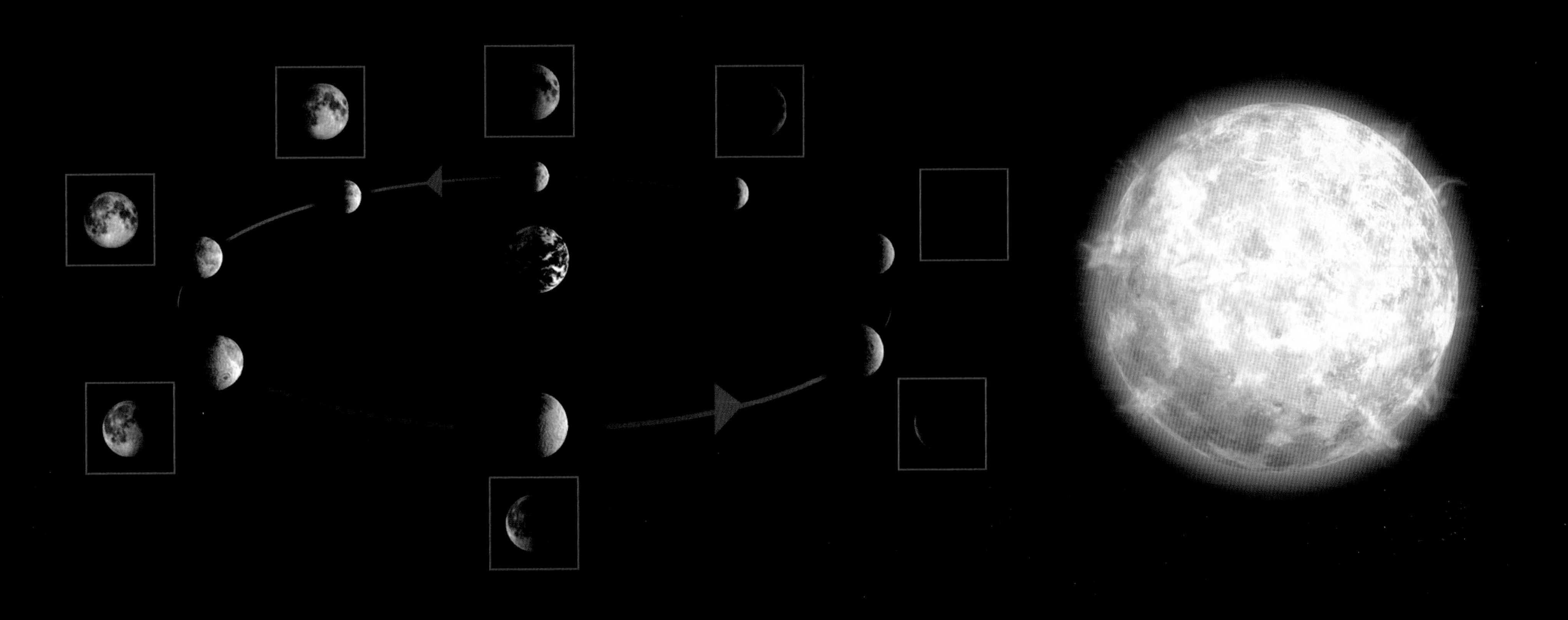

The Moon is the most rewarding object of all for amateur astronomers – a whole other world on Earth's cosmic doorstep, on average just 400,000 km away.

To the naked eye, the Moon's disc is roughly half a degree across in the sky. This changes slightly because our satellite's orbit is not a perfect circle, but is hardly enough to be noticed – the more obvious effect of the Moon appearing larger when close to the horizon is actually an optical illusion.

As our satellite orbits Earth, we see changing amounts of its daylit side, creating the familiar cycle of phases. But even as the Moon waxes and wanes, the features on it remain fixed: the Moon's rotation and orbital periods are locked together so that it rotates once for each orbit and always keeps the same face turned to Earth. This phenomenon, a result of tidal forces between Earth and Moon, is common in satellites across the solar system.

The Moon's orbit is tilted at about 5 degrees to the ecliptic, which gives it a complex cycle of movement in the sky. Although it follows a similar cycle of movement to the Sun, sometimes this additional tilt increases the angle between the Moon and the celestial equator, and sometimes it subtracts from it. The end result is a cycle of motion that sees Earth, Sun and Moon return to the same relative positions just once every 18 years – a cycle which also governs the pattern of eclipses of the Sun and Moon (see p.27)

Naked-eye observers can easily tell that the near side of the Moon is made up of two different types of terrain. Darker patches called maria (from the Latin for seas) fill large, roughly circular basins amid a brighter landscape of 'highlands'. In ancient times, observers thought they saw figures in the maria, such as the 'Man in the Moon' or the Chinese 'Rabbit in the Moon'.

Binoculars will begin to hint at the Moon's magnificent array of surface features. They will show many of the Moon's larger craters, and reveal that these are more plentiful in the highlands, and comparatively sparse on the lunar seas. Binoculars should also show the prominent bright 'rays' that spray out from several of the larger impact craters, and will show the terminator – the dividing line between the Moon's night and day, as a jagged line broken by the shadows of deep craters and high mountains.

Telescopes will show the Moon in all its glory – revealing the highlands as a bright landscape pulverized by countless meteor impacts, and the maria as rolling lowland plains of solidified lava, peppered with the occasional crater. Elsewhere there are mountain ranges, and lava tubes running from extinct volcanoes. Many of these features are best seen when they are near the terminator, and the long shadows cast by the rising or setting Sun help to reveal the surface relief. With a good telescope, the detail is almost endless, and our satellite alone is well worthy of a lifetime's amateur study.

Above: As the Moon orbits Earth and Earth spins round the Sun, we see different amounts of our satellite's sunlit face (shown in the red insets). The Moon's phases grow or wax from 'new', when the sunlit side is completely hidden, through crescent to 'first quarter', then through a bloated 'gibbous' phase to full. They then wane through gibbous, last quarter and 'decrescent' back to new.

Above: The Moon orbits Earth every 27.3 days, on a path that is tilted at about 5 degrees from the ecliptic – the general plane of Earth's orbit. Its distance from Earth varies between 363,000 and 406,000 kilometres.

Moon atlas

The near side

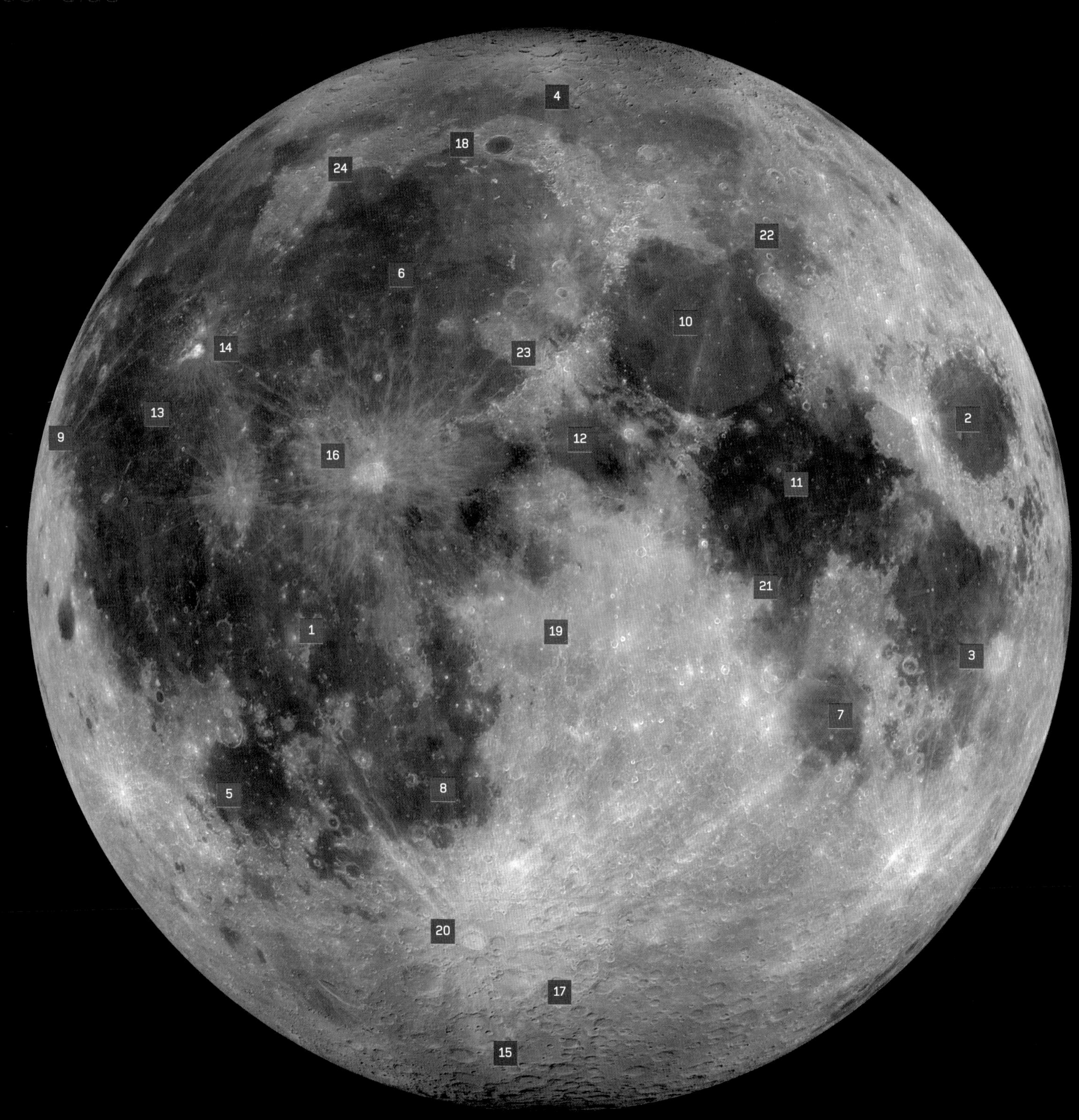

Seas

1. Mare Cognitum
2. Mare Crisium
3. Mare Fertilitatis
4. Mare Frigoris
5. Mare Humorum
6. Mare Imbrium
7. Mare Nectaris
8. Mare Nubium
9. Mare Orientale
10. Mare Serentitatis
11. Mare Tranquilitatis
12. Mare Vaporum
13. Oceanus Procellarum

Craters

14. Aristarchus
15. Clavius
16. Copernicus
17. Magninus
18. Plato
19. Ptolemaeus
20. Tycho

Other features

21. Apollo 11 landing site
22. Lacus Somniorum
23. Lunar Appenine Mountains
24. Sinus Iridium

Moon atlas

The far side

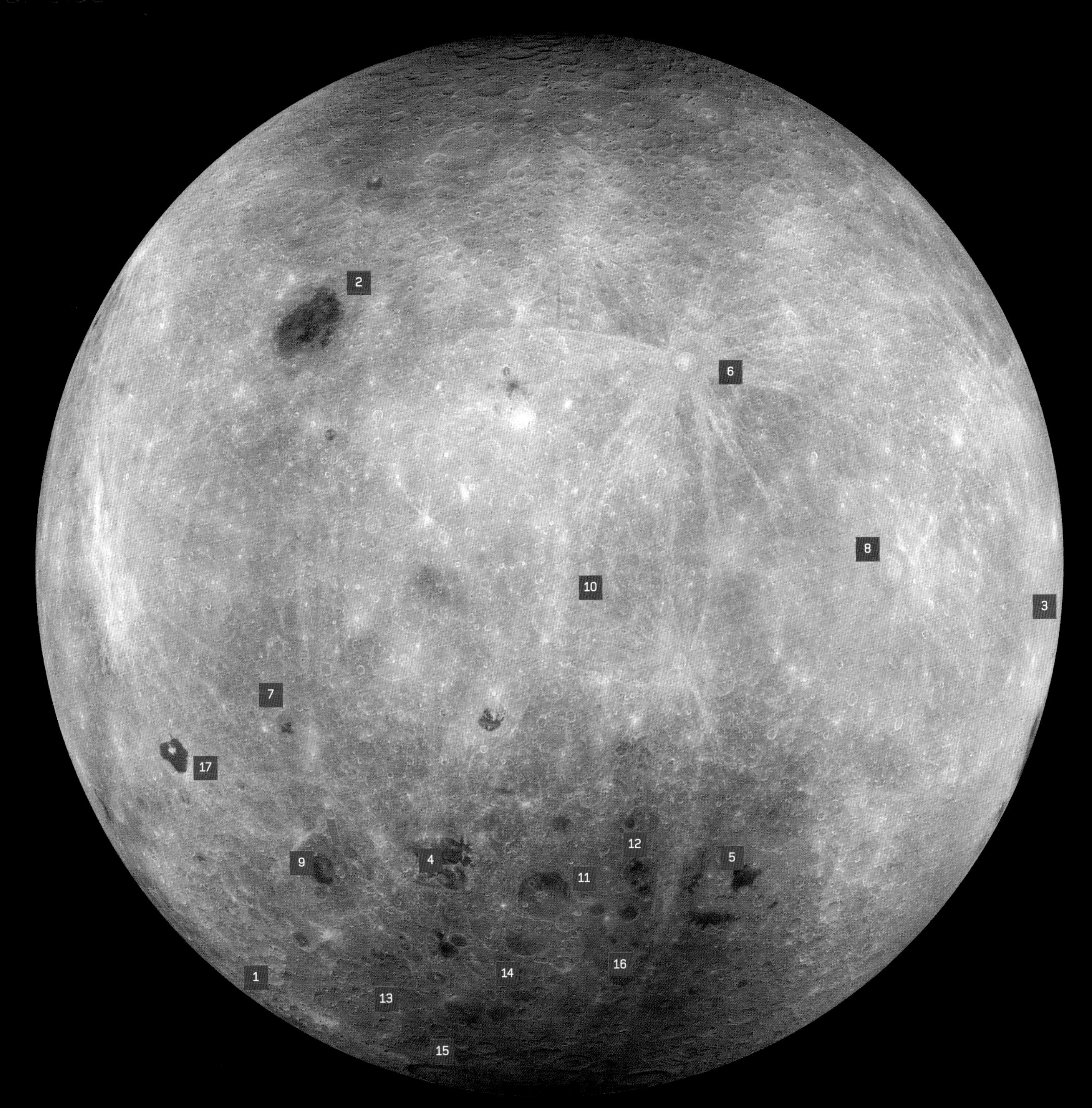

Seas

1 Mare Australe
2 Mare Moscoviense
3 Mare Orientale
4 Mare Ingenii

Craters

5 Apollo
6 Cockcroft
7 Gagarin
8 Hertzsprung
9 Jules Verne
10 Korolev
11 Leibnitz
12 Oppenheimer
13 Planck
14 Poincaré
15 Schrödinger
16 South Pole-Aitken Basin
17 Tsiolkovskii

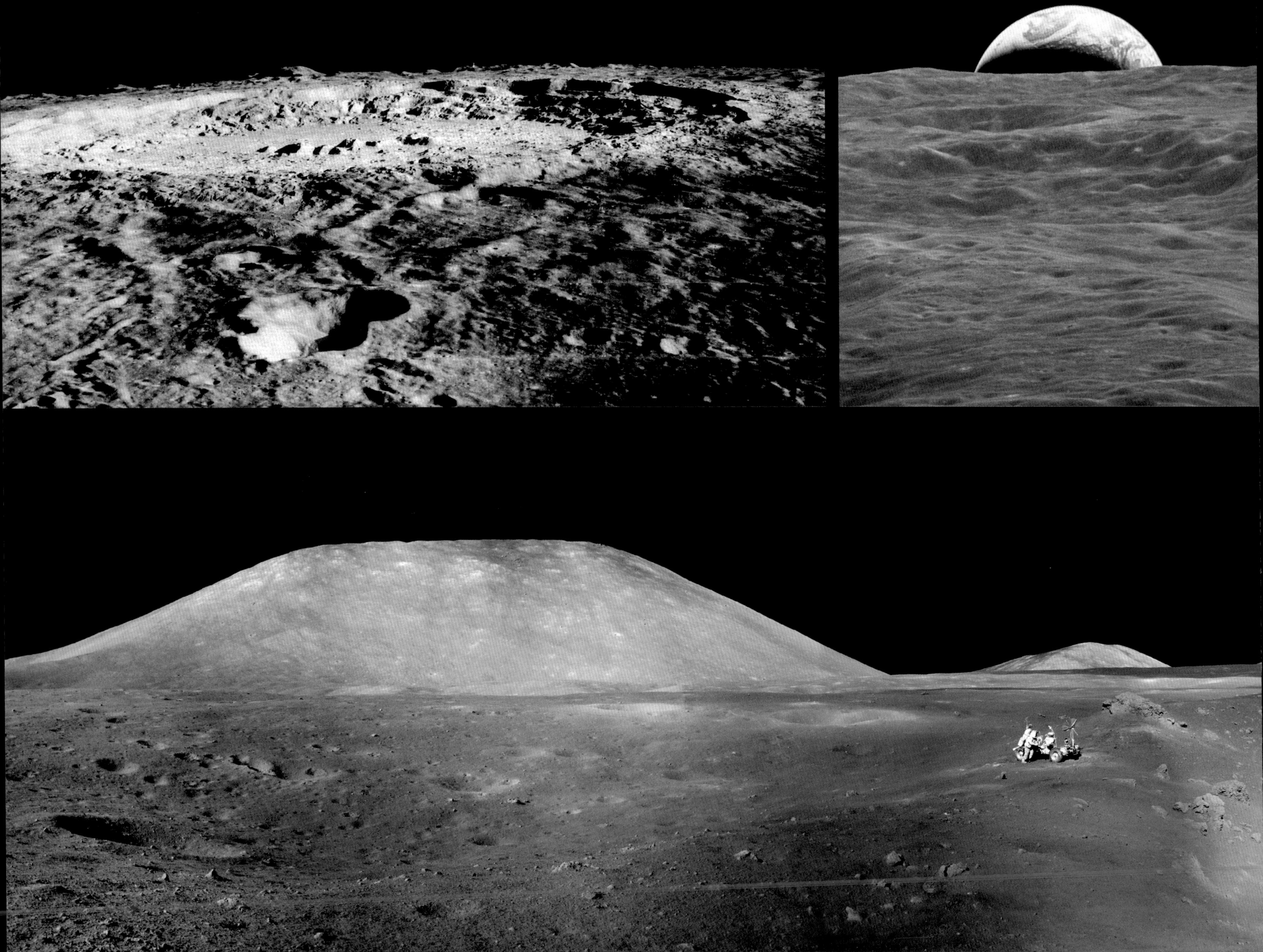

Far left: The bright nearside crater Copernicus is a typical young lunar crater, 90 km wide and about 4 km deep. With no protective atmosphere, the Moon has been bombarded with meteorites throughout its history, although the peak of bombardment tailed off about 3.9 billion years ago.

Left: From the Moon, the Earth passes through a similar set of phases every 28 days. But our planet looms much larger in lunar skies, and never changes its position – the Apollo astronauts only saw 'Earthrise' because they themselves were in orbit around our satellite.

Right: The lunar south pole conceals the largest known crater in the solar system. The South Pole-Aitken Basin is a huge dent in the crust, 2,500 km across and 13 km deep. However, it lies largely on the far side of the Moon, and is disguised by the many craters that have formed inside it.

Below: The Moon is the only other world where humans have explored. This stunning vista of the Taurus-Littrow Valley on the edge of the Sea of Serenity captures geologist Harrison Schmitt alongside the Lunar Roving Vehicle.

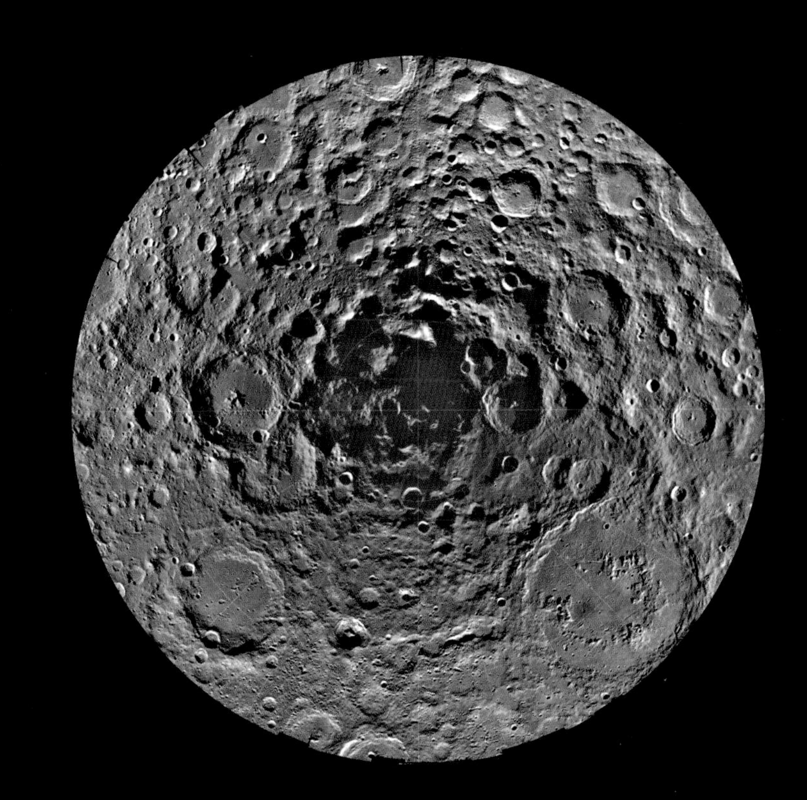

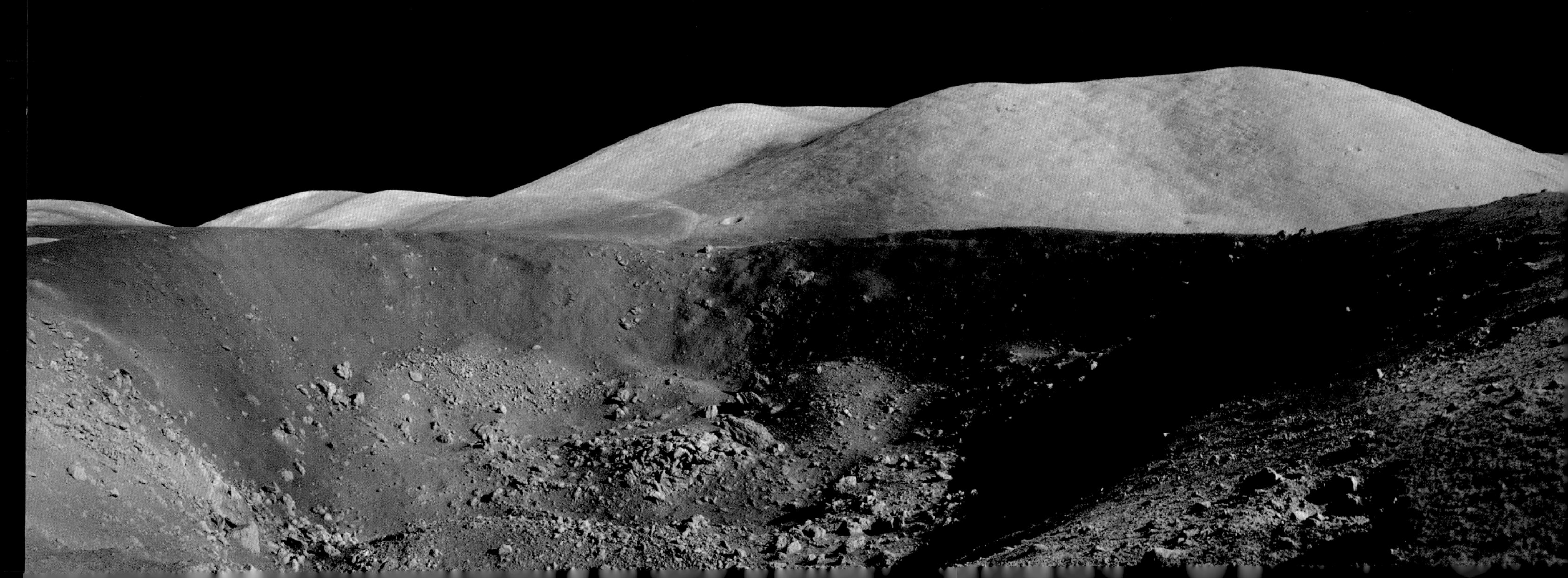

Right: Only one spaceprobe has so far visited Mercury, although others are planned or are already on their way. Mariner 10 flew past the planet three times in 1974 and 1975, but only photographed about half of the surface. Its images revealed a cratered world, superficially resembling the Moon, but with some important differences.

Opposite: Mercury's most impressive feature is the Caloris Basin, an impact crater 1,300 kilometres across. This mosaic of Mariner 10 images shows some of the mountain ranges and rays of ejected material that surround the crater's central plain.

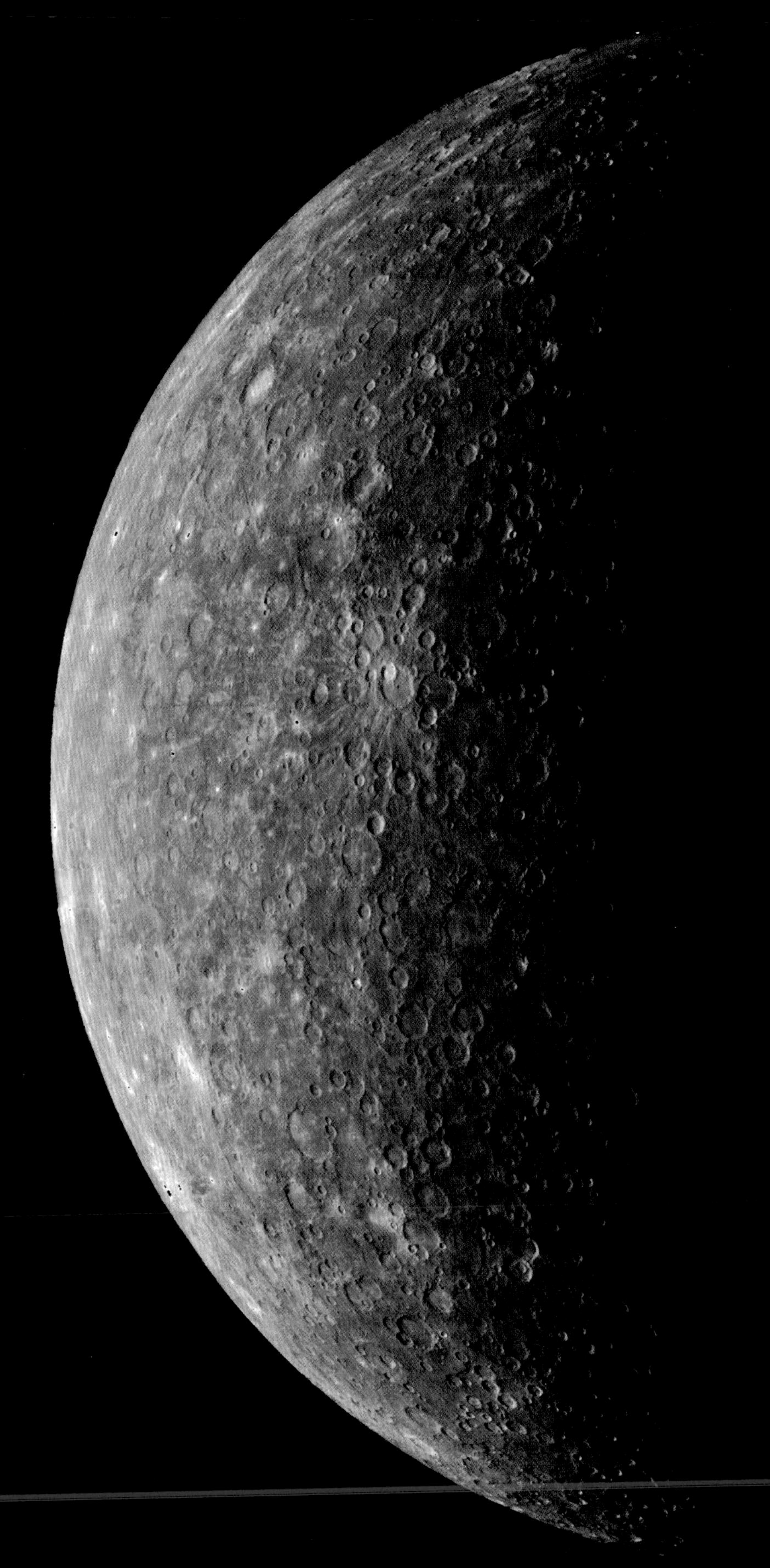

Observing Venus

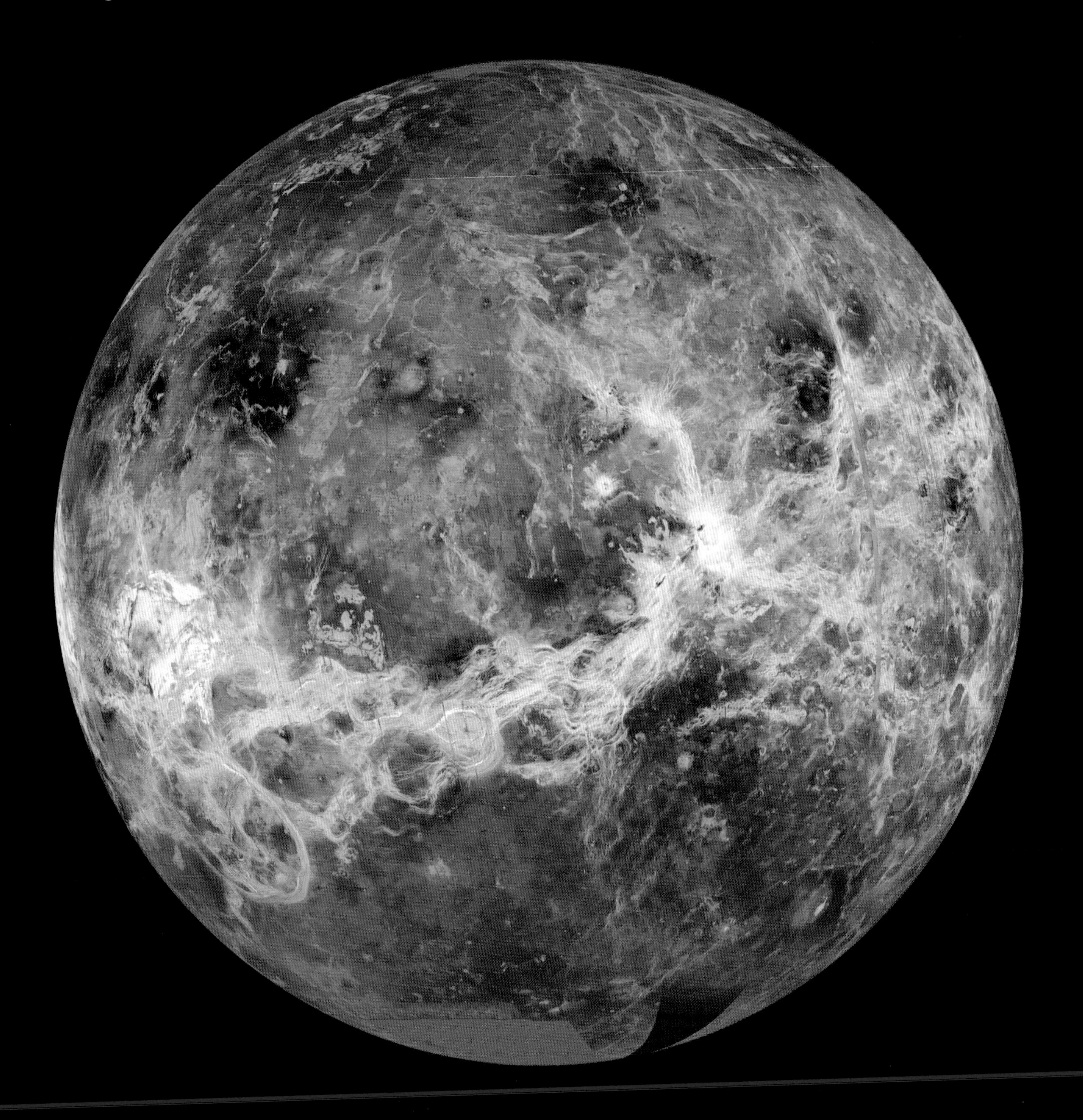

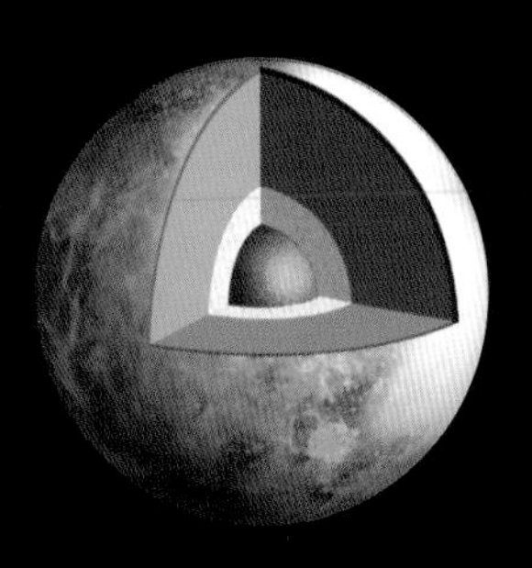

With a diameter 95 percent of Earth's, Venus is a near-twin of our own planet. Its internal structure is similar, with a rocky crust, outer and inner mantle, and a core that has probably solidified at least partially.

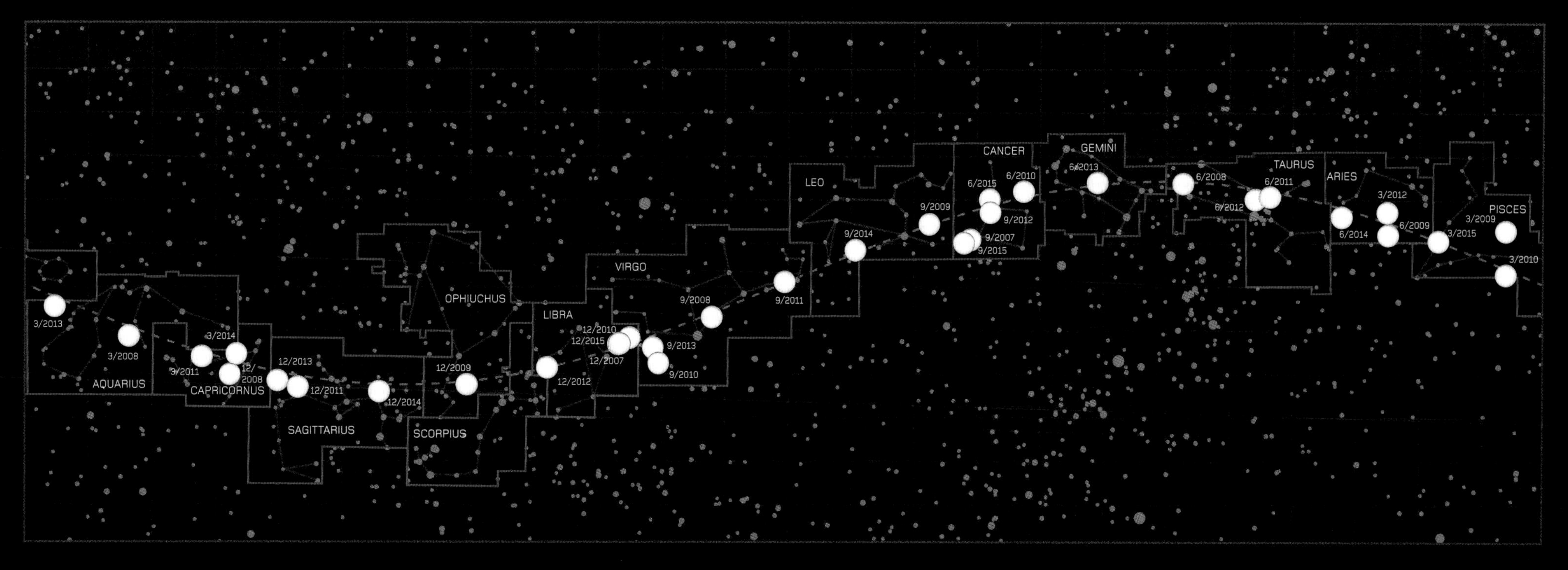

The closest planet to Earth, Venus is an unmistakable sight, the brightest object in the sky after the Sun and Moon, with a magnitude of between -3.8 and -4.6. It was known to the ancients as both Hesper and Lucifer – the Evening and Morning Stars, although they knew full well that both objects were in fact the same planet. Unfortunately for astronomers, Venus is shrouded in a dense, opaque atmosphere, which keeps many of the planet's secrets hidden from view.

Like Mercury, Venus is an 'inferior' planet orbiting closer to the Sun than Earth. However, since its orbit is much larger than Mercury's (circling the Sun every 225 days), Venus is not permanently lost in twilight. It is still 'locked' to the Sun and condemned to loop perpetually around it, but it can reach 'elongations' of up to 47 degrees, allowing it to be seen against a truly dark sky.

Through binoculars or a telescope, Venus is a dazzling sight. The smallest instrument should be enough to distinguish the planet's moonlike phases. Even if binoculars cannot resolve the true shape of the planet's sunlit side, most people can tell the small, almost full sphere of light visible around superior conjunction from the elongated crescent arc around inferior conjunction. The smallest telescope makes the phases unmistakable – they were first identified by Italian astronomer Galileo Galilei in the 16th century, and were a crucial piece of evidence that the Sun-centred theory of the solar system was correct.

When the planet is a narrow crescent, still relatively close to the Sun and comparatively dim, telescopic observers may want to look for the 'ashen light' – a curious glow that many have seen suffusing the night side of the planet. The light resembles the 'Earthshine' of reflected light seen on the night side of the crescent Moon, but since Venus has no companion to reflect light onto it, its origin is still a mystery.

Venus's dense atmosphere is a good reflector of sunlight, partly explaining its dominance in our skies, but it hides any surface features from view. Occasionally, amateur astronomers have seen wisps of darker cloud in the atmosphere – perhaps connected to the chevron-like cloud structures revealed in ultraviolet images from spaceprobes, but it has been left to robotic landers and orbiting missions equipped with radar mapping instruments to reveal the hellish world beneath the clouds.

At their inferior conjunctions, the tilted orbits of Venus and Mercury mean that the planets rarely fall into an exact straight line between Sun and Earth, but occasionally they do, and a planet crosses the face of the Sun. Such 'transits' of Venus and Mercury can be studied by amateurs using the method described on p. 27 to project the Sun's image. The next such events will happen in 2012 (for Venus) and 2016 (for Mercury).

Above: This Venus locator chart tracks the motions of our nearest neighbour in space against the background stars and constellations of the zodiac over the coming years.

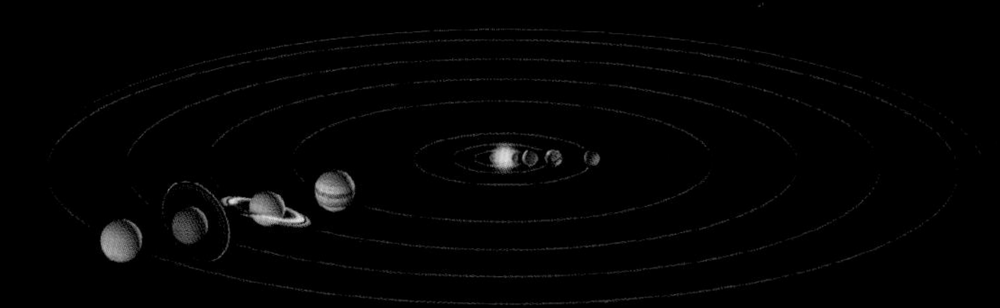

Above: Venus has a near-circular orbit 108 million km from the Sun. Strangely, the planet is effectively upside down, rotating in the opposite direction from normal, with a period of 243 Earth days – longer than its year.

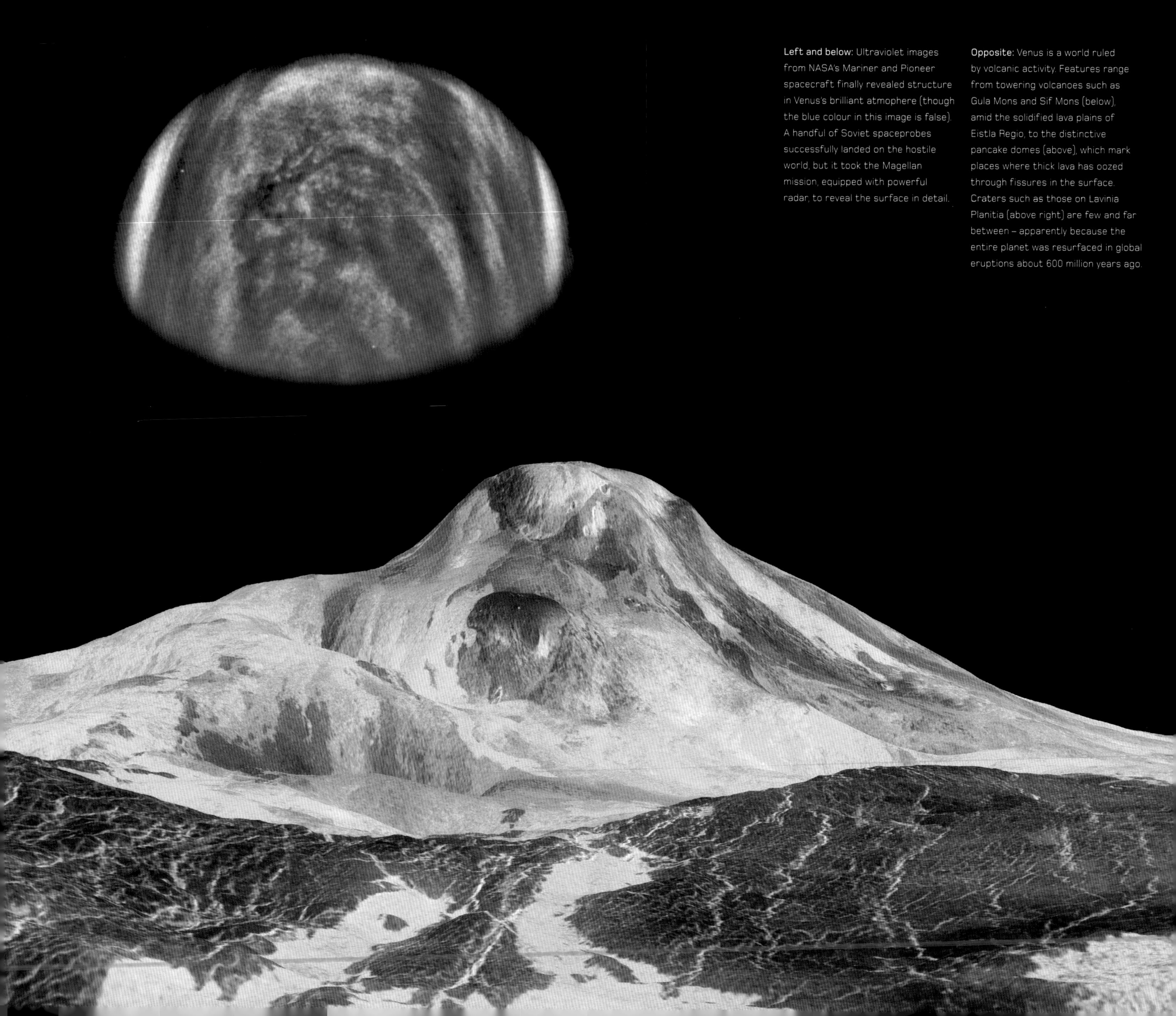

Left and below: Ultraviolet images from NASA's Mariner and Pioneer spacecraft finally revealed structure in Venus's brilliant atmophere (though the blue colour in this image is false). A handful of Soviet spaceprobes successfully landed on the hostile world, but it took the Magellan mission, equipped with powerful radar, to reveal the surface in detail.

Opposite: Venus is a world ruled by volcanic activity. Features range from towering volcanoes such as Gula Mons and Sif Mons (below), amid the solidified lava plains of Eistla Regio, to the distinctive pancake domes (above), which mark places where thick lava has oozed through fissures in the surface. Craters such as those on Lavinia Planitia (above right) are few and far between – apparently because the entire planet was resurfaced in global eruptions about 600 million years ago.

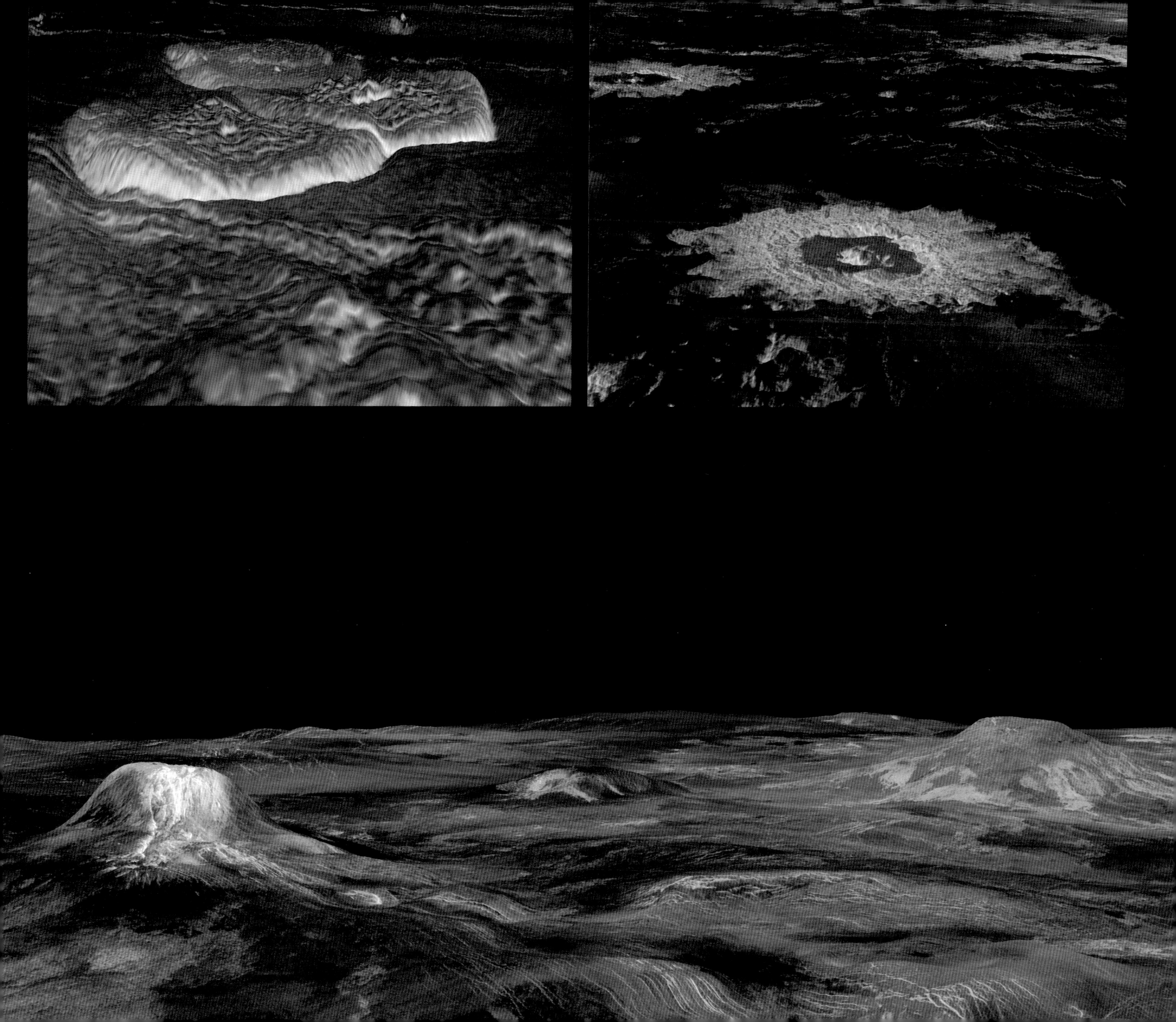

Observing Mars

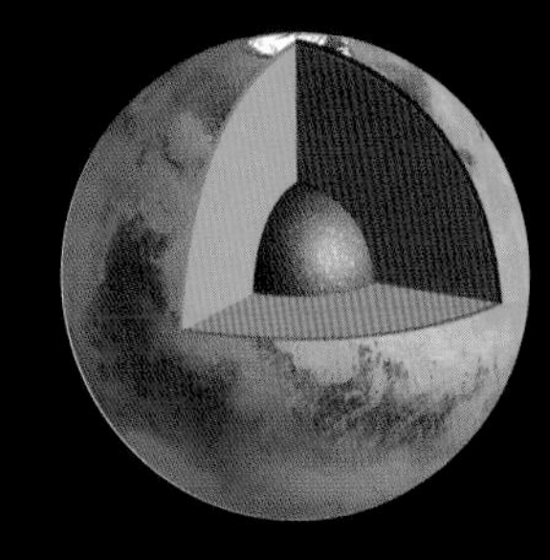

Because Mars is smaller than the Earth, it has cooled down more quickly from its formation, and so has a simpler structure, with a deep mantle surrounding a large, solidified core. The crust is thin below the northern plains, and thick beneath the cratered southern highlands.

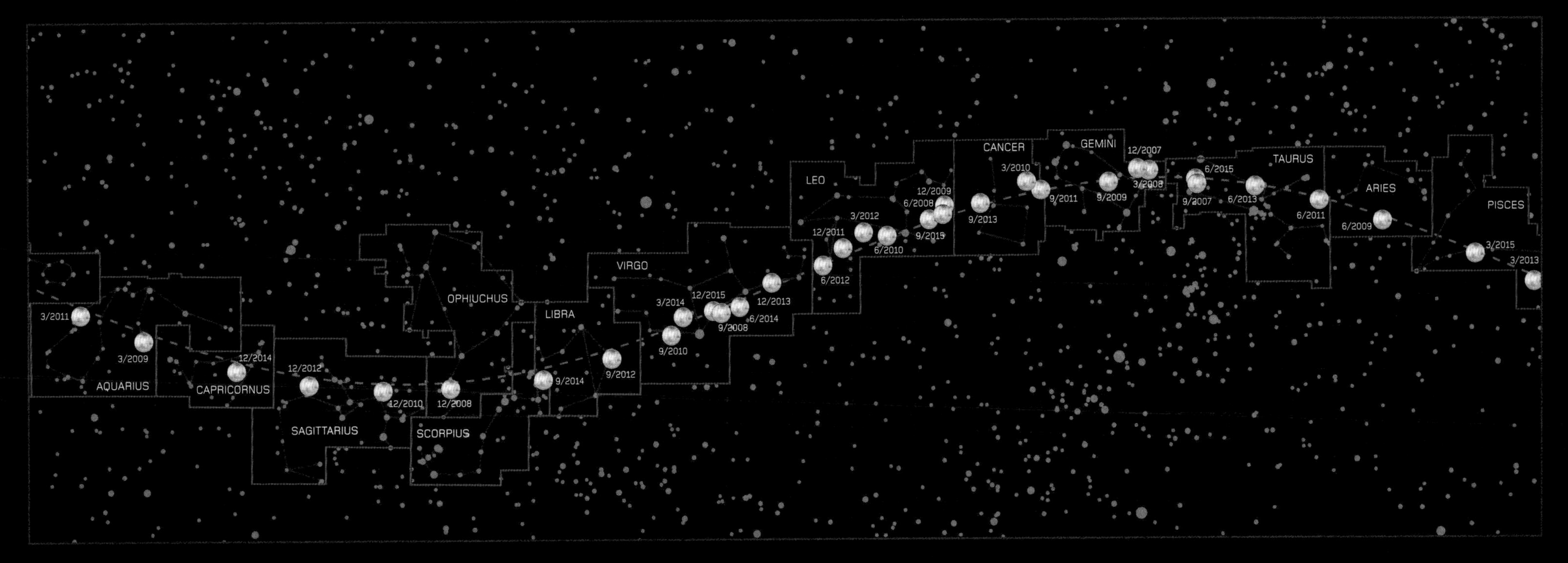

Above: This locator chart for Mars records the position of the Red Planet over the coming years as it moves through the constellations of the zodiac. Gaps in the sequence indicate periods when Mars is in conjunction and cannot be seen from Earth.

Mars is a fascinating target for amateur astronomers. Once seen as a possible home for intelligent aliens, then as a barren desert planet, the pendulum has swung again to reveal a planet with a rich past and a potential haven for life.

To the naked-eye observer, Mars's rich orange-red colour makes it unique among the planets – although budding stargazers should be careful not to confuse it with red stars such as Antares, Betelgeuse or Arcturus. The planet orbits the Sun every 1.88 Earth years, so it moves fairly rapidly around the sky. It is also the first 'superior' planet, with an orbit further from the Sun than Earth's, and this means that it circles the entire sky, moving between a position beyond the Sun at 'conjunction', to 'opposition', when Earth and Mars line up on the same side of the Sun and Mars lies exactly on the opposite side of the sky.

The combined motions of Earth and Mars mean that oppositions occur every 26 months, and as the faster-moving Earth 'overtakes' Mars, for a few months the planet appears to loop backwards in its path around the sky. This phenomenon, called 'retrograde motion', affects all the superior planets, but is at its most pronounced for Mars.

Binoculars will show Mars as a tiny red disc, and observant viewers will see how it changes its size between distant conjunction and nearby opposition. Small telescopes, meanwhile, will start to reveal some of the planet's surface features, the most obvious of which are the bright spots at its north and south poles. These are ice caps that grow and shrink with the changing seasons, and thanks to spaceprobes we know they are made from a combination of familiar water ice and frozen carbon dioxide or 'dry ice'.

Somewhat larger telescopes will show a variety of light and dark patches on the planet's surface, and the contrast between these can be exaggerated using coloured filters. Among the most prominent are the dark triangular plain of Syrtis Major, and the bright basin of Hellas Planitia – the remnant of an enormous impact in the planet's ancient history.

Mars's orbit is quite elliptical, and some oppositions bring it closer to Earth than others (most recently in 2003 and 2006). These close oppositions are an ideal chance for observing the planet in detail, but unfortunately they often coincide with the most violent Martian weather – huge seasonal dust storms that whip red sand high into the thin Martian atmosphere, from where it may take months to settle. A small consolation for frustrated observers is the interest to be gained from tracking the development of weather on another planet.

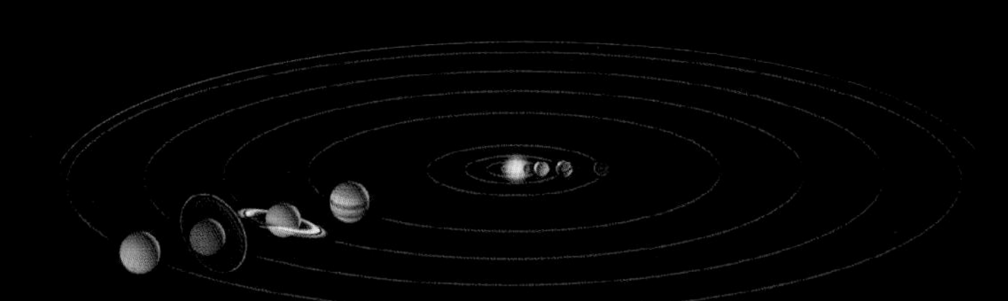

Above: Mars orbits beyond Earth, in a quite eccentric path that ranges between 206 and 250 million kilometres from the Sun. Beyond it lies a substantial gap before the giant planet Jupiter, where most of the asteroids lurk.

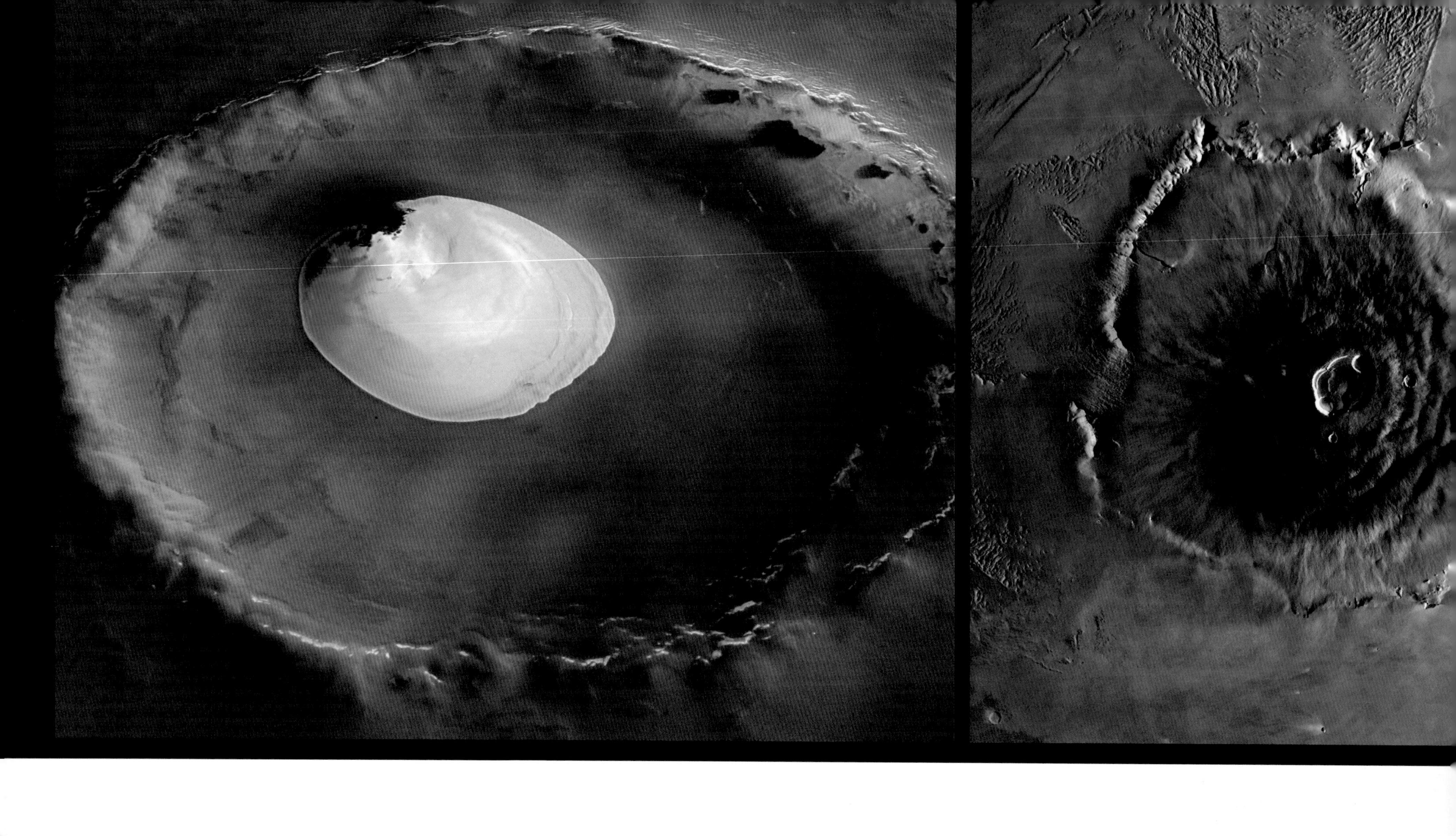

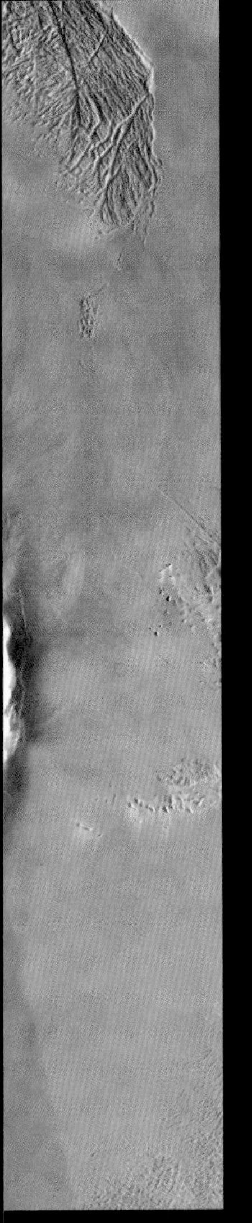

Opposite left: Early spaceprobe images depicted Mars as a lifeless dry desert, but more recent discoveries have cast the planet in a new light, with a wet, hospitable past and plentiful water still locked away as ice. This unusual ice-filled crater close to the north pole may be a relic of a time when there was much more ice on the surface.

Opposite right: Mars is famous for its enormous volcanoes, of which the grandest is Olympus Mons, 27 km high and more than 500 km across. In this picture, the central caldera itself is 56 km wide. However, there is no evidence that the giant Martian volcanoes are still active.

Below: Robots sent from Earth have explored the Martian surface since the 1970s. One of the most recent is the Mars Exploration Rover *Spirit*, which landed in 2004 and made this panoramic view while exploring Gusev Crater.

Observing Jupiter

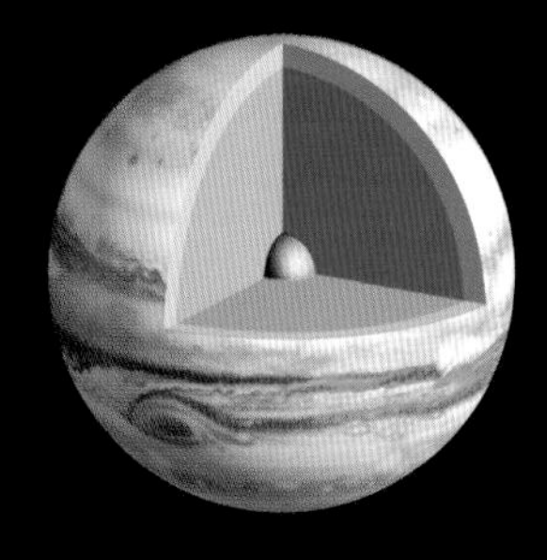

Jupiter's interior is dominated by hydrogen gas, though the upper layers are made colourful by impurities. Within a few thousand kilometres of the surface, the gas is compressed into liquid hydrogen, which eventually breaks down into an electricity-conducting metallic form deep inside the planet. Jupiter's solid core is very small, and some suspect it does not exist at all.

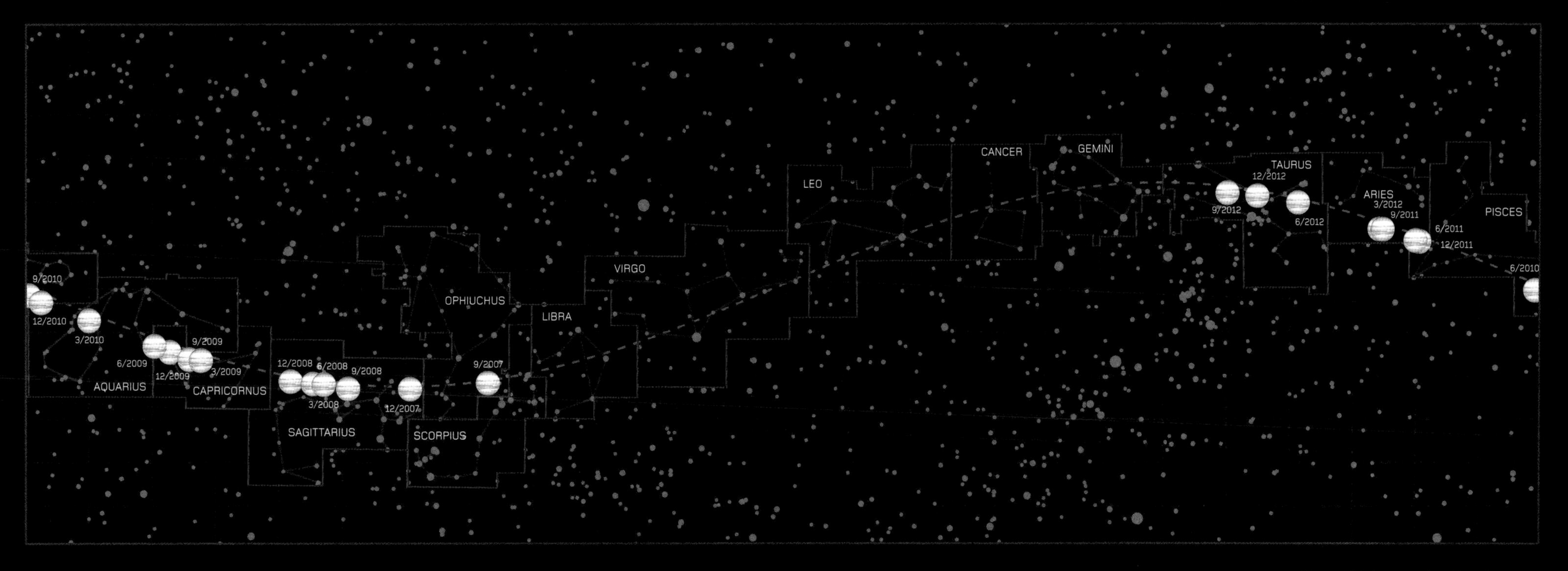

The giant of our solar system, Jupiter is so large that, despite its location beyond the asteroid belt, it is outshone only by the Sun, Moon and Venus. With its constantly changing cloud patterns and shifting system of satellites, it is perhaps the most rewarding planet for observers with any type of optical aid.

Even to the naked eye, Jupiter is impressive. Its disc can be up to 1/30 the diameter of a Full Moon, and almost perceptible without assistance. Its brightest moons, meanwhile, are on the limit of naked-eye visibility, although for most (but not all) observers they get lost in the glare from the planet itself.

Binoculars will reveal Jupiter's disc and the four bright moons that orbit around it (the largest in a system of at least 63). In order from the planet, these are Io, Europa, Ganymede and Callisto. Innermost Io orbits the planet in 42.5 hours, while outer Callisto takes 17 days to make its journey around Jupiter. As a result, the moons are constantly shifting position, and it is best to refer to a monthly magazine, software or an online chart to get your bearings. The moons frequently vanish as they pass across the face of Jupiter or are 'occulted' behind it, and during the occultation events, it is interesting to note how the moons usually disappear when they encounter or leave Jupiter's shadow, before or after they pass behind the planet itself.

Small telescopes will reveal more – the satellites cast their own shadows back onto Jupiter as they pass across its face, and they can be resolved into tiny discs in their own right. And Jupiter itself is transformed by a small telescope. The planet's disc becomes an obvious oval, crossed by light and dark-coloured bands. Jupiter's shape is a result of its gassy composition and rapid rotation (despite being the largest planet in the solar system, it spins in just 10 hours, and so bulges noticeably at the equator). Its striped appearance, meanwhile, is a sign of the weather systems that wrack its surface. Light bands are known as 'zones', and dark ones as 'belts'. They are formed by different types of cloud at different levels in Jupiter's deep atmosphere, corresponding to areas of high and low pressure just like those on Earth. The major difference is that the forces created by the planet's rapid rotation 'stretch' the belts and zones until they wrap all the way around Jupiter.

Larger telescopes reveal still more detail – bands of different intensities and colours, delicate swirls along the boundaries between belts and zones, and huge oval storms floating in the midst of it all. The largest of these storms is the famous Great Red Spot – observed since at least 1665, and usually observable through even small instruments (though its size and intensity vary unpredictably).

Above: This locator chart tracks the movements of the giant planet Jupiter over the coming years, showing its position against the background constellations of the zodiac. Gaps in the sequence indicate times when Jupiter is behind the Sun, and out of sight from Earth.

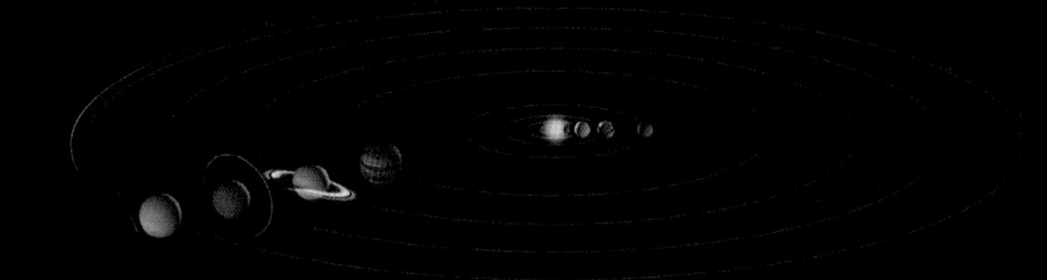

Above: Jupiter lies well beyond the asteroid belt, in an orbit that takes it between 741 and 817 million kilometres from the Sun. It has more mass than all the other planets combined, and an equatorial diameter of 143,000 km.

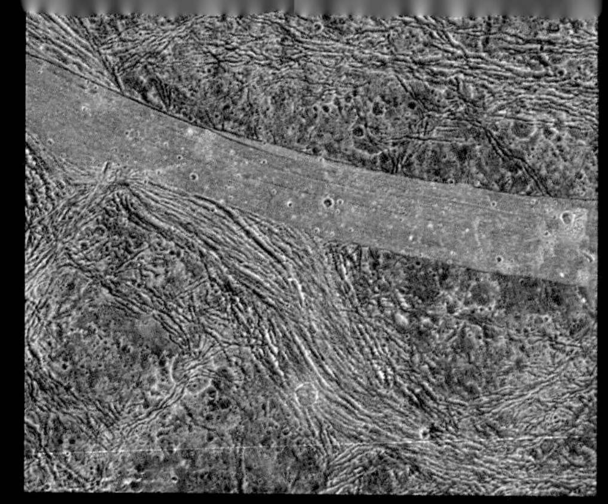
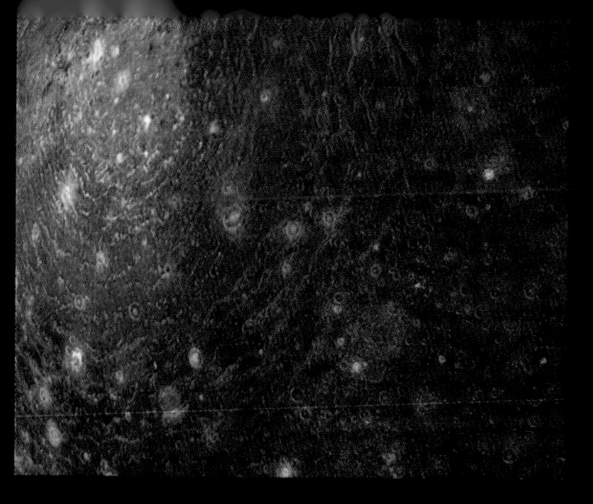

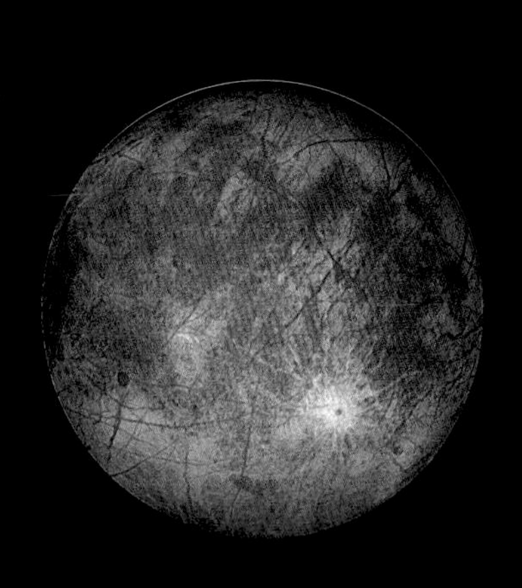

Above: Jupiter's major satellites are a colourful and varied quartet, shown here in their correct scale and order from the planet. Io is a world of active volcanoes, Europa a ball of ice with a concealed ocean, Ganymede the largest satellite in the solar system, with a complex past, and Callisto an unchanging iceball, target for countless impacts.

Opposite: The Saturn-bound Cassini spaceprobe captured this spectacular image of Io against Jupiter's cloud bands as it passed the giant planet in 2000.

Observing Saturn

Saturn's outer haze conceals a chemical make-up and internal structure similar to Jupiter's with hydrogen and helium rapidly turning from gas to liquid, and an inner ocean of liquid metallic hydrogen around a solid core of rock and ice.

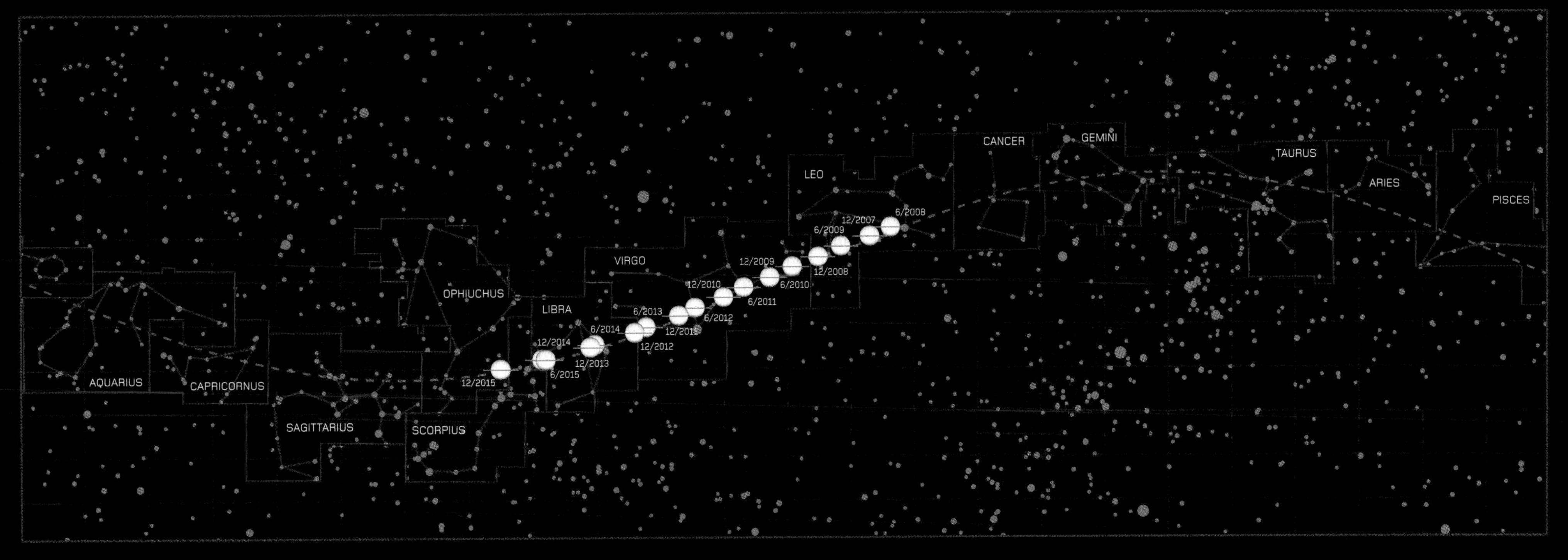

Above: This Saturn locator chart tracks the movement of the ringed planet against the constellations of the zodiac in the coming years.

Saturn marks the edge of the naked-eye solar system, and is also the most distant world that most amateurs can observe in detail. A gas giant and close cousin of Jupiter, its cloud bands seem muted and disappointing compared to its colourful neighbour, largely due to an overlying haze of white ammonia crystals.

However, Saturn's system of rings more than makes up for this. Today, we know that all the giant planets have rings, but Saturn's are in a class of their own, and the only ones easily visible from Earth.

Good binoculars may hint at something 'wrong' with the shape of the planet, and a small telescope will usually show the rings as handle-like extensions to either side of the planet, with space visible through the gaps on either side of Saturn's disc. Slightly larger telescopes will start to reveal structure in the rings – the bright A and B rings are separated by a dark circle called the Cassini Division, while the narrower Encke Division splits the A ring itself in two. On the inner edge of the B ring lies the semi-transparent C or 'Crepe' ring, and the even more elusive D ring stretches down from there almost to the planet itself. The ring system is easier to see at some times than at others – because Saturn, like Earth, is tilted on its axis, we see the rings from different angles at different points in Saturn's 30-year orbit.

Beyond the rings circle the largest members of Saturn's huge satellite family. Easiest to see is giant Titan, which appears as an 8th magnitude star and can be easily tracked. Rhea and Iapetus are the next brightest satellites, although Iapetus's strange surface (see over) causes it to change brightness from one side of its orbit to the other.

Above: Saturn orbits the Sun between 1.51 and 1.35 billion kilometres out, and is the most remote planet that can be seen with the unaided eye.

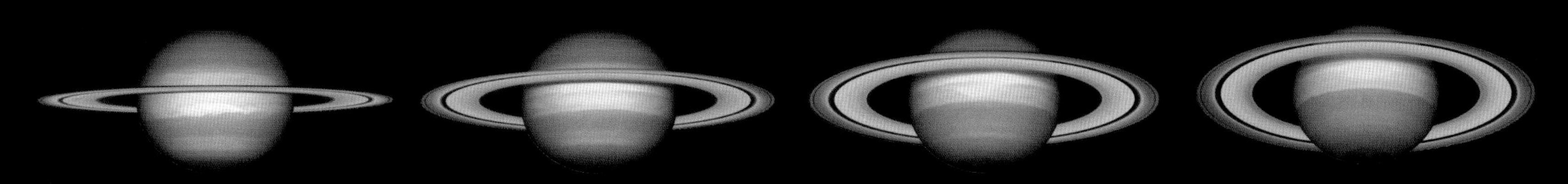

Left: A sequence of photographs from the Hubble Space Telescope tracks the changes to Saturn's appearance as its rings 'opened' towards Earth between the years 1996 and 2000.

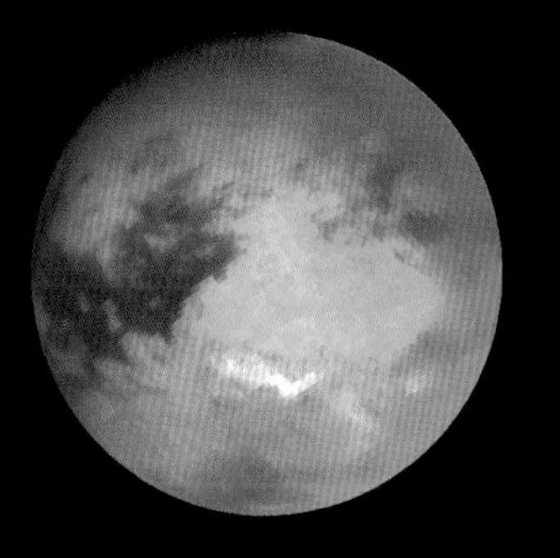

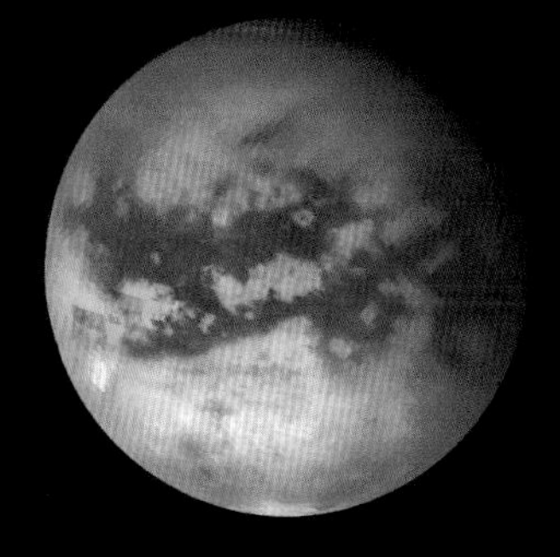

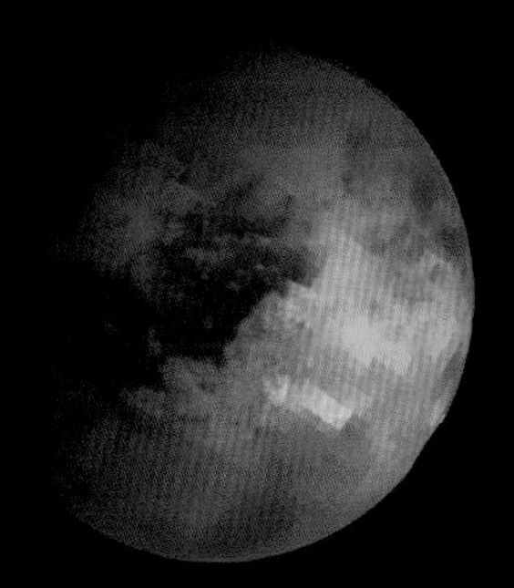

Opposite: The Cassini probe that entered orbit around Saturn in 2004 has sent back many spectacular image of the interplay between Saturn, its rings and its satellites, such as this beautiful portrait of the small moon Mimas against the ring-striped, twilight bulk of Saturn.

Left: Cassini went equipped with cameras capable of looking through Titan's atmosphere to the surface below. They revealed a startlingly Earth-like world of erosion, lakes and hills, where liquid methane plays the same role that water does on Earth.

Below: Saturn's three largest satellites are a mixed bunch – Rhea (left) is a heavily cratered iceball, Titan (centre) the only moon in the solar system with a substantial atmosphere, and Iapetus (right) has a bizarre surface with one bright hemisphere and one dark.

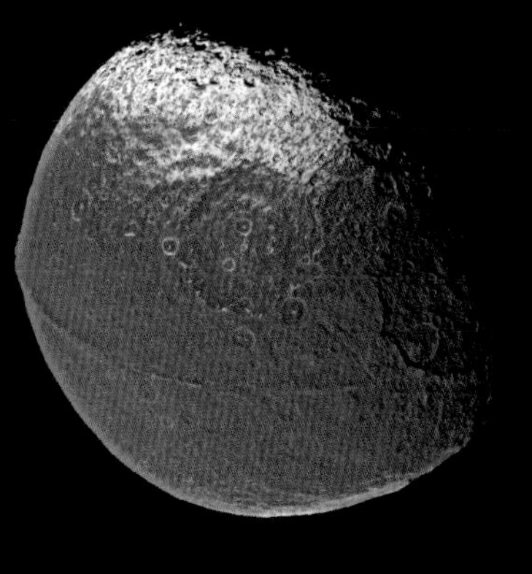

Observing Uranus

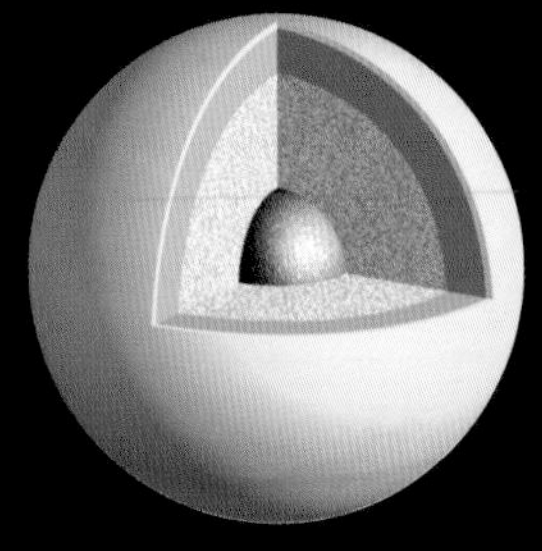

While Uranus's outer layers of gas and liquid hydrogen and methane resemble those of the large inner giants, its mantle is made up of slushy chemical ices, swirling around a large core of rock and ice.

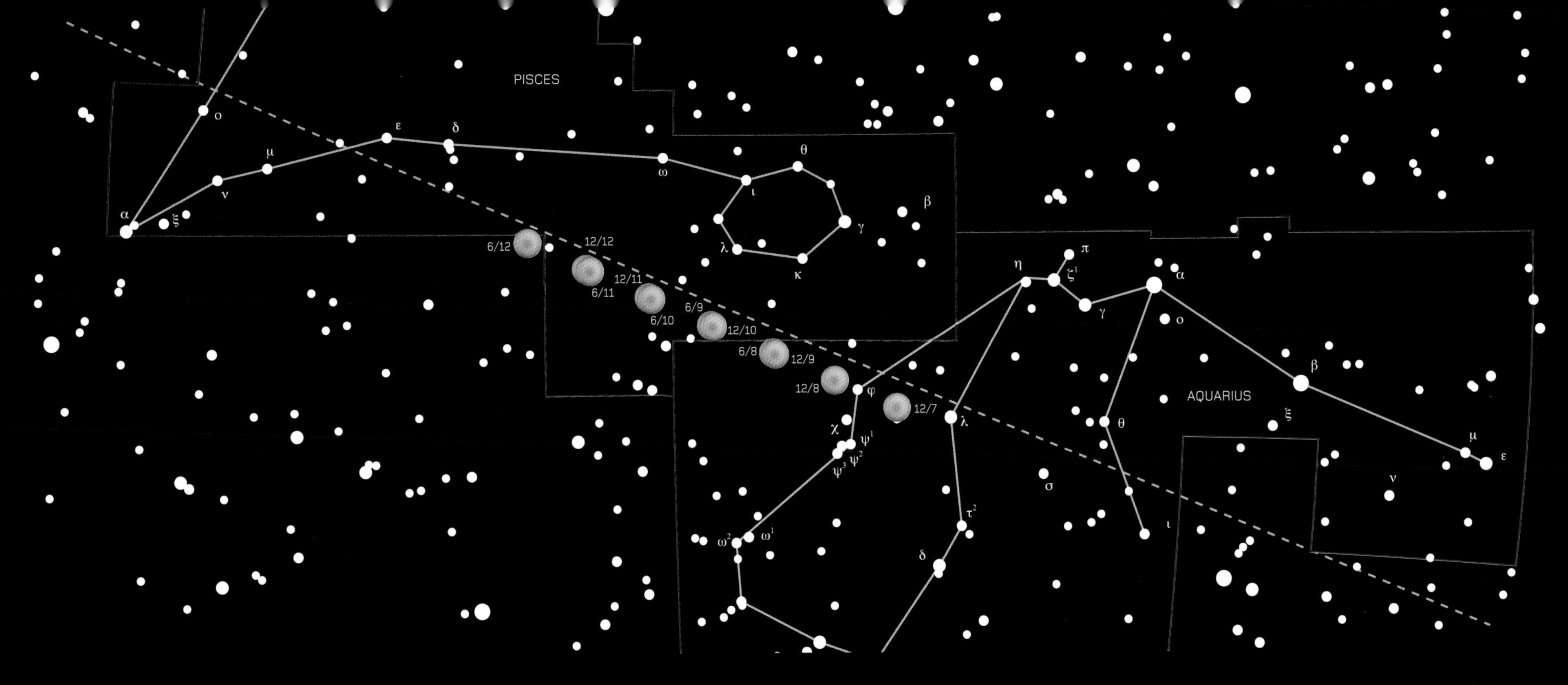

Above: This locator chart for Uranus tracks its motion through the zodiac constellations Aquarius and Pisces between now and 2012.

Left: This Hubble Space Telescope image of Uranus offers the clearest view ever obtained of the strangely tilted planet, its rings and its major satellites.

The first of the 'modern' planets, Uranus was discovered accidentally by William Herschel in 1781. When astronomers looked back through the records, they found that it had been sighted but ignored before – a testament to how insignificant it can appear even through a telescope.

In fact, Uranus lies on the edge of naked-eye visibility at magnitude 5.8, so it can be seen without optical aid by those with good eyesight who know where to look. Binoculars easily show it as an unremarkable pale green 'star', which only reveals itself as a planet by its slow drift against the background sky, seen in carefully recorded observations several nights apart.

Uranus is an 'ice giant', with a thick cloak of gas surrounding a slushy interior of various partially frozen chemicals. Thanks to the Voyager 2 spaceprobe, we know that its surface is featureless through much of its long, 88-year orbit – but even when it does become active, it reveals its surface bands and storms only to the most powerful Earth-based telescopes. However, a decent-sized telescope may show the planet's brightest moons – Titania and Oberon, both around magnitude 14 – and reveal the planet's strangest secret. While most satellites loop back and forth parallel to the ecliptic as their parent planet moves along its path through space, the moons of Uranus loop 'up and down' – almost at right angles to the planet's motion through space. Like most moons, they are orbiting over the planet's equator – their strange appearance is a result of the entire Uranian system being 'tipped over' at 98° from the vertical: Uranus effectively rolls around its orbit on one side. This bizarre arrangement, most likely the result of a cosmic collision early in the planet's history, was confirmed in 1977 when astronomers discovered a system of rings around the planet, tilted at this same extreme angle.

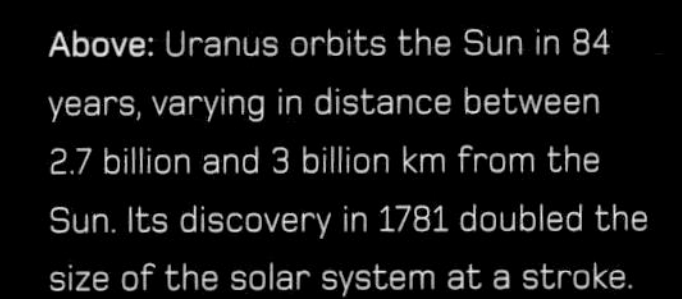

Above: Uranus orbits the Sun in 84 years, varying in distance between 2.7 billion and 3 billion km from the Sun. Its discovery in 1781 doubled the size of the solar system at a stroke.

Observing Neptune

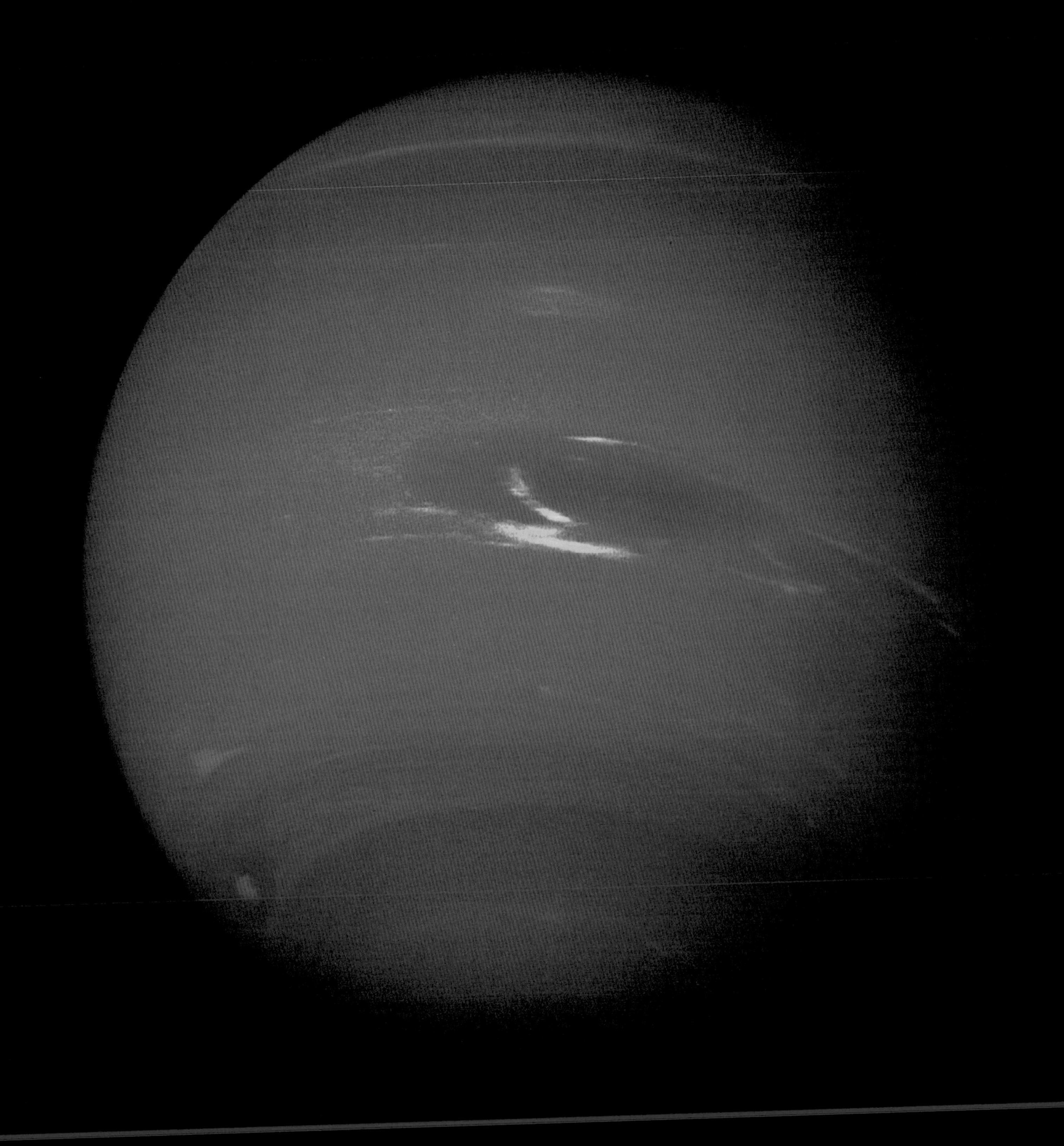

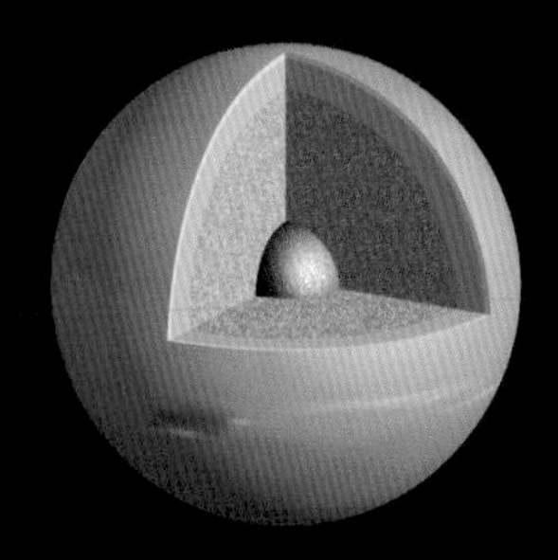

Neptune's interior is similar to that of Uranus, with an outer atmosphere of liquid hydrogen and helium giving way to a mantle of slushy ices around a rock/ice core. The planet is bluer than Uranus because the upper atmosphere contains more methane, which absorbs red light.

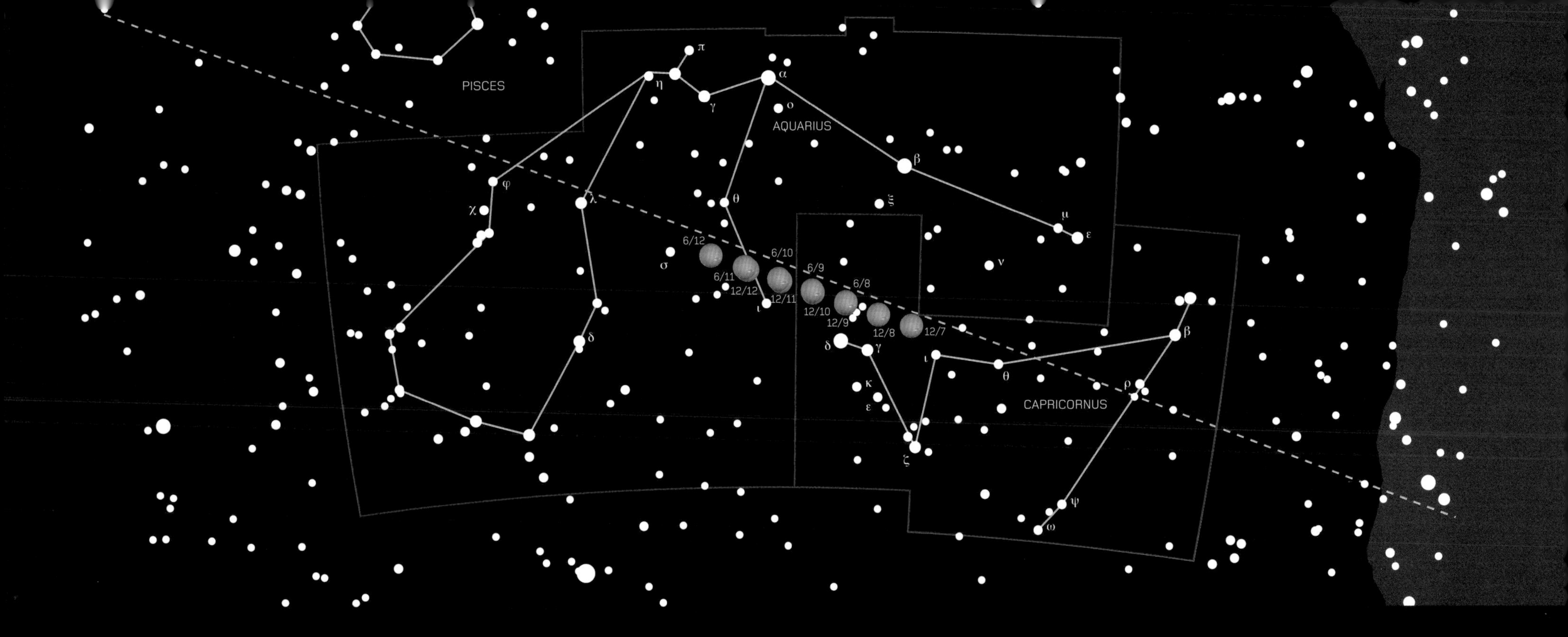

Neptune is the most challenging planet to observe – at magnitude 7.9 it is well beyond naked-eye visibility and requires either binoculars or a small telescope. Discovered in 1846, it was the first planet to be found by a deliberate search, after French mathematician Urbain Le Verrier predicted its existence from the way its gravity disturbed the orbit of Uranus.

Neptune's 165-year circuit of the Sun means that it moves very slowly against the background stars, and is easily overlooked by those who are not looking for it – Italian astronomer Galileo Galilei actually recorded it when he was observing Jupiter with a basic telescope as early as 1612. It is presently moving between the constellations of Capricornus and Aquarius, but a detailed finder chart showing the many faint stars in this area of the sky is a necessity before searching for it.

Although the Voyager flyby of 1989 revealed that Neptune is a fascinating world of raging storms and some of the most powerful winds in the solar system, the planet is a featureless blue disc to Earth-based telescopes. Amateurs with fairly large telescopes, however, may try to locate Neptune's largest satellite, Triton, which looks like a 14th magnitude star, but orbits the planet in a little under six days. Curiously, Triton orbits Neptune in the 'wrong direction' compared to the planet's rotation (although the direction of its orbit is impossible to detect without professional equipment). This suggests that the moon is probably a captured 'dwarf planet' – a member of the Kuiper Belt similar to Pluto. Triton's arrival in the Neptune system may have sent other satellites spinning out of it, explaining the lack of other large moons.

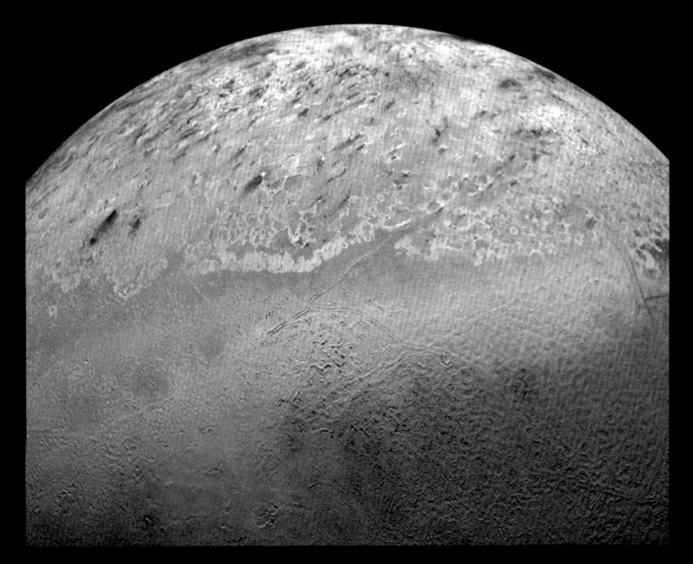

Above: This locator chart for Neptune tracks its motion through the zodiac constellations Capricornus and Aquarius between now and 2012.

Left: The Voyager 2 spaceprobe revealed that Neptune's single major satellite, Triton, was a facinating and active world. The streaks in the upper half of this image are plumes of dust released in icy geysers of liquid nitrogen.

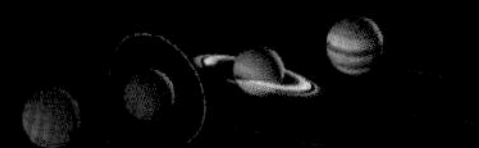

Above: The outermost planet, Neptune orbits between 4.45 and 4.54 billion kilometres from the Sun – about 30 times the Earth-Sun distance. Beyond it lies the broad Kuiper Belt of icy dwarf planets.

B o o t e s
Arctur
Krone
Gemma
Berg Mänalus
J u n g f r a u
Vindemiatrix
S c h l a n g e
Quadrant
Jagdhunde
210
225
XIV
XV

3 The constellations

An apparently unchanging backdrop to the movements of the planets, the stars, galaxies and nebulae of the more distant Universe are just as fascinating as our nearer neighbours in space. These celestial splendours are divided among 88 constellations that vary from the majestic to the insignificant. The pages that follow chart each area of the sky and highlight its most intriguing objects.

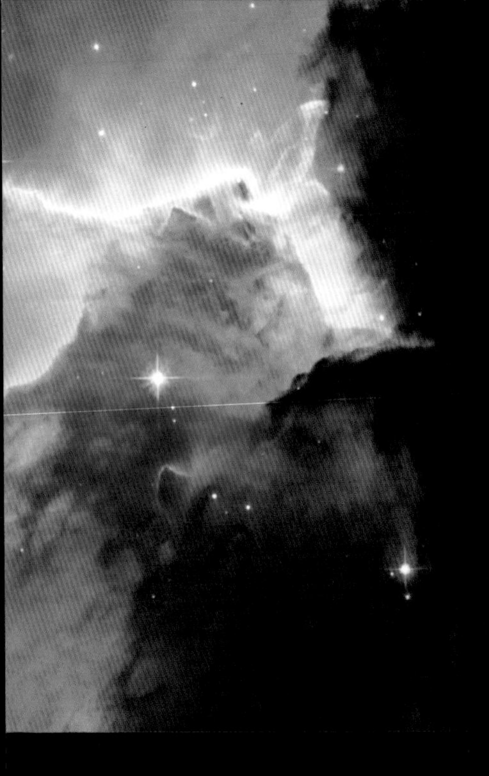

Introducing the stars

Aside from a handful of bright, nearby planets and other members of the solar system, most of the points of light in the night sky are stars and objects associated with them – luminous balls of exploding gas like the Sun, so unimaginably remote that we measure their distance in terms of how long their light takes to reach us. A 'light year', the most widely used measure of distances beyond the solar system, is equivalent to 9.5 million million (9.5 trillion) kilometres, and even the closest star is more than four light years away.

Cataloguing the sky

Roughly 6,000 stars are bright enough to be visible to the naked eye, so about 3,000 are typically visible at any one time, and numbers increase exponentially when any sort of optical aid is used. In order to make sense of such huge numbers of stars, astronomers long ago divided the sky into constellations – star patterns made by linking together the brighter stars. Today, there are 88 recognized constellations, of which 48 derive from a book called the *Almagest*, written by the Egyptian-Greek astronomer Ptolemy of Alexandria in the 2nd century AD. European exploration of the southern hemisphere saw the addition of many new constellations in the far southern sky, while the invention of the telescope (and the increasing importance of comparatively faint naked-eye stars) encouraged some astronomers to fill in the 'gaps' in the northern hemisphere.

In 1603, German astronomer Johannes Bayer introduced a new system for cataloguing stars that avoided the need to memorize their proper names (which are often daunting words derived from Arabic). The theory was simply to name the brightest star in a constellation Alpha, the next brightest Beta, then Gamma, Delta, Epsilon, and so on through the Greek alphabet. This system should allow for the clear identification of the 24 brightest stars in any constellation before the Greek letters run out, but in reality Bayer's catalogues, and the work of those who followed him, were full of contradictions and mis-orderings of the stars. However, once the designations were in use, they stuck, and no one dared to suggest a complete re-ordering with the confusion that would doubtless ensue.

Since 1712, Bayer's designations have been supplemented by so-called Flamsteed numbers, named after their inventor, British Astronomer Royal Sir John Flamsteed. These numbers list all the reasonably bright stars in a constellation in order of increasing right ascension, and are often used where no Bayer letter exists. Additional confusion comes from the fact that the constellations were originally only lines drawn between specific stars, and the boundaries between them were not finally settled until 1930.

While the Bayer and Flamsteed numbers and the constellation names encompass most of the significant naked-eye stars, there are many other objects in the sky that are not stars – these include nebulae, star clusters and distant galaxies. Two major catalogues list the brightest of these so-called 'deep sky' objects – the Messier catalogue (first published by French astronomer Charles Messier in 1774), which lists the brightest objects with their 'M' numbers, and the New General Catalogue (NGC), compiled by J.L.E. Dreyer in the 1880s.

Stellar evolution

Stars are not perpetual – they are born from collapsing clouds of gas in star-forming nebulae, shine for most of their lives through nuclear fusion of hydrogen, helium and perhaps other elements, and eventually dissipate, scattering the enriched material they have processed back across space to form the next stellar generation. However, they live and die on geological timescales of millions to billions of years, so it is very rare that we can actually observe stars in the act of a major change. Instead, astronomers have pieced together the story of their life cycles, and the relationships between different types of stars, from a series of stellar 'snapshots' scattered across the sky.

In the early 1900s, Swede Ejnar Hertzsprung and American Henry Norris Russell independently had the idea of comparing the properties of stars in the sky on a graph. In particular, they looked for relationships between a star's true luminosity and its colour, and the statistical distribution of

different classes of star. Taking into account the natural bias towards brighter stars and closer ones, they discovered something remarkable – nearly all stars follow a trend that goes from faint, cool red stars to bright, hot blue ones. The number of stars to which this rule applied was so overwhelming that it became clear that nearly all stars spend almost all of their lives obeying it. The major exceptions were a large group of bright red, orange and yellow stars (giants) and immensely brilliant stars of many colours called supergiants.

This trend soon became known as the 'main sequence' of stellar evolution, and when it became clear that it also related to a star's mass (with lower mass stars being the cool red ones and higher mass stars being the hot blue ones), it revealed the key to stellar evolution. Any individual star spends most of its life in more or less one place on the main sequence, predetermined by its mass. Our Sun for instance (a fairly average star with low mass and a middling surface temperature of about 5,500°C) sits neatly in between the hot and cool extremes.

It is only when the star exhausts its main fuel supply of hydrogen that it begins to change, brightening to become a giant, but at the same time normally swelling so much that its surface cools and it turns red or orange. The star may stabilize itself again for a time, but eventually it will drain its core completely of potential fuel, and destroy itself. For most stars, this is a process of casting off the outer shells in a so-called 'planetary' nebula and revealing a hot, dense core known as a white dwarf. For a few massive stars, the end is far more violent – a sudden collapse and rebound that tears the star apart in a supernova blast of nuclear reactions that can sometimes outshine an entire galaxy, and can leave a strange superdense object – a neutron star or a black hole – where the star once shone.

Beyond the Milky Way

All the individual stars we can see with the unaided eye are members of our own galaxy, but there are as many galaxies beyond our own as there are stars in the Milky Way. Galaxies vary in size from dwarfs perhaps ten thousand light years across, to giants a couple of hundred thousand light years wide. They have shapes ranging from complex elegant spirals (of which the Milky Way is an example), through elongated balls of stars, to chaotic clouds of stars and gas.

The closest galaxies, including our own, are members of a small cluster called the Local Group. They vary from small dwarfs tens of thousands of light years away (some of which are colliding with the Milky Way), to the giant spiral Andromeda Galaxy almost 3 million light years away. Galaxies cluster together on every possible scale, and our Local Group is bound to the much more distant Virgo Cluster, which contains several hundred large galaxies and is centred about 60 million light years away.

Only the very closest galaxies are within range of binoculars, though a small telescope will reveal many more. However, larger instruments and long-exposure photography are needed to reveal the detailed structure of these vast stellar empires – to the observer at the eyepiece, they generally appear as fuzzy, nebulous objects, and it was less than 100 years ago that astronomers proved beyond doubt that the so-called 'spiral nebulae' were indeed galaxies beyond our own.

Ursa Minor

The Lesser Bear

Resembling a smaller and fainter version of the famous Plough or Big Dipper, Ursa Minor is the most northerly of the constellations, permanently on show throughout the northern hemisphere, but forever hidden from residents of the south.

This distinctive pattern has been interpreted in several ways throughout history. Occasionally it has been seen as a wing extending from Draco, the Dragon. It was also seen as a dog, accompanying Boötes on his journey around the pole. The most popular interpretation, however, sees Ursa Minor as Arcas, son of the beautiful Callisto who was turned into the greater bear by the jealous goddess Juno. According to this myth, Callisto wandered alone in the woods for many years, until one day Arcas went hunting there. Overjoyed to see her son, Callisto ran towards him, only for Arcas to recoil in horror and draw his bow to kill her. At the fatal moment, Jupiter intervened, casting them both into the sky as the celestial bears. (Another version of the story sees Arcas transformed into Boötes.)

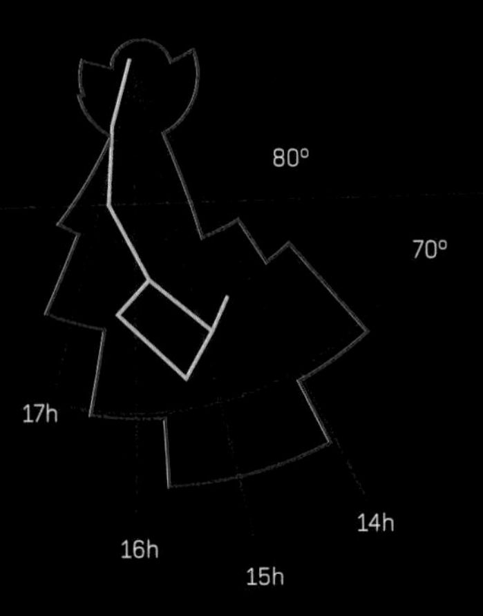

Polaris

Alpha (α) Ursae Minoris

2h31m, +89°15′	Magnitude: 2.0
Multiple star	Distance: 2,400 light yrs

Polaris is the one (more or less) fixed point in the skies of the northern hemisphere – the star about which all others appear to revolve thanks to Earth's daily rotation. By a lucky chance, this moderately bright star happens to lie within a degree of the north celestial pole. In the course of each day it describes a small circle around the pole itself, but its movement is so slow as to be imperceptible. Polaris has borne this name for many centuries, but thanks to the slow and steady drift of precession, its time as pole star is (in astronomical terms) comparatively brief. Ancient Greek astronomers said that the pole itself lay in the empty space within a quadrangle of stars (with Polaris or 'Cynosura' at one corner), but by early medieval times, the pole had drifted close enough for Polaris to earn its name, and become a vital aid to navigation (it was often known as the 'Navigator's Star').

Aside from its location, Polaris is interesting in its own right – not least because it has been well observed for centuries. There is good evidence that the star has brightened by a whole magnitude since classical times, contradicting many of our ideas about stellar evolution. The increase in luminosity is probably linked to the fact that Polaris was, until recently, a variable star, similar to Delta Cephei. Around 1900, the star's brightness varied by around 0.15 magnitude with a period of just under four days, but the variations gradually grew smaller until by 1994 they had stopped completely.

Polaris is also a multiple star system. The main star is a bright yellow supergiant, almost 450 times as luminous as the Sun. It has a yellow sunlike companion of magnitude 8.2, easily seen through a good amateur telescope, and it is also orbited by a much smaller and fainter dwarf star.

Epsilon (ε) Ursae Minoris

16h45m, +82°02′	Magnitude: 4.2
Binary star	Distance: 340 light yrs

This star in the little bear's 'tail' normally shines at a steady magnitude 4.23 most of the time, but dips in brightness by less than 0.1 magnitude every 39 days. This effect is caused because the visible star, roughly 200 times as luminous as the Sun, has a faint dwarf companion in orbit around it on a plane that happens to more-or-less line up with Earth. Every 39 days, the dwarf partially eclipses the bright primary star, blocking out a little of its light.

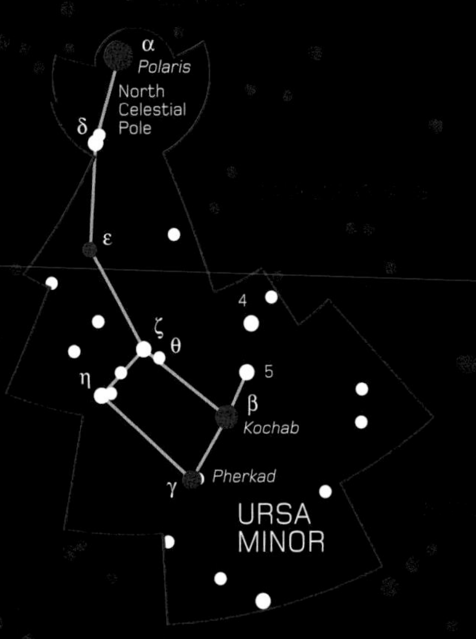

Pherkad

Gamma (γ) Ursae Minoris

15h20m, +71°50′	Magnitude: 3.05
Variable star	Distance: 540 light yrs

The third star of Ursa Minor, Pherkad's name means 'calf'. Along with Kochab, it forms a pair known as the 'guardians of the pole'. Northern skywatchers can use the square of stars in the pan of the 'little dipper' to familiarize themselves with the magnitude system, since they happen to have magnitudes of almost exactly 2, 3, 4 and 5. However, Pherkad's light is not exactly constant – it varies by about 0.03 magnitude in a period of around 3 hours (too little to detect with the naked eye) as its surface pulsates.

Kochab

Beta (β) Ursae Minoris

14h50m, +75°09′	Magnitude: 2.1
Giant star	Distance: 165 light yrs

Just slightly fainter than Polaris, the orange star Kochab was actually the pole star around 1000 BC. This may account for its name, which seems to derive simply from the Arabic for 'The Star'. Kochab itself is noticeably orange, and in reality about 190 times as luminous as our Sun – it is a star that has entered the giant phase of its life, increasing in brightness and ballooning in radius so that its outer surface cools and reddens. It is also slightly unstable and variable, and has a faint companion of magnitude 11.4.

Draco

The Dragon

One of the largest constellations in the sky, Draco winds around the north celestial pole, largely enclosing the polar constellation of Ursa Minor. Mythologically, it is said to represent the dragon that guarded the golden apples in the orchard of the Hesperides, fought by the hero Hercules as one of his twelve tasks – Hercules himself is often depicted as kneeling on the dragon's head, preparing to strike the death blow.

Despite its size, Draco contains few bright stars, and, lying well away from the Milky Way, it is also fairly lacking in deep sky objects. And even though it offers unobstructed views across intergalactic space, there are no nearby galaxies in this direction to enliven the view.

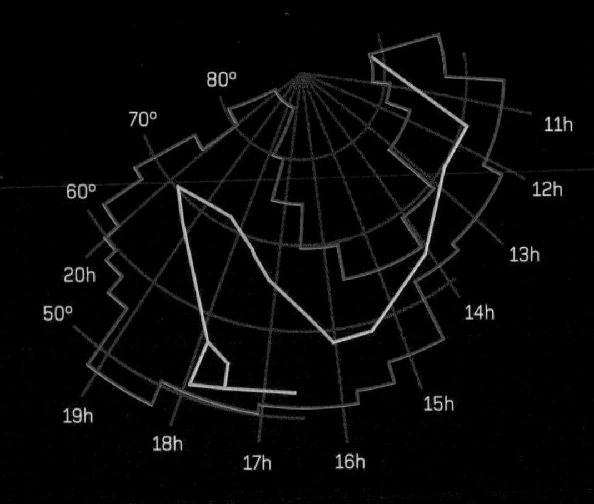

Etamin

Gamma (γ) Draconis

17h56m, +51°29′	Magnitude: 2.2
Giant star	Distance: 150 light yrs

Etamin is the brightest star in Draco, an orange giant currently about 150 light years from Earth. Its name comes from the Arabic word for serpent, but another old name for it is the Zenith Star – so-called because its rotation around the pole carried it directly overhead from the latitude of London. While attempting to measure the precise location of this star in 1728, British astronomer James Bradley discovered an effect called the aberration of starlight, by which the motion of the Earth around the Sun affects the angle of incoming starlight, and hence the apparent position of stars in the sky.

Etamin is approaching Earth at a rate of about 28 km per second, and will pass within 28 light years of us in about 1.5 million years. At this time, assuming the stars remain unchanged, it will rival Sirius as the brightest star in the sky, shining around magnitude -1.5.

Eldasich

Iota (ι) Draconis

15h25m, +58°58′	Magnitude: 3.3
Extrasolar planetary system	Distance: 98 light yrs

Eldasich is a red giant about 45 times as luminous as the Sun, swollen to a diameter of more than 200 million km as it uses up the last of its nuclear fuel and heads towards the end of its life. It is slightly variable, but is best known for its companion, Iota Draconis b – the first planet to be discovered orbiting a giant star. The planet has a mass more than eight times that of Jupiter, and orbits its star in a highly eccentric orbit, every 511 days.

Arrakis

Mu (μ) Draconis

17h5m, +54°28′	Magnitude: 4.9
Binary star	Distance: 90 light yrs

Arrakis, whose name in Arabic means 'dancer', is an attractive binary system, best split with a small-to-medium-sized telescope and high magnification. It consists of twin yellow-white stars in a 480-year orbit around each other. Each star has an individual magnitude of 5.7, suggesting they are more than three times as luminous as the Sun.

39 Draconis

18h24m, +58°48′	Magnitude: 5.0
Multiple star	Distance: 190 light yrs

This complex multiple system is thought to contain eight members, but most are so closely bound in orbit around each other that, from a distance of almost 200 light years, they are impossible to distinguish. From Earth, there are three distinct members – a magnitude 5.0 blue primary with a magnitude 7.4 yellow neighbour (easily seen in binoculars) and a closer magnitude 8.0 companion that requires a telescope.

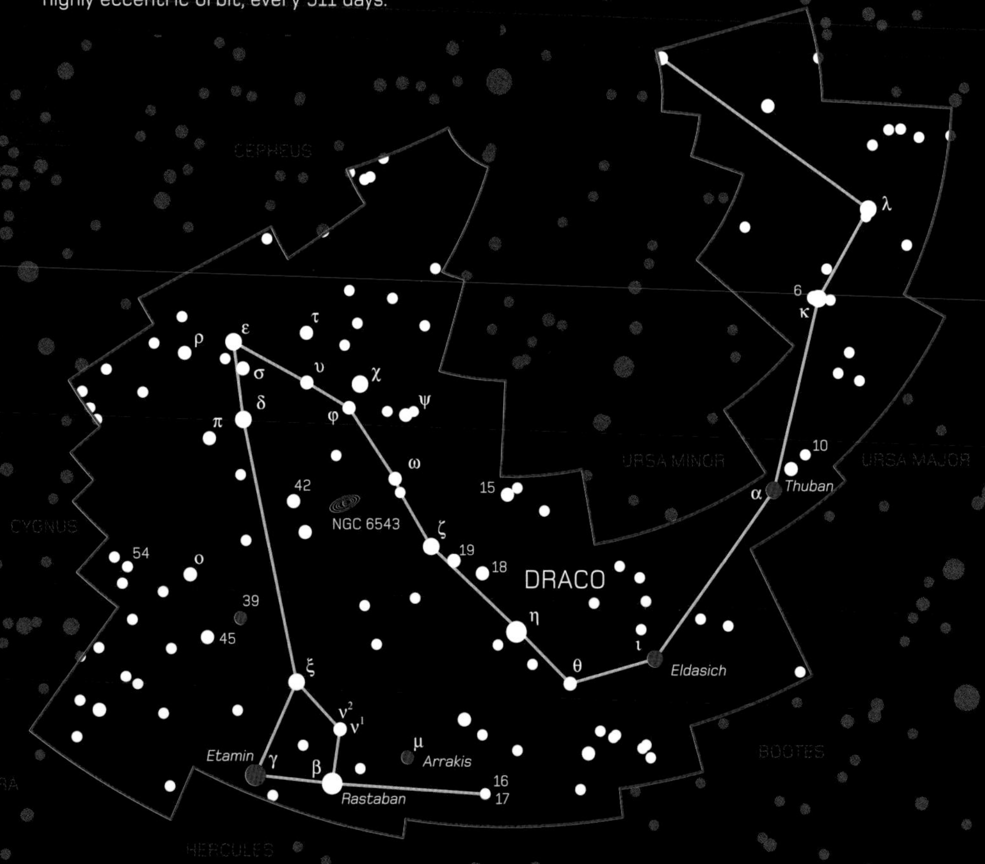

Thuban

Alpha (α) Draconis

14h4m, +64°22′	Magnitude: 3.6
Binary star	Distance: 310 light yrs

Although it bears the Greek designation alpha, Thuban is an average, fairly inconspicuous star of magnitude 3.5, outshone by half a dozen other stars in the long and winding constellation. It is a yellow-white giant star with the luminosity of more than 270 Suns, and is a spectroscopic binary, with a companion star that cannot be seen, but whose influence becomes clear when Thuban's spectrum is analysed. However, Thuban's chief claim to fame is that this was the pole star for ancient Egyptians. It came closest to the north celestial pole around 2700 BC, and some archaeologists claim that shafts within the pyramids, built around this time, were deliberately aligned with Thuban's precise position.

Cat's Eye Nebula

NGC 6543

17h58m, +66°38′	Magnitude: 8.1
Planetary nebula	Distance: 3,600 light yrs

The beautiful Cat's Eye Nebula is Draco's deep-sky highlight, and indeed one of the finest planetary nebulae in the sky. Discovered by William Herschel in 1786, it is formed by the glowing outer layers of a giant star nearing the end of its life and blowing most of its material away into space.

Cepheus & Camelopardalis

King Cepheus and the Giraffe

These two indistinct constellations of far northern skies complete the circle around the pole largely formed by Draco. Cepheus is an ancient pattern, representing the king of Ethiopia – husband of nearby Cassiopeia and a key player in the story of Perseus and Andromeda. Camelopardalis was inserted into a vacant patch of sky in the early 1600s by Dutch astronomer and theologian Petrus Plancius, who invented several other (now defunct) constellations based on Biblical characters and events. Although the word camelopardalis is Latin for giraffe, Plancius intended his new pattern to represent the beast of burden that carried Rebecca into Canaan to marry Isaac – so it seems likely Plancius got his zoology or his translation mixed up at some stage.

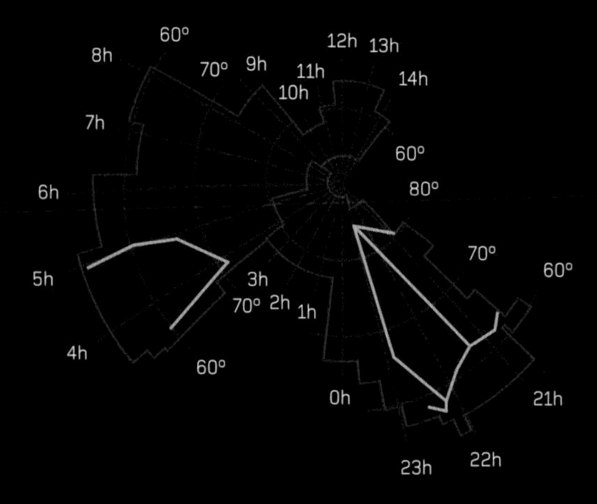

Z Camelopardalis

8h25m, +73°07′ | Magnitude: 12 (var)
Cataclysmic variable | Distance: 530 light yrs

This faint and obscure star for medium-sized telescopes is actually a remarkable 'dwarf nova' system – the first of its type to be discovered. Every few weeks, its brightness rises sharply from around magnitude 13 to magnitude 10 or 11, before slowly tailing off again. It seems to be a nova-like system, in which a small but dense white dwarf star is pulling material off an orange giant companion and into an accretion disc around it. The disc occasionally becomes unstable, dumping large amounts of hot gas onto the surface of the white dwarf, where it creates a nuclear firestorm. In contrast, the supply of material onto the surface of a normal nova is more gradual, allowing it to build up to much greater densities before an eruption is triggered.

NGC 2403

7h37m, +65°36′ | Magnitude: 8.4
Spiral galaxy | Distance: 12M light yrs

This attractive face-on spiral galaxy, discovered by William Herschel in 1788, is part of the group associated with M81 in Ursa Major, and lies 12 million light years away. Its face-on orientation means that its light is fairly well spread out, so those without larger telescopes are best using binoculars or a wide-field, low-magnification eyepiece.

Alfirk

Beta (β) Cephei
21h29m, +70°34′ | Magnitude: 3.2
Variable star | Distance: 595 light yrs

Another prototype variable, Beta is a hot blue-white star that varies in brightness by around 0.1 magnitude in 4.57 hours. However, one characteristic of Beta Cephei stars is that this main oscillation period is overlaid with numerous others – some longer and some shorter, so that when the brightness is plotted over time, an extremely complex 'light curve' emerges.

The Garnet Star

Mu (μ) Cephei
21h44m, +58°47′ | Magnitude: 4.0
Variable star | Distance: 5,000 light yrs

One of the reddest stars in the sky, Mu Cephei, like Delta and Beta, is the prototype for an entire class of variables – some of the most extreme stars known. It is a red supergiant, producing the same energy as 350,000 Suns (although much of this is emitted as invisible infrared radiation), and is thought to have a diameter of about 4.5 billion km – so big that, if it took the place of our Sun, it would engulf the orbit of Jupiter.
It is only its great distance that stops Mu from being one of the brightest stars in the sky, but despite its great brilliance, the star's surface is extremely cool – around 3,200°C. Mu's brightness fluctuates unpredictably by up to a magnitude, though there seems to be an underlying cycle of a little over two years. It is almost certainly in the final stages of its life, squandering the last vestiges of nuclear fuel in its core, on its way to a dramatic supernova explosion.

Beta (β) Camelopardalis

5h3m, +60°27′ | Magnitude: 4.0
Multiple star | Distance: 1,000 light yrs

The primary star in this system is a yellow supergiant, roughly 3,000 times as luminous as the Sun, and seven times its mass. This star has consumed its main hydrogen reserves in a few tens of millions of years, and has now swollen to enormous size. With an unstable surface, it is prone to outbursts – in 1967 it leapt a whole magnitude for a few minutes, perhaps due to enormous stellar flares.

Small telescopes reveal a companion star of magnitude 8.6, and larger instruments will show that this is itself a double.

Delta (δ) Cephei

22h29m +58°25′ | Magnitude: 4.7 (var)
Variable star | Distance: 980 light yrs

Inconspicuous to the naked eye, Delta Cephei is in fact the prototype of an important class of variable stars, which act as valuable measurement tools for astronomers. Delta is a yellow supergiant star a little less than a thousand light years away, and on average 2,000 times more luminous than the Sun. However, its magnitude as seen from Earth varies between magnitudes 3.6 and 4.3 in a precise cycle of 5.37 days. The variation in brightness is related to pulsations in the star's diameter, and astronomers have proven that the period of a 'Cepheid' variable's oscillations is linked with its average luminosity.

This means that, from a star's cycle of variations and its magnitude as seen from Earth, astronomers can get a good idea of its true distance. This was the technique used by Edwin Hubble in the 1920s to calculate the distance to the so-called 'spiral nebulae' and prove that they were in fact galaxies millions of light years from our own.

Cassiopeia

Queen Cassiopeia

The distinctive W-shape of Cassiopeia circles the north celestial pole directly opposite the equally unmistakable Plough in Ursa Major. It is highest in the sky on northern summer nights, and marks the northernmost limit of the Milky Way, where the rich starfields contain many attractive clusters and nebulae.

In mythology, Cassiopeia was the vain queen of Ethiopia, wife of King Cepheus (the constellation depicts her sitting on her throne). Bragging about the beauty of her daughter Andromeda, she incurred the wrath of Hera, queen of the gods, who sent a terrible sea monster, Cepheus, to ravage the land until Andromeda was sacrificed. Only the intervention of the hero Perseus saved the day.

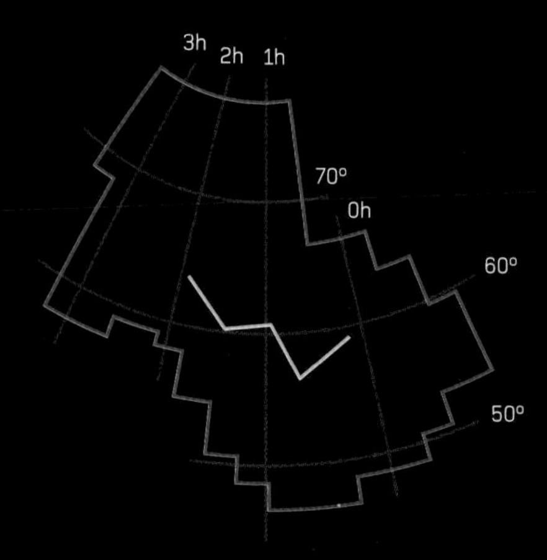

NGC 457

1h19m, +58°20′ | **Magnitude: 6.4**
Open cluster | **Distance: 9,000 light yrs**

This group of about 100 stars is one of the brightest non-Messier clusters in Cassiopeia. Through binoculars or a small telescope on low power, the strands of stars trace out a figure that inspires the object's nickname of the 'ET Cluster'.

Caph

Beta (β) Cassiopeiae
0h9m, +59°09′ | **Magnitude: 2.3**
Giant star | **Distance: 54 light yrs**

Beta Cassiopeiae's name commemorates the former Arab interpretation of the constellation – it means 'the stained palm', recalling Arab identification of the distinctive W as a hand marked with a henna tattoo. Caph is a yellow-white giant star, and we are seeing it in the comparatively brief period where it has exhausted the primary fuel source of hydrogen at its core, but has not yet swollen to become a red giant.

Iota (ι) Cassiopeiae

2h29m, +67°24′ | **Magnitude: 4.5**
Multiple star | **Distance: 141 light yrs**

Through the smallest telescopes, magnitude 4.5 Iota Cassiopeiae reveals a magnitude 8.4 companion. A slightly larger instrument will also reveal a closer companion of magnitude 6.9, orbiting the primary every 840 years. The primary itself is also a spectroscopic binary in a 52-year orbit, and a variable, changing in brightness by about 0.1 magnitude over 1.74 days.

Gamma (γ) Cassiopeiae

0h47m, +60°43′ | **Magnitude: 2.15 (var)**
Variable star | **Distance: 610 light yrs**

Gamma Cassiopeiae is one of the brightest and most notable stars of the northern sky without a proper name. As the central star in the 'W' of Cassiopeia, it is prominent enough, but it is also an unpredictable variable star, that can outshine its neighbours Alpha and Beta.

Gamma's brightness at a distance of more than 600 light years reveals that it is an incredibly luminous star – shining with the light of more than 40,000 Suns. It is the prototype for an entire class of variable stars, all of which have similar characteristics – they are blue-white type 'B' stars weighing more than 10 times as much as our Sun, and spin so rapidly that they fling material off at their equator to form an orbiting ring called a 'decretion disc' that produces bright 'emission lines' in the star's spectrum. These stars are therefore classed 'Be' stars.

The star is still not fully understood – not least because it generates X-rays. These might be associated with an unseen companion star that orbits it in 203.6 days. The companion has about the same mass as the Sun, so it might be a normal star, but it could also be a dense white dwarf that could produce X-rays while pulling material onto its surface from the decretion disc.

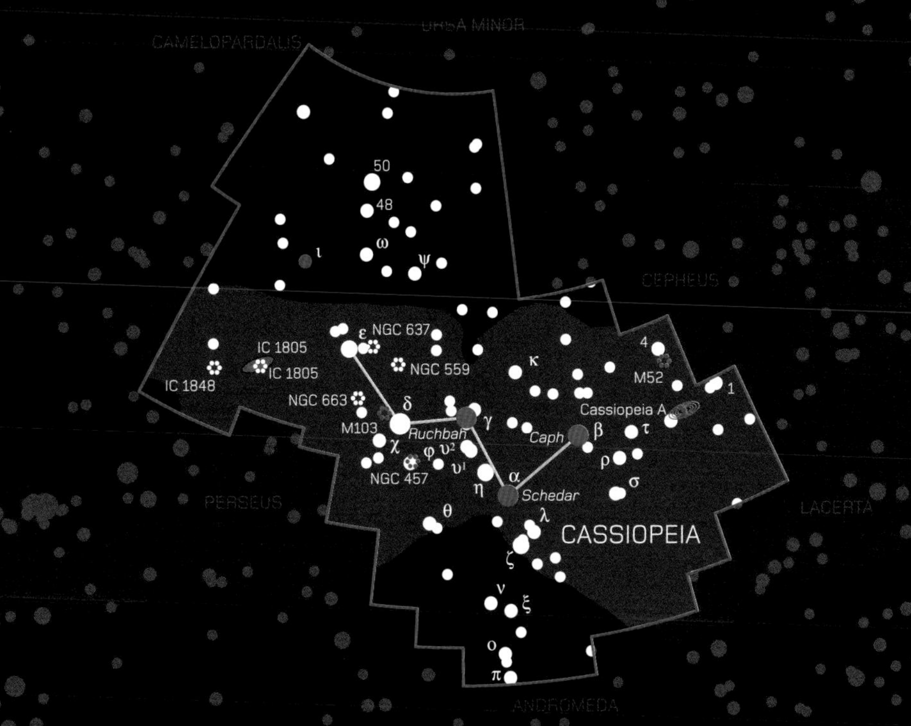

Schedar

Alpha (α) Cassiopeiae
40h31m, +56°32′ | **Magnitude: 2.25**
Giant star | **Distance: 230 light yrs**

With a name that indicates its location at Cassiopeia's breast, Schedar is a swollen yellow-orange giant more than 500 times as luminous as the Sun. Up until the 19th century it was considered a variable star, though it is certainly not today. A nearby magnitude 8.9 star is obvious in a small telescope, but this is a chance alignment, not a true binary companion.

M103

NGC 581
1h33m, +60°42′ | **Magnitude: 7.4**
Open cluster | **Distance: 8,500 light yrs**

This late addition to Messier's catalogue was first observed by French astronomer Pierre Mechain in 1781. It is a loose open cluster of stars about a quarter the diameter of the Full Moon, easily spotted with binoculars. However, the brightest star in the field, Struve 131, is not actually a member.

M52

NGC 7654
23h24m, +61°35′ | **Magnitude: 7.3**
Open cluster | **Distance: 5,000 light yrs**

This rich open cluster, discovered by Messier in 1774, is thought to have about 200 members distributed across a roughly spherical region about 20 light years across. Binoculars will show it as a faint nebulous disc about half the diameter of the full Moon.

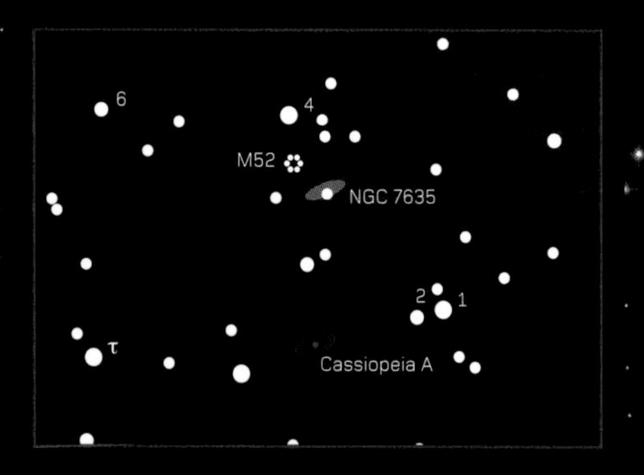

Cassiopeia A

23h23m, +58°48′
Supernova remnant
Magnitude: –
Distance: 10,000 light yrs

Although extremely faint in visible light, Cassiopeia A is one of the most powerful radio sources in the sky. It is the expanding cloud of gas created by a supernova explosion about 350 years ago, which for some reason went unobserved on Earth – perhaps obscured by interstellar dust. The outer edge of the gas shell has an astounding temperature of 30 million °C, and this makes it a strong source of high-energy X-rays.

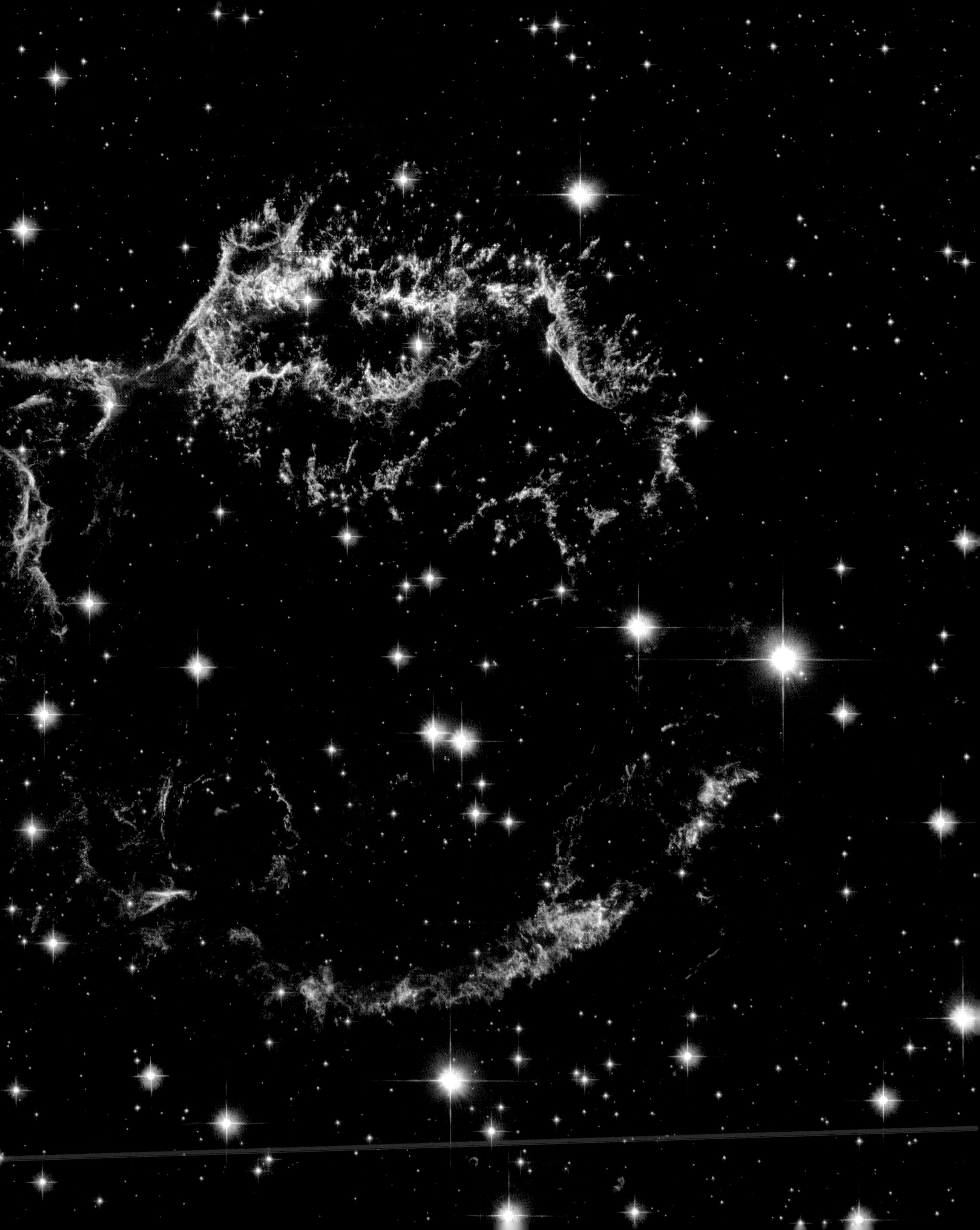

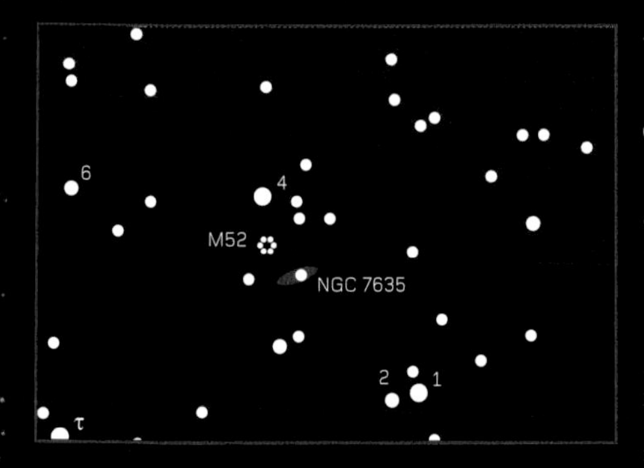

The Bubble Nebula

NGC 7635

23h21m, +61°12′
Emission nebula
Magnitude: 8.7 (central star)
Distance: 11,000 light yrs

The spectacular overlapping shells of the Bubble Nebula are blown into shape by the hot star at their centre, catalogued as SAO 20575. This stellar heavyweight has the mass of about 15 Suns, making it highly luminous and creating fierce stellar winds that blow from its surface. These streams of particles billow out through the surrounding gas cloud that gave birth to the star itself a few million years ago, blowing it into bubbles and causing it to glow.

Au
The C

The br
host t
bright
patter
king Er
story –
the st
suckle
his mo
The tig
Epsilon
'the kid
in one i
depicti
two ba
Nor
stars k
a cat o
on the
Johann
that on

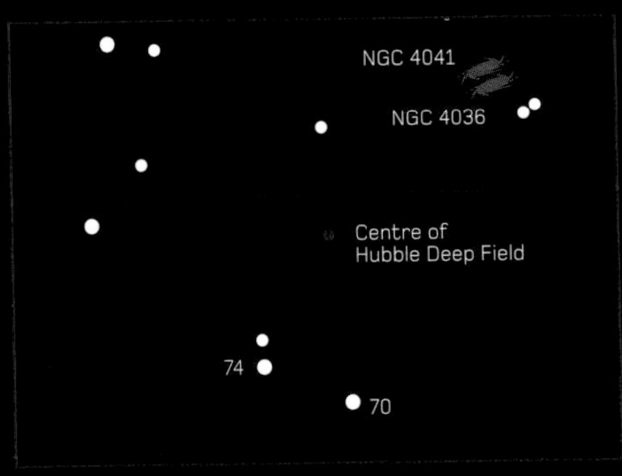

Hubble Deep Field

12h37m, +62°13'
Field of galaxies

Magnitude: –
Distance: Up to 10,500 million light yrs

In 1995, the Hubble Space Telescope turned its powerful gaze onto a small region of 'empty' space in Ursa Major. Staring into the depths of the Universe for a total of ten days, the telescope captured light from the most remote objects yet seen – galaxies so distant that their light left on its journey to Earth more than 10 billion years ago, and we are seeing them in the early stages of their evolution.

9h

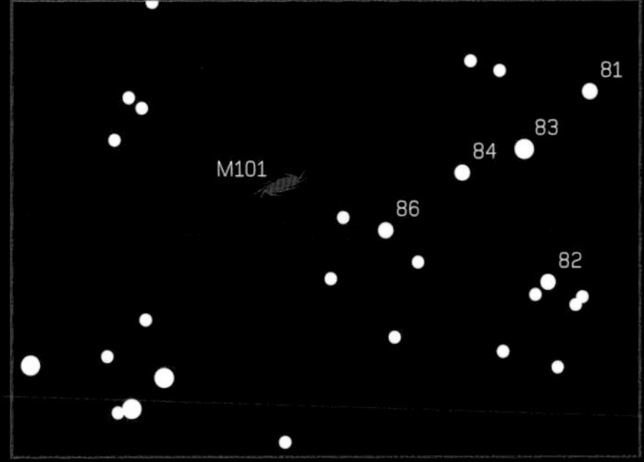

Pinwheel Galaxy

M101 NGC 5457

14h3m, +54°21'	Magnitude: 7.9
Spiral galaxy	Distance: 27M light yrs

The glorious Pinwheel Galaxy displays its impressive spiral arms to the Hubble Space Telescope in this superb image – a composite of 51 individual pictures. With a diameter of 170,000 light years, the Pinwheel is roughly twice the diameter of the Milky Way, and is extremely rich in star-forming regions.

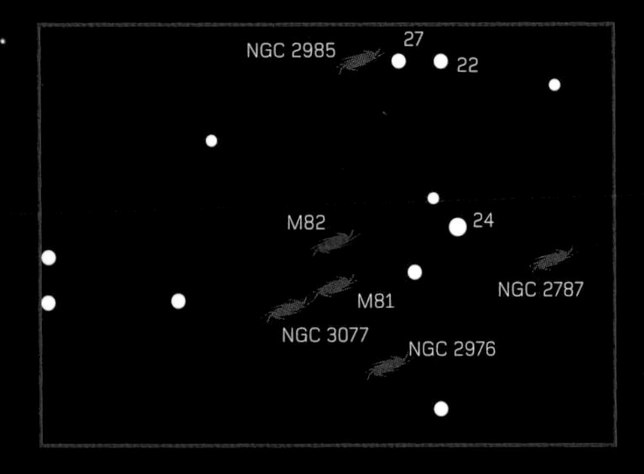

The Cigar Galaxy

M82

9h56m, +69°41′ | Magnitude: 8.4
Irregular galaxy | Distance: 12M light yrs

For a long time, M82 was described as an 'exploding galaxy', such is its ferocious appearance in a variety of different radiations. It is surrounded by huge lobes of radio-emitting gas, and is undergoing a great wave of star formation, probably as a result of its recent encounter with M81. Today, M82 is classed as a 'starburst' galaxy.

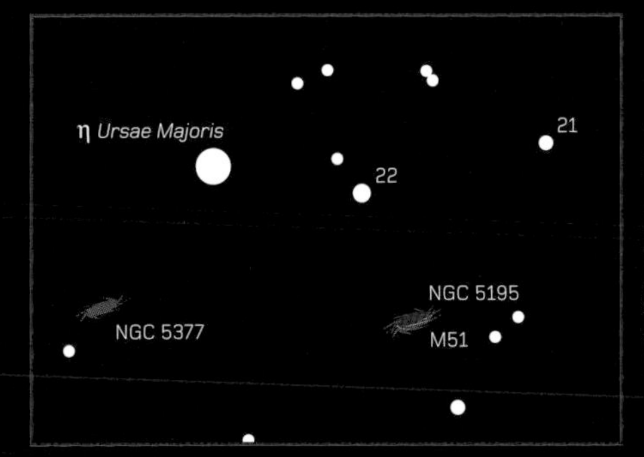

Whirlpool Galaxy

M51

13h30m, +47°12′	Magnitude: 8.4
Spiral galaxy	Distance: 37M light yrs

The beautiful Whirlpool Galaxy of Canes Venatici displays one of the finest spiral structures in the entire sky – in fact it was the first 'spiral nebula' to be recognized, by Irish astronomer Lord Rosse in 1845. The arms are rich in star-forming regions and clusters of bright young stars, suggesting a wave of starbirth intensified by M51's interaction with nearby NGC 5195.

Boötes & Corona Borealis

The Herdsman and the Northern Crown

The distinctive kite-shape of Boötes lies eastward of Ursa Major, its base marked by the brilliant red Arcturus, one of the brightest stars in the sky. It is usually said to represent the celestial herdsman, Boötes, forever driving away the bears Ursa Major and Minor, with assistance from his hunting dogs (nearby Canes Venatici). However, in another tale, it is Arcas, the doomed son of Callisto, the Great Bear. Although the constellation lies well away from the Milky Way, it contains several interesting objects, including one of the best-understood planetary systems beyond our own.

Close to Boötes lies the unmistakable arc of the Northern Crown. Although its stars are not the brightest, they form a distinct circlet that is easy to spot – they represent the crown worn by Ariadne at her wedding to the god of wine Bacchus.

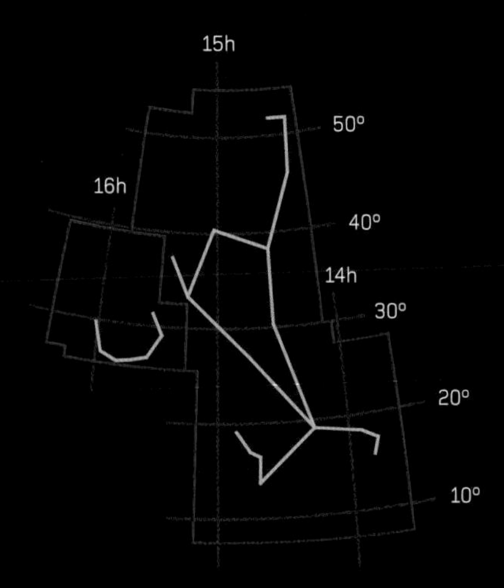

Arcturus

Alpha (α) Boötis

14h16m, +19°10′	Magnitude: -0.04
Giant star	Distance: 37 light yrs

Arcturus offers a glimpse of our Sun's far future. At a distance of 37 light years from Earth, it is a star with 1.5 times the mass of our own, that has already lived a long hydrogen-burning life on the 'main sequence' of stellar evolution, and has swollen to become a red giant with a diameter of about 15 million km – or 24 times greater than the Sun. It is the nearest red giant to Earth, and it is thought that it may now be contracting again as a new wave of nuclear fusion reactions spread through its core utilizing the helium waste left behind by hydrogen fusion. Eventually, though, it will exhaust this fuel supply too, swelling once again to a red giant and shaking off its outer layers in a planetary nebula.

Arcturus has some intriguing mysteries – it moves at a different speed from most of the surrounding stars, and has fewer traces of heavier elements than stars like the Sun. This all fits in with the idea that it was born early in our galaxy's history, but some astronomers have suggested an alternative explanation – that Arcturus and several stars associated with it are interlopers from another galaxy that was absorbed by our own billions of years ago.

Izar

Epsilon (ε) Boötis

14h45m, +27°04′	Magnitude: 2.4
Multiple star	Distance: 210 light yrs

This attractive contrasting double was nicknamed Pulcherrima (meaning 'most beautiful') by its discoverer, Friedrich Struve. It requires a moderate-sized telescope to resolve, and a high magnification reveals an orange and blue-white pair of stars at magnitudes 2.7 and 5.1 respectively. Both stars are giants, with the masses of four and two Suns respectively.

Tau (τ) Boötis

13h47m, +17°27′	Magnitude: 4.5
Extrasolar planetary system	Distance: 51 light yrs

This average white star at the constellation's southern end is best known for its planetary companion – a so-called 'hot Jupiter' class world that was detected through perturbations in the spectrum of its parent star. Tau's planet has at least 3.9 times the mass of Jupiter, but orbits its sun in just 3.3 days, virtually skimming the star's surface. Hot Jupiters are a bizarre type of world, and still not fully understood, but they are an important sign that not all solar systems are like our own.

In addition, Tau has a distant red dwarf companion, orbiting far beyond any other planets on a highly elliptical path. Considering that multiple stars far outnumber singletons like the Sun in our galaxy, it's encouraging to extrasolar planet hunters that solar systems can form in such disruptive conditions.

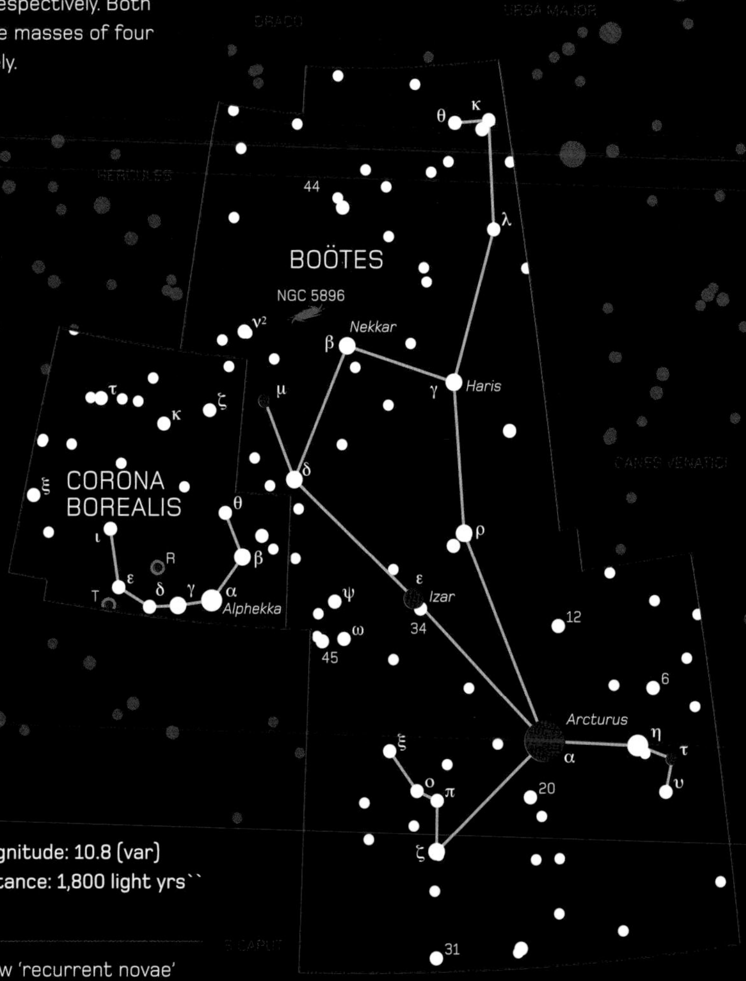

Alkalurops

Mu (μ) Boötis

15h25m, +37°23′	Magnitude: 4.3
Multiple star	Distance: 120 light yrs

To the naked eye, only the blue-white primary star of this multiple system is visible, but it is worth investigating further. Binoculars are enough to show that Alkalurops A (the name means 'club') has a well-separated companion of magnitude 6.5, and a small telescope will show that this companion is itself a double star, consisting of two sunlike yellow dwarfs of magnitudes 7.0 and 7.6. These twins orbit each other every 260 years, but take about 125,000 years to waltz with their brighter companion. There is good evidence that the primary, too, is a double star, with an unseen companion that circles it in about 300 days.

The Blaze Star

T Coronae Borealis

16h0m, +25°55′	Magnitude: 10.8 (var)
Cataclysmic variable	Distance: 1,800 light yrs``

This is one of the very few 'recurrent novae' visible to amateurs. Normally visible through a small telescope as a magnitude 11 star, it is a system that contains a white dwarf star orbiting around a swollen red giant from which it pulls away material. When the material accumulating on the surface of the dwarf becomes hot and dense enough, it burns away in a burst of nuclear fusion that can reach magnitude 2.0 as seen from Earth. Such outbursts were recorded in 1866 and 1946, and another could happen at any time.

R Coronae Borealis

15h49m, +28°09′	Magnitude: 5.9 (var)
Variable star	Distance: 6,000 light yrs

This bizarre variable star behaves in precisely the opposite way to its neighbour the Blaze Star. A distant yellow supergiant that normally shines at around sixth magnitude, it sometimes drops, unpredictably and suddenly, beyond even the range of small telescopes, perhaps as low as magnitude 15, taking many months to regain its former brightness. Astronomers know that the star's light is being blocked by a cloud of carbon, but they still don't fully understand what's going on – the best theory is that carbon is blown away from the star's upper atmosphere, cooling to a point where it can condense into dark obscuring clouds. As the pressure of the star's radiation pushes them further away, the clouds part and the star recovers its brightness.

Hercules

Hercules

Hercules is one of the larger constellations, but the extended chains of stars that supposedly mark the limbs of the Greek demigod can be hard to spot. The easiest way to find the constellation in northern skies is to look for brilliant white Vega in neighbouring Lyra, and then identify the lopsided square of stars called the 'Keystone', to its west. The Keystone represents Hercules' body, and his limbs can be traced outward from its four corners.

This pattern of a kneeling hero crouching to despatch a vanquished dragon is actually thought to pre-date the Hercules myth – it may have originated a thousand or more years earlier, as a depiction of the Sumerian warrior Gilgamesh, hero of history's first epic. Whatever its origins, though, the Greeks ensured that it now commemorates the son of Zeus and the mortal princess Alcmene.

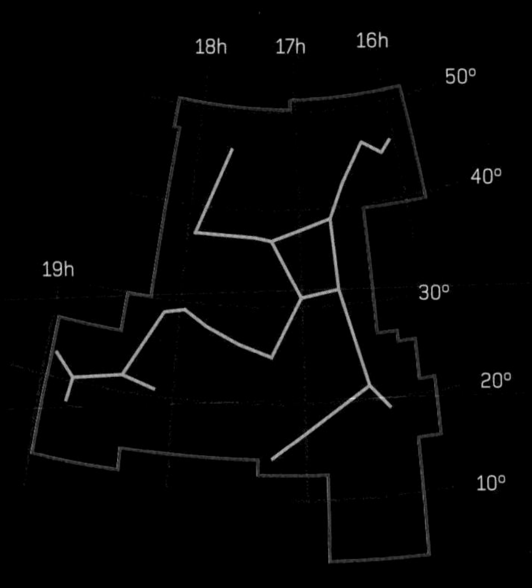

Rasalgethi

Alpha (α) Herculis

17h15m, +14°23′	Magnitude: 3.5
Multiple star	Distance: 380 light yrs

The name of Hercules' brightest star means 'the kneeler's head', indicating that the constellation is upside-down as seen from the northern hemisphere. A small telescope reveals that the brighter red star has a greenish-white companion of magnitude 5.4.

Rasalgethi A is a red giant about 300 million km across – the same diameter as Earth's orbit around the Sun. It is surrounded by an extended envelope of gas extending to 50 times this diameter, and is one of the few stars that have had their diameter measured directly. It is also variable, fluctuating erratically between magnitudes 3 and 4.

Rasalgethi B, which orbits the giant in about 3,600 years, is itself a double star system consisting of a yellow giant and a yellow-white dwarf star, indivisible with even the largest telescope. Its greenish tint may be an illusion created by contrast with its red neighbour.

The 'Turtle in Space'

NGC 6210

16h45m, +23°49′	Magnitude: 9.0
Planetary nebula	Distance: 6,000 light yrs

This compact but relatively bright planetary nebula resembles a swimming turtle only through the largest telescopes. Smaller telescopes will find it as a greenish, star-like blob, which medium-sized ones will transform into a faint disc with a magnitude 12.7 central star.

M92

17h17s, +43°08′	Magnitude: 6.4
Globular cluster	Distance: 27,000 light yrs

Hovering at the limit of naked-eye visibility, M92 is Hercules' other impressive globular cluster. Binoculars will find it easily, but it is smaller and dense, so larger telescopes are needed to resolve any of its stars. It was discovered by Johann Elert Bode in 1777, and independently rediscovered by Messier in 1781.

DQ Herculis

18h8m, +45°52′	Magnitude: 14.2
Cataclysmic variable	Distance: 1,800 light yrs

In 1934, one of the brightest novae of modern times erupted in Hercules, gradually building in intensity until it reached a peak of magnitude 1.3, before fading over the following months. The so-called progenitor system, DQ Herculis was, and still is, an apparently insignificant star of magnitude 14, but it proved vital to explaining the concept of novae.

We now know that the DQ system consists of a small red dwarf star, orbited by a dense white dwarf (the burnt-out remnant of a much more massive star). The white dwarf is close enough to its companion to pull away material which builds up in a flattened accretion disc around it. However, the magnetic field tugs material out of the disc and channels it down onto the white dwarf's magnetic poles. In 1934, it built up to a point where the pressure and temperature on the surface were equivalent to those in the heart of a star like the Sun, allowing the huge envelope of material around the dwarf to burn away in a burst of nuclear fusion.

M13

NGC 6205

16h42m, +36°28′	Magnitude: 5.8
Globular cluster	Distance: 25,100 light yrs

This spectacular globular cluster, the best visible from many northern latitudes, is about 150 light years across, and contains perhaps a million stars. It was discovered by Edmond Halley in 1714, who noted that it could just be seen with the naked eye on a dark, moonless night. Binoculars show it as a circular patch of light about half the size of the Full Moon, while small telescopes can reveal some of the individual stars around the cluster's ragged outer edges.

M13 was chosen as the target for the radio message to aliens transmitted in 1974 from the Arecibo radio telescope in Puerto Rico. In the unlikely event that an extraterrestrial civilization receives the message, we might expect an answer in 50,000 years.

Zeta (ζ) Herculis

16h41m, +31°36′	Magnitude: 2.9
Binary star	Distance: 35 light yrs

An attractive double, Zeta is a good test of amateur instruments. The primary yellow star, seven times brighter than our Sun, is orbited every 34.5 years by an orange companion of magnitude 5.5. The system's orientation causes the separation between the stars to constantly change. At the most recent close approach in 2002, they were too close to separate with amateur telescopes, but as they draw apart to a maximum separation around 2018, they will grow steadily easier to distinguish.

Lyra, Vulpecula & Sagitta

The Lyre, the Fox and the Arrow

Three compact constellations flank the southern limits of Cygnus. Of these, Lyra is by far the most prominent, thanks to the presence of brilliant white Vega, one of the brightest stars in the sky. Lyra represents a lyre, the ancient musical instrument played by, amongst others, Orpheus during his journey into the underworld of the Greek mythological afterlife. Lying just a little way from the rich Milky Way starfields of Cygnus, Lyra also contains several interesting stars and deep sky objects.

Sagitta, the arrow, is small and faint, but still readily identified thanks to its distinctive pattern. Curiously, the arrow is far away from the archer Sagittarius – instead it seems to have been fired by Hercules in the general direction of Cygnus, the swan, and Aquila, the eagle.

Vulpecula is the most obscure of these three constellations, with no distinct pattern and no stars above fourth magnitude. It represents a fox fleeing with a goose (marked by Alpha Vulpeculae, also known as Anser).

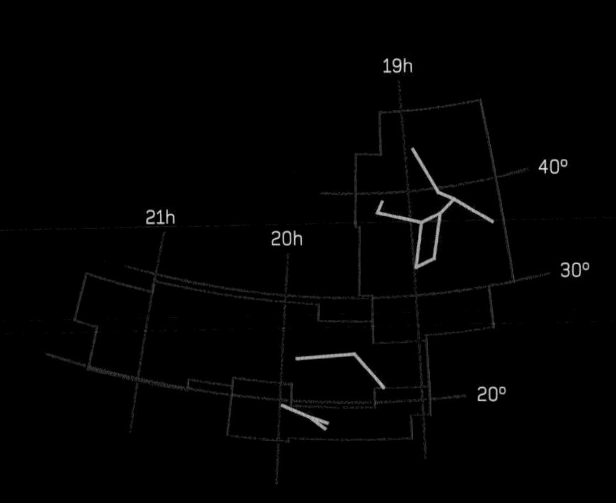

Vega

Alpha (α) Lyrae

18h37m, +38°47′ | Magnitude: 0.03
Main sequence star | Distance: 25 light yrs

The fifth brightest star in the sky, Vega is an unmistakable sight on northern summer nights, where it forms the northeast apex of the northern 'summer triangle'. Like many bright stars, Vega is not really that spectacular – it just happens to lie on our cosmic doorstep, at a distance of just 25 light years. Vega is a normal white star that shines by fusing hydrogen into helium at its core. Because it has about 2.3 times as much mass as the Sun, it is about 40 times more luminous, and has a surface temperature of about 9,200°C. More massive stars live fast and die young – Vega is about one-tenth of the Sun's age, and already halfway through its hydrogen fuel supply. Its youth also means that it spins far more rapidly than the Sun – rotating in about 12 hours and bulging out significantly around its equator as a result. It is also still surrounded by a debris disc of gas and dust left over from its formation. The debris may be in the process of forming planets.

Sheliak

Beta (β) Lyrae

18h50m 33°22′ | Magnitude: 3.5 (var)
Binary star | Distance: 880 light yrs

Binoculars reveal that Beta Lyrae is a multiple system, with a white primary component and a blue companion of magnitude 7.2. The primary, however, is a remarkable system in its own right. It varies in brightness between magnitudes 3.4 and 4.6, in a cycle that lasts 19.4 days, and is the prototype for a class of variables known as 'contact binaries'. It has two components, a blue-white star and a white giant, that are so close together that they are both distorted into elliptical shapes, and material pulled off them forms an obscuring disc around the system. As the stars pass behind one another, they cause primary and secondary minima (brightness dips) at intervals of 12.9 and 6.5 days – but their distorted shapes mean that the surface they present to Earth is constantly changing, causing their brightness to vary continuously.

Brocchi's Cluster

Collinder 399

19h25m, +20°11′ | Magnitude: 3.6
Optical cluster | Distance: -

Also known as the Coathanger (though this shape is best identified by looking at the cluster 'upside-down'), Brocchi's Cluster is a group of 40 stars in Vulpecula that was first recorded by the Arab astronomer al Sufi in the mid-10th century. Unusually, though, this prominent cluster (best studied with binoculars) seems to be the result of a chance alignment rather than a true group of stars in space – measurements of its different members have produced wildly varying distances.

Epsilon (ε) Lyrae

18h44m, 39°40′ | Magnitude: 4.6
Multiple star | Distance: 160 light yrs

This famous multiple star system is a great target for binoculars or a small telescope. Binoculars split it easily into two components of magnitudes 4.7 and 4.6, but a telescope will show that each of these stars is itself double, making this a stunning quadruple system.

The northern pair, Epsilon 1, has magnitudes 4.7 and 6.1, and the stars waltz about their common centre of gravity every 1,200 years. The southern pair, Epsilon 2, has magnitudes 5.1 and 5.5, and an orbital period of 585 years. Epsilon 1 and Epsilon 2 are separated by about 0.16 light years, but they are gravitationally bound together in an orbit that takes hundreds of thousands of years to complete.

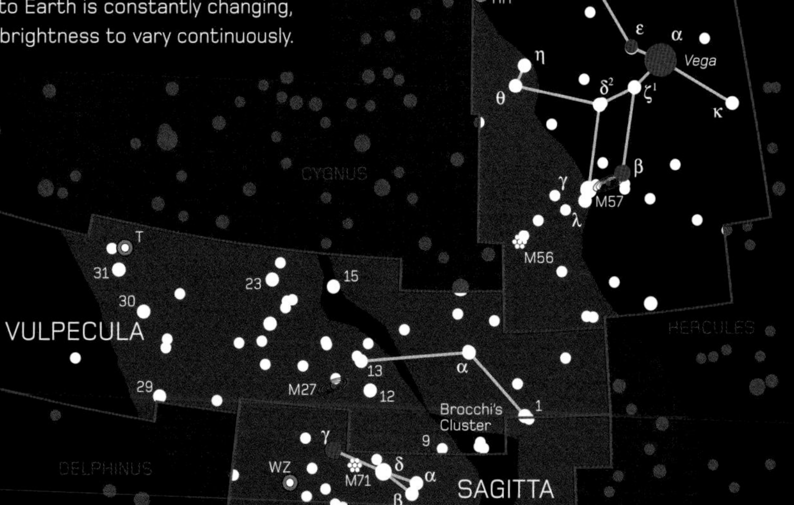

RR Lyrae

19h26m, +42°47′ | Magnitude: 7.5 (var)
Variable star | Distance: 855 light yrs

Varying in brightness by about a magnitude in a cycle that repeats every 13 hours, RR Lyrae is the prototype for a common type of variable star that has proved invaluable in measuring the shape of our galaxy. These stars are yellow-white giants nearing the end of their lives, somewhat older, less massive and fainter than the cepheid variables such as Delta Cephei. Their pulsations follow a characteristic pattern that makes them easy to spot across great distances (though they are too faint to see in other galaxies), and each star's period is related directly to its luminosity. This makes it possible to estimate the true distance of RR Lyrae stars across the galaxy, and map its structure. The stars are also vital to estimates of the distance to globular clusters, where they are quite common.

Gamma (γ) Sagittae

19h59m, +19°30′ | Magnitude: 3.5
Giant star | Distance: 275 light yrs

Sagitta's brightest star is one of the few extremely cool, red class M stars visible to the naked eye. It has a mass of 2.5 Suns, and is a giant in the final stages of its evolution, having exhausted both hydrogen and helium fuel supplies in its core. Its brightness fluctuates slightly already, and it will probably become a long-period, Mira-type variable before finally shedding its outer layers as a planetary nebula in the next few million years.

Dumbbell Nebula

M27

20h0m, +22°43′ | Magnitude: 7.4
Planetary nebula | Distance: 1,250 light yrs

Vulpecula's Dumbbell Nebula was the first planetary nebula to be discovered, by Charles Messier in 1764. Visible through binoculars on a dark night, its high contrast relative to the background sky makes it the best object of its kind for small telescopes.

The nebula gets its distinctive appearance because the star at its heart seems to be flinging off material mostly around its equator – since we see it almost edge-on, the bubble of gas appears to have a bar across its centre. Measurements of the Dumbbell's expansion have allowed astronomers to find its age, and it is a surprisingly young feature, about 4,000 years old. This suggests that planetary nebulae are a very brief phase in the life story of a typical star.

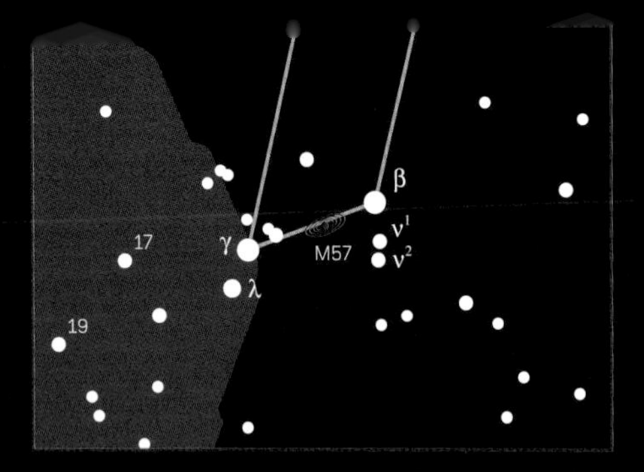

Ring Nebula

M57

18h54m, +33°02′
Planetary nebula

Magnitude: 8.8
Distance: 2,300 light yrs

Probably the most famous of all planetary nebulae, Lyra's Ring Nebula was for a long time thought to be a spherical shell of gas, with a ring-like appearance caused by the thickness of the shell along its edges as seen from Earth. Now, however, it seems that the nebula is a true ring – an expanding doughnut of gas – or perhaps even a funnel-like structure that we happen to see end-on.

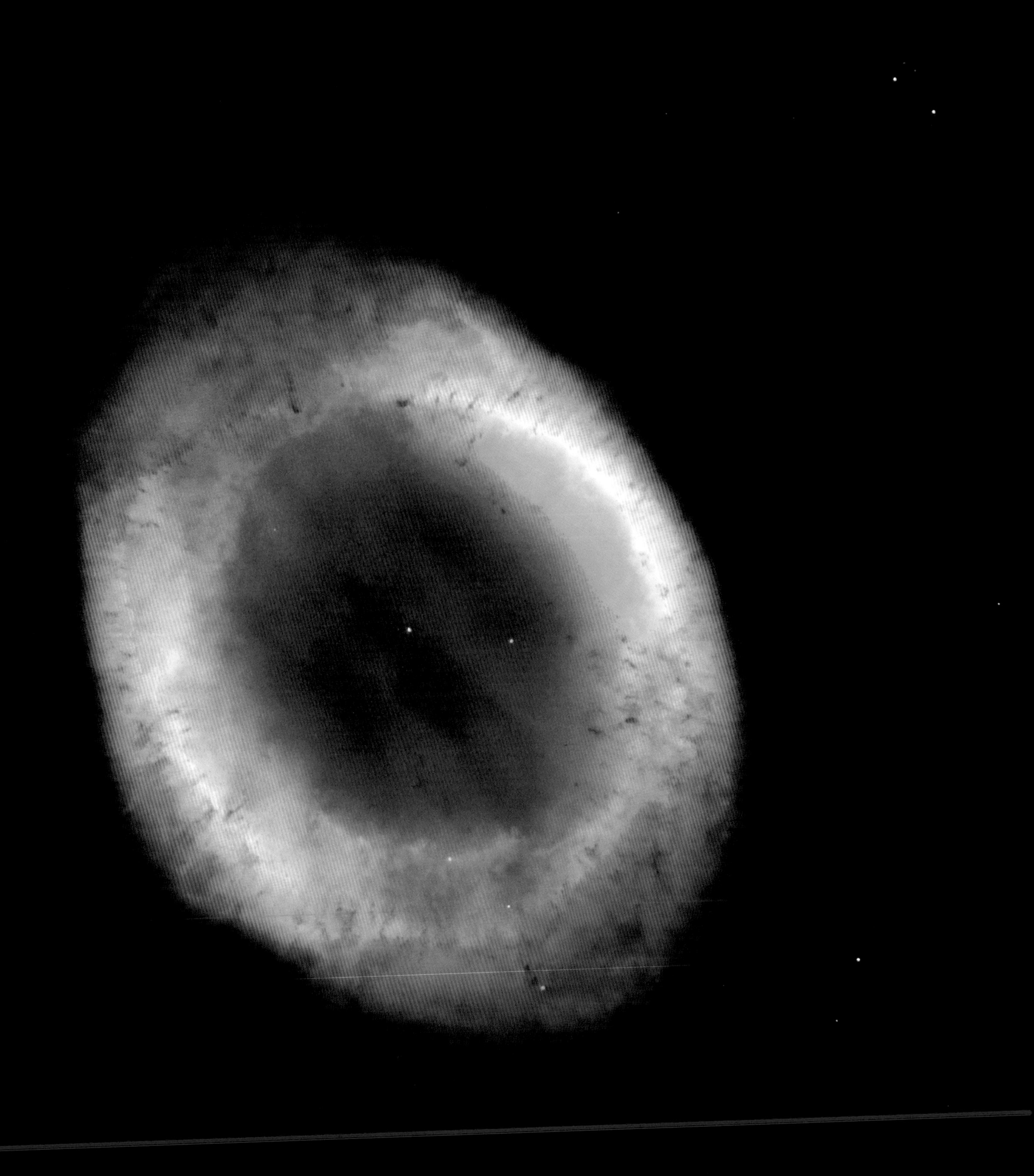

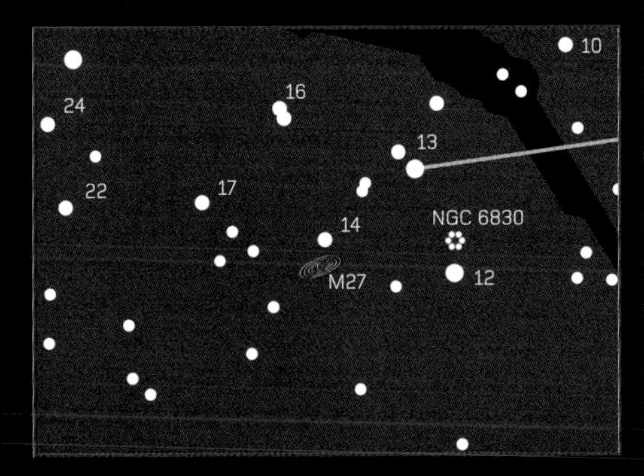

Dumbbell Nebula

M27

20h0m, +22°43′	Magnitude: 7.4
Planetary nebula	Distance: 1,250 light yrs

The first planetary nebula to be discovered, Vulpecula's M27 is one of the brightest and most exquisite objects of its type. Measurements of its expansion indicate that the outer shell of gas was blown off its central star about 4,000 years ago.

Cygnus

The Swan

Easily picked out, even against the dense starfields of the northern Milky Way, it's little wonder Cygnus is also sometimes known as the Northern Cross. Widely separated cultures around the world have transformed this bright cruciform pattern into a creature of some sort – either a bird or a dragon – flying down the Milky Way. Today, it is almost universally recognized as a swan. Some say it is Zeus, king of the Greek gods, assuming the form of a swan to seduce Leda. Others see it as the musician Orpheus, transformed into a bird after his death and placed in the sky alongside his instrument, the lyre. A third legend sees Cygnus as a friend of Phaeton, who stole the Sun god's chariot and fell into the mythical river Eridanus. Cygnus repeatedly plunged into the river in an attempt to rescue his friend, and Zeus transformed him into a swan to aid his efforts.

Stretched across the plane of our galaxy, Cygnus is rich in fascinating objects, ranging from nearby double stars to distant giants, and from black holes to remote active galaxies, far beyond the Milky Way.

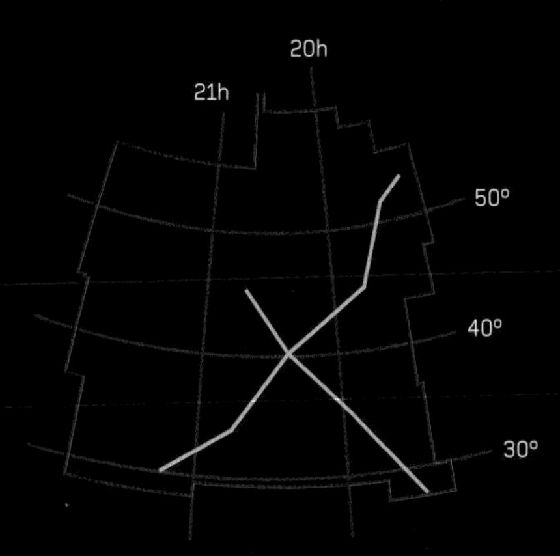

Cygnus A

19h59m, +40°44′ | **Magnitude: 15.0**
Active galaxy | **Distance: 600M light yrs**

This distant, misshapen galaxy is visible only through larger telescopes, but for radio astronomers it is one of the brightest and largest features in the sky. Two radio-emitting jets emerge from the core of the visible galaxy, 120,000 light years across, and billow out into two vast lobes of gas that extend to a diameter of more than half a million light years.

Cygnus A is among the nearest and most dazzling active galaxies, visible not just through radio waves, but also through X-ray emitting 'hot spots' that form where the expanding lobes meet gas in the surrounding space. Astronomers think that the galaxy at the heart of Cygnus A has recently swallowed another, and the fresh supplies of gas and dust in this cosmic snack have woken the dormant black hole at the heart of Cygnus, which is now spitting out jets from its poles as it feasts.

Cygnus Rift

20h20m, +35° | **Magnitude: -**
Dark nebula | **Distance: 2,300 light yrs**

Easily spotted with the naked eye on a moderately dark night, the Cygnus Rift is a dark gap in the Milky Way, running close to the constellation's central spine. In reality, it is not a true gap, merely a dark cloud of obscuring dust that blocks our view of the more distant star clouds beyond, so we only see the handful of stars that lie in front of it as seen from our point of view.

Albireo

Beta (β) Cygni
19h31m, +27°58′ | **Magnitude: 3.1**
Binary star | **Distance: 385 light yrs**

Widely acknowledged as the most beautiful double star in the northern skies, Albireo reveals its binary nature to the smallest telescope, or even good binoculars. The system is dominated by a yellow giant of magnitude 3.1, orbited by a magnitude 5.1 blue-green companion that forms a striking contrast.

61 Cygni

21h7m, +38°45′ | **Magnitude: 5.0**
Double star | **Distance: 11.4 light yrs**

One of the closest stars to Earth, binoculars reveal that this faint magnitude 5.0 stars is in fact a pair of near-twin orange star of magnitude 5.2 and 6.0 The two stars orbit each other every 650 years, and have masses of about 0.5 and 0.4 Suns respectively, and there is evidence from their spectra that each has a smaller, unseen companion.

61 Cygni is an important star to the history of astronomy, since it was the first to have its distance measured directly. In 1838, German astronomer Friedrich Wilhelm Bessel successfully worked out the system's parallax – the amount by which the stars appear to move in the sky when they are viewed from opposite sides of Earth's orbit. Parallax gets smaller over greater distances, and Bessel found that 61 Cygni has a comparatively huge parallax of just over 0.3 seconds of arc (roughly 1/12,000 of a degree). Knowing this and the distance between opposite sides of Earth's orbit, he calculated that 61 Cygni was some 10.4 light years away – an impressively accurate measurement for the time. Parallax has since proved a vital tool in revealing the true properties of the stars.

North America Nebula

NGC 7000

20h59m, +44°20′ | **Magnitude: -**
Star-forming nebula | **Distance: 1,600 light yrs**

The North America Nebula is among the largest of the sky's 'emission nebulae' – huge gas clouds that glow as their atoms are excited by the fierce radiation of the young stars being born inside them. It is such a diffuse object that it disappears when looked at through small telescopes, and although some claim it can be seen with the naked eye, it has proved impossible to assign an overall magnitude to it. NGC 7000 is best located with binoculars or larger instruments with a wider field of view, but long-exposure photographs are needed to reveal its structure and distinctive shape.

P Cygni

20h18m, +38°02′ | **Magnitude: 4.8 (var)**
Variable star | **Distance: 5,500 light yrs**

This unpredictable variable star is thought to lie between 5,000 and 6,000 light years away – much too far away to show measurable parallax. It is a blue supergiant – a truly enormous star, many times the mass of the Sun, which has exhausted the supply of hydrogen fuel in its core, and is now evolving into an even more luminous red supergiant. As it does so, it fluctuates wildly in brightness – around 1600, it reached magnitude 3, and it has shown bursts of brightness on several other occasions. If current distance estimates are accurate, then it is also one of the most luminous stars known, pumping out the same energy as about 700,000 Suns.

Cygnus X-1

19h58m, +35°12′ | **Magnitude: 8.95**
Black hole | **Distance: 8,200 light yrs**

In visible light, this bizarre object appears to be a monstrous ninth-magnitude blue supergiant star, with the mass of perhaps 20 or more Suns. However, X-ray telescopes show it as a strong X-ray source, flickering about 1,000 times a second. The X-rays do not come from the star itself, but from a massive, invisible object in a 5.6-day orbit around it. Astronomers believe that this is a black hole – the collapsed core of an even more massive star. As it drags material away from its companion, it pulls it to its doom and heats it to extreme temperatures that are capable of emitting X-rays.

Deneb

Alpha (α) Cygni
20h41m, +45°17′ | **Magnitude: 1.25**
Giant star | **Distance: 3,000 light yrs**

While many of the sky's brightest stars benefit from their proximity to Earth, Deneb is a truly brilliant monster, about 3,000 light years away and blazing with the luminosity of 200,000 Suns. It may be the brightest star of its type in the entire galaxy – a white supergiant that started life just a few million years ago with twenty times the mass of the Sun. During its life as a main-sequence (hydrogen-burning) star, it was so hot and blue that even now, when it has grown larger than Earth's orbit round the Sun, and cooled substantially, it is still a hot white star with a surface temperature of more than 8,000°C. In a few million years at most, Deneb will end its life in a dramatic supernova explosion.

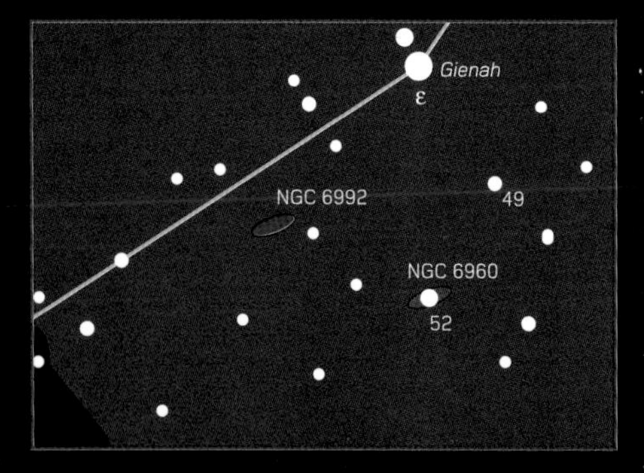

Veil Nebula

NGC 6960, 6979, 6992, 6995

20h51m,+30°40′	Magnitude: 5.0
Supernova remnant	Distance: 2,600 light yrs

The Veil Nebula marks the place where the shredded remains of a long-dead star are colliding with a slower-moving, cooler gas cloud, and being re-heated until they glow in different colours – oxygen emits blue light, hydrogen green and sulphur red. The Veil (the brightest part of a larger ring-like structure called the Cygnus Loop) is thought to have formed in a supernova explosion 15,000 years ago.

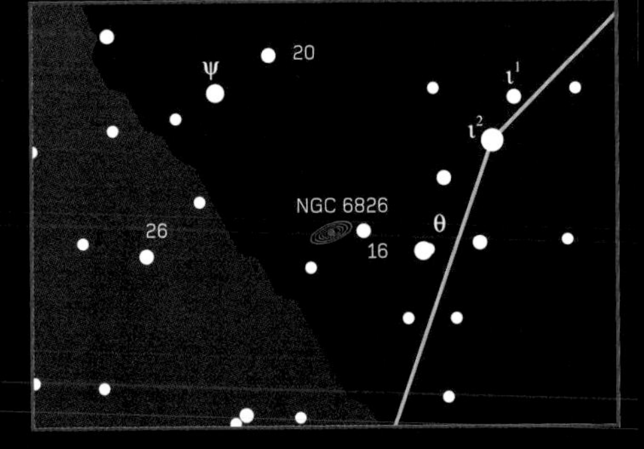

The Blinking Nebula

NGC 6826

19h45m, +50°31′ | Magnitude: 10.0
Planetary nebula | Distance: 2,200 light yrs

This planetary nebula gets its name from its habit of disappearing when looked at directly – it demonstrates how the light sensitive cells in the centre of the eye are less receptive than those around the edge, so that 'averted' vision can reveal more than a direct stare. The nebula itself shows an expanding gas shell, with a pair of bright objects (most likely previously emitted gas clouds) flanking the central star.

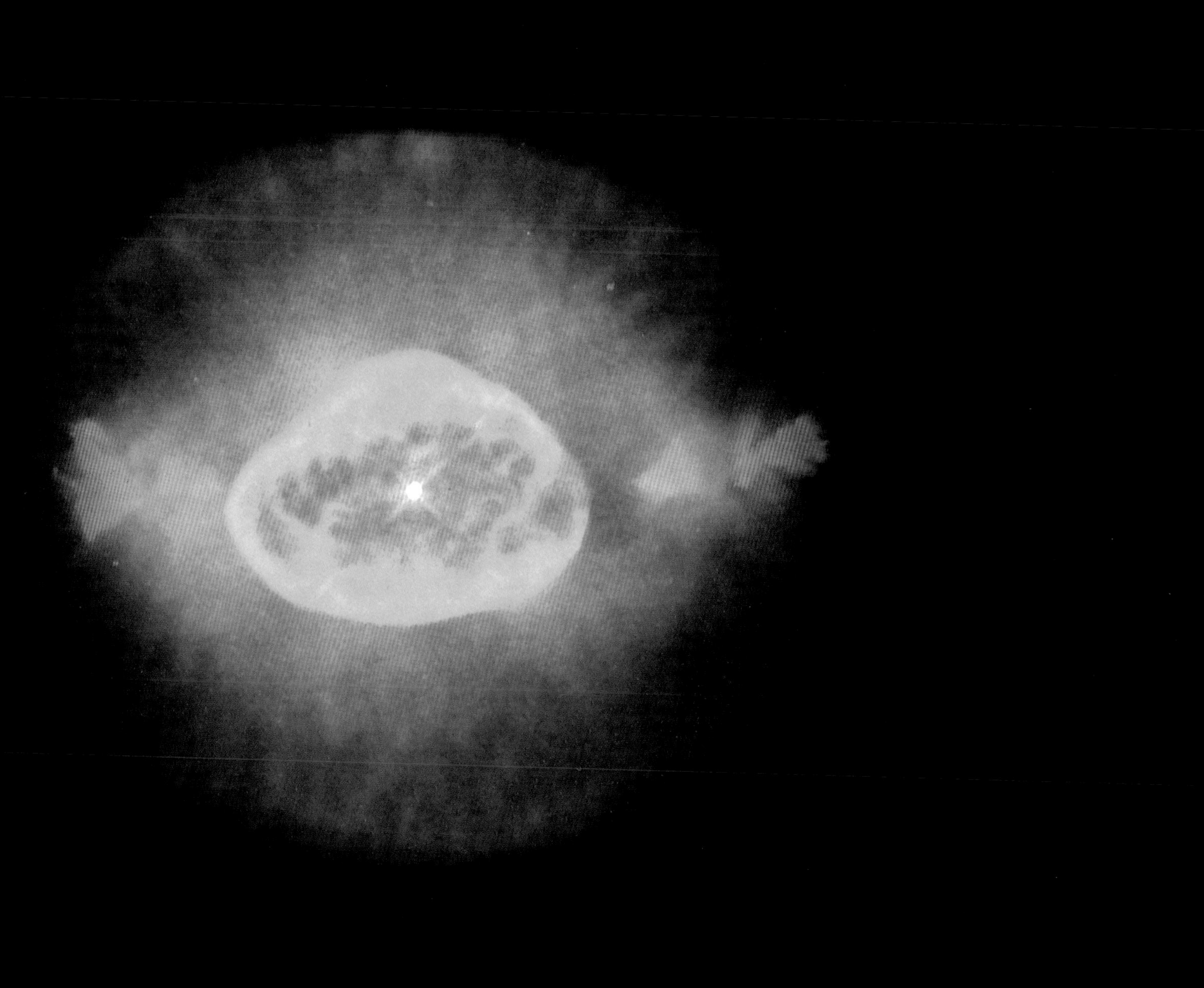

Andromeda & Lacerta

The Princess and the Lizard

The fairly indistinct branching shape of Andromeda is still easy to locate in the sky because its brightest star is also the northeast corner of the Square of Pegasus. In mythology, Andromeda was the beautiful daughter of Cepheus and Cassiopeia. Her mother's boasts about Andromeda's beauty enraged the habitually jealous Juno, queen of the gods, who sent a monster, Cetus, to ravage their kingdom until Andromeda was sacrificed. Fortunately, the hero Perseus, armed with the head of Medusa, arrived just in time to save the princess.

While Andromeda is an ancient constellation, its western neighbour Lacerta is a later invention, added to the sky by Polish astronomer Johannes Hevelius in 1687. Although the lizard's stars are faint, their shape is quite distinct, and easy to pick out in a dark sky.

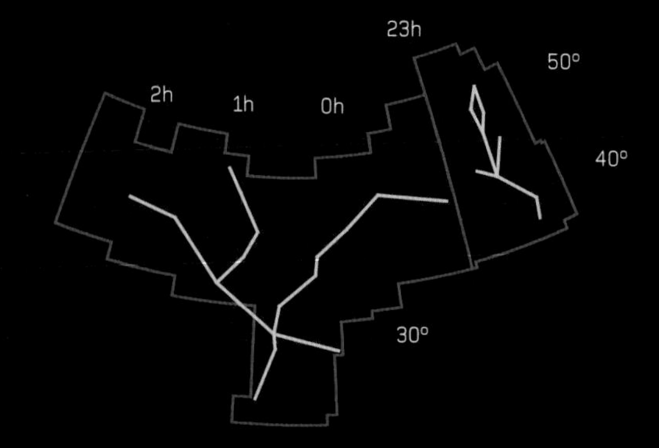

The Andromeda Galaxy

M31

0h43m, +41°16′	Magnitude: 3.4
Spiral galaxy	Distance: 2.9M light yrs

The most spectacular and fascinating object in this area of the sky, M31 is the closest major galaxy to our own, and the most distant object visible from Earth with the naked eye. Shining with the light of a fourth magnitude star, its fuzzy glow is easy to locate on a dark, Moonless night. Binoculars will show the elliptical shape of the galaxy's nucleus more clearly, but small telescope observers should stick to a low magnification to keep a wide field of view – the spiral extends over three degrees and is best seen in contrast with the dark background sky. With patience and a well-adapted eye, it is possible to trace the hazy outer edge of the galaxy, but larger telescopes or long-exposure photography are needed to reveal the dark dust lanes that twist around the galaxy and define its spiral arms – because we see M31 almost edge-on, its spiral structure tends to blur into a hazy, elongated ellipse.

The great spiral has been known for centuries – Persian astronomer Al Sufi, writing in AD 964, called it the 'little cloud'. However, it was only in the early 20th century that Edwin Hubble proved indisputably that it, and other 'spiral nebulae' like it, are galaxies far beyond our own, rather than star-clouds in a halo around the Milky Way. He did this by identifying faint Cepheid variables in M31 (a type of variable star whose period is linked to its intrinsic brightness – see Delta Cephei), and proving that they lay at least 2 million light years away.

Today, we know that Andromeda is the largest member of the Local Group (our small cluster of galaxies). Its disc is twice the diameter of the Milky Way's, some 200,000 light years across, but it is actually less massive than our galaxy, because the Milky Way contains much greater amounts of invisible but weighty 'dark matter'. At the heart of M31 lies an intriguing mystery – the galaxy seems to have two distinct cores (invisible to the optical observer, but revealed by image processing). It's possible that the Andromeda Galaxy has swallowed up a smaller neighbour in the past few tens of millions of years.

M32

NGC 221

0h43m, +40°52′	Magnitude: 8.1
Elliptical galaxy	Distance: 2.8M light yrs

The great spiral of M31 is accompanied by two satellite galaxies – M32 and NGC 205. Each is an elliptical ball of stars about 10,000 light years across and containing many millions of stars. M32 is the easier of the two to locate with a small telescope – it appears as an eighth-magnitude glow embedded in the spiral's outer haze about half a degree south of the central nucleus. In reality, it is 'above' the plane of the spiral, and lies in front of it as seen from our perspective.

Almach

Gamma (γ) Andromedae

2h4m, +42°20′	Magnitude: 2.3
Multiple star	Distance: 355 light yrs

Gamma is a beautiful triple star for telescopes of any size. The two principal components are a yellow star of magnitude 2.3 and a blue one of magnitude 4.8, easily split in small telescopes. The blue star has a fainter, magnitude 6.1 companion that presents more of a challenge – it orbits its parent every 61 years, and is currently nearing a close alignment that will make it indivisible with even larger telescopes around 2012.

Alpheratz

Alpha (α) Andromedae

0h8m, +29°05′	Magnitude: 2.1
Binary star	Distance: 97 light yrs

Andromeda's brightest star sits on the northeastern corner of the Square of Pegasus, and is also sometimes known as Delta Pegasi. Alpheratz is a blue-white star of magnitude 2.1, though it is also a variable similar to Cor Caroli (see Canes Venatici), so it undergoes rapid fluctuations of less than 0.1 magnitude. Analysis of the star's spectrum has revealed that it is a spectroscopic binary, consisting of two components orbiting each other in about 97 days, so close together that they cannot be distinguished with even the most powerful telescope.

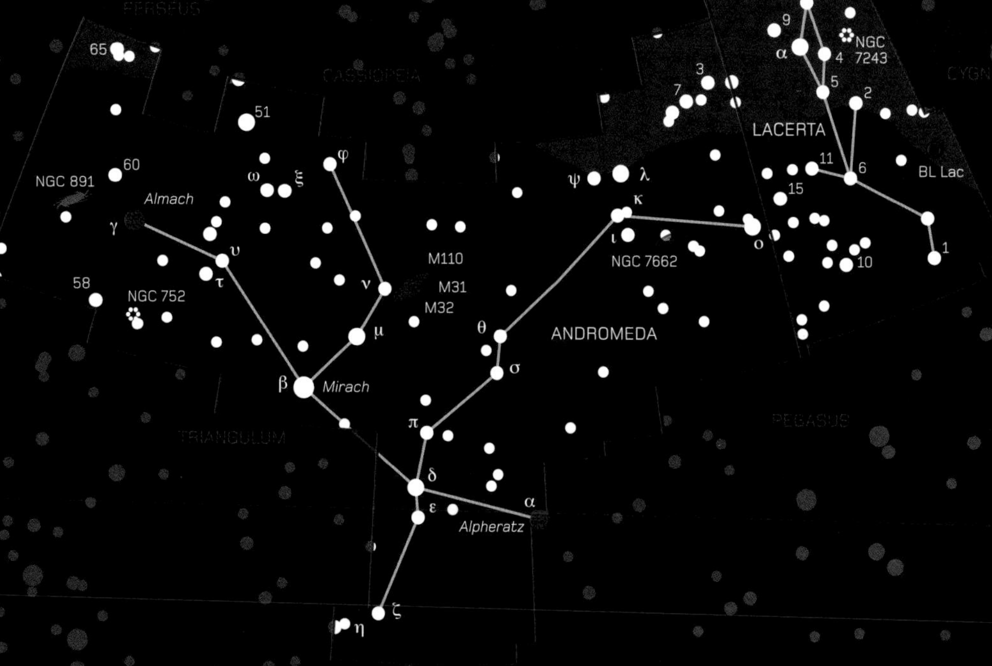

Blue Snowball Nebula

NGC 7662

23h26m, +42°33′	Magnitude: 9
Nebula	Distance: 2,200 light yrs

This attractive planetary nebula is one of the easiest to spot with a small telescope, appearing as a blue-green 'star' of magnitude 9 that resolves into a fuzzy disc at higher magnifications. The nebula's source – the central star – is only visible through larger telescopes, and is variable between magnitudes 12 and 16.

BL Lacertae

22h2m, +42°17′	Magnitude: 15 (var)
Active galaxy	Distance: 1Bn light yrs

Through a medium-sized telescope, this object appears to be a highly variable starlike point of light, ranging unpredictably between magnitude 12.4 and 17.2 over a matter of days or even hours. At first, astronomers assumed 'BL Lac' was a variable star, and named it accordingly. But in 1969 BL Lac was found to be a strong radio source, and in the 1970s, its position turned out to coincide with an extremely faint and distant elliptical galaxy.

BL Lacertae is now known to be the first in a new class of galaxies known as 'blazars'. A blazar is similar to a quasar – a galaxy with a supermassive black hole at its centre that is feeding hungrily on gas, dust and stars, and spitting out jets of high-energy particles from its poles as it spins. Most quasars can be identified because of these jets, but a blazar happens to be tilted so that the jets line up with the direction of the Earth – in effect, we are staring down the throat of a distant active galaxy.

Perseus

Perseus

This constellation, representing the hero at the core of a mythological story that spans much of the northern heavens, contains several bright and notable stars, but is still not the most easily identified, thanks in part to its position against the rich starfields of the northern Milky Way. However, it is a particularly rewarding hunting ground for amateur astronomers, since it contains many dense 'open' clusters of recently formed stars. Most notable among these is the famous 'double cluster' of NGC 869 and 884.

Perseus was the exiled Greek prince of Argos, favoured by Athena, goddess of wisdom. He confronted and slew the gorgon Medusa before flying to rescue the princess Andromeda from the sea monster Cetus. In his hand, Perseus still holds the head of Medusa, with which he turned the monster to stone. The Gorgon's evil eye is marked by the distinctive star Algol, the 'Winking Demon'.

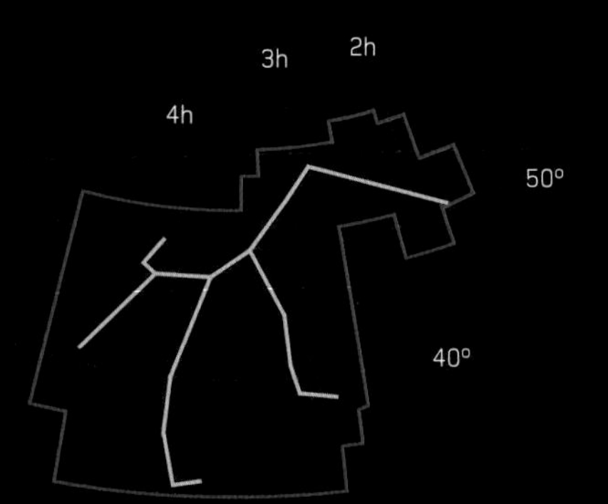

Algol

Beta (β) Persei

3h8m, 40°57′	Magnitude: 2.1 (var)
Binary star	Distance: 93 light yrs

One of the most famous stars in the sky, Algol was among the very first variable stars to be discovered. The earliest definite record of its changing light was made by Italian astronomer Geminiano Montanari in 1670, but it may have been noted centuries before that by Arab astronomers who first gave this star the name Algol, from a phrase that means 'the winking demon'.

While some variable stars pulsate steadily between maximum and minimum brightness, Algol 'winks'on and off fairly abruptly, dipping from its usual magnitude of 2.1 to a minimum of 3.4 for ten hours in a cycle of 2 days and 21 hours. Its changes are easy to track with the naked eye.

British astronomer John Goodricke was the first to correctly explain its behaviour in 1783 – it is an 'eclipsing binary' consisting of two stars with differing brightness that pass in front of each other with each orbit as seen from Earth.

Most of Algol's light comes from a hot blue-white star with 3.5 times the Sun's mass and 100 times its luminosity in visible light. The fainter component in the eclipse is actually the larger star – a yellow-orange giant with about 80 percent of the Sun's mass and about four times its luminosity. This runs against many of our ideas about stellar evolution, which says that the more massive star should evolve and reach giant status much faster – the only explanation is that the giant star must have lost a lot of its mass in its recent past – probably stripped away from its tenuous atmosphere by the gravitational tug of its nearby companion.

In addition, there is at least one other star involved in the system, orbiting the central pair in 600 days and not directly involved in the eclipse. Its gravity still causes the system to 'wobble', so that the eclipse period varies in a 32-year cycle.

Mirfak

Alpha (α) Persei

3h24m, +49°52′	Magnitude: 1.8
Giant star	Distance: 590 light yrs

The brightest star in Perseus is this white supergiant, 590 light years away and shining with the energy of more than 5,000 Suns. Mirfak, whose name means 'elbow', is the brightest and most evolved member of a cluster of hot blue and white stars that surrounds it. This cluster, called Melotte 20, is about 50 million years old, and Mirfak is the first of its stars to evolve away from the main sequence into a supergiant. This is largely because of its great mass – about eight times that of the Sun. The other stars must be less massive, since they are still in their main sequence, hydrogen-burning lifetimes. Seen through binoculars or a small telescope, Mirfak is surrounded by a beautiful field of blue and white stars.

M34

NGC 1039

2h42m, +42°47′	Magnitude: 5.5
Open cluster	Distance: 1,400 light yrs

Far older than either member of the famous Double Cluster, M34 contains about 80 stars visible to amateur instruments, with about 20 bright members within the reach of binoculars and small telescopes, distributed in a rough 'X' shape. M34 is thought to be about 180 million years old, and its stars have clearly drifted apart, so they are now scattered over a volume of space perhaps 15 light years across.

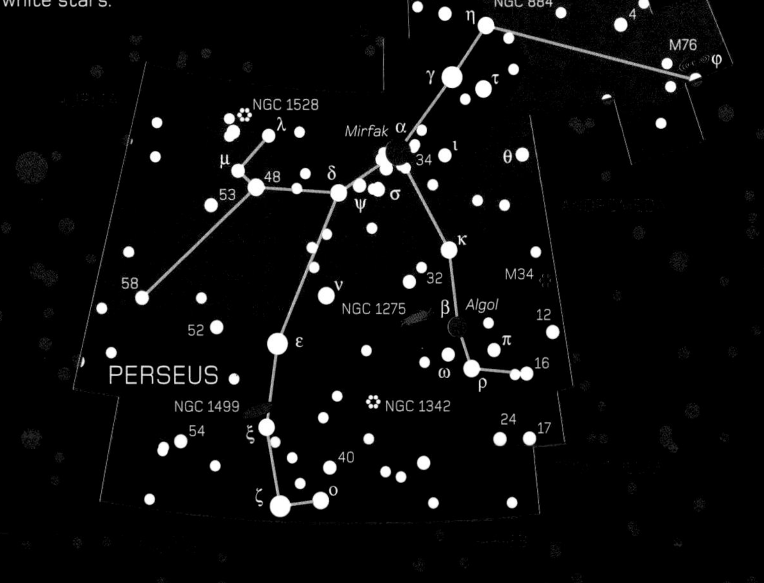

The Little Dumbbell Nebula

M76

1h42m, +51°34′	Magnitude: 10.1
Planetary nebula	Distance: 8,200 light yrs

Discovered by French astronomer Pierre Méchain in 1780, the deceptive structure of this nebula puzzled astronomers for many decades. Some thought that it was a 'spiral nebula' similar to the Andromeda Galaxy, and reported seeing stars in its 'arms'. Others believed, rightly, that it was entirely gaseous, but saw it as bipolar – two lobes of gas emerging from a central star. It was not until 1918 that the Little Dumbbell (so named from its resemblance to Vulpecula's brighter Dumbbell Nebula) was correctly explained as a planetary nebula – the expanding bubble of gas shed by a dying star. The nebula's unusual shape arises because of a slow-moving, dense ring of material expanding around the star's equator. We see this ring edge-on, so it forms the nebula's narrow central 'bar', while the faster-moving gas escaping above and below the ring forms the faint spherical shell once interpreted as spiral arms.

The Double Cluster

NGC 869/884

2h21m, +57°08′	Magnitude: 4.3, 4.4
Open clusters	Distance: 7,100 light yrs 7,400 light yrs

This famous pair of star clusters in northern Perseus is a stunning sight through binoculars or a small telescope with a wide field and low magnification. To the naked eye, it appears as a pair of fuzzy 'stars', and each cluster actually has its own stellar designation too – 'Chi Persei' for NGC 884 and 'h Persei' for NGC 869. The clusters are at different distances and not bound to one another by gravity, but they do share a common origin in a group of stars and clusters with similar age known as the Perseus OB 1 association. NGC 869 seems to be the older of the two clusters – it is about 19 million years old, while NGC 884 is 12.5 million years old. The brightest stars of each cluster are a mix of brilliant blue-white main sequence stars and more massive red and orange giants that are already nearing the end of their lives.

Pisces, Aries & Triangulum

The Fish, the Ram and the Triangle

East of the great Square of Pegasus and Andromeda lies a relatively sparse region of sky that contains three constellations – two large and well known as signs of the zodiac, the third small but with an intriguing secret of its own.

Pisces is the sign of the fishes, consisting of two long lines of faint stars meeting at its luminary Alrescha. Today, it is usually depicted as a pair of fish bound together at their tails, but in classical times the fish swam freely, and represented Aphrodite and her son Cupid fleeing from the approaching sea monster Typhon (nearby Cetus).

Aries, the Ram, covers the line of the ecliptic between Pisces and Taurus, and is marked by three moderately bright stars on its western edge. This is the animal that bore the Golden Fleece sought by Jason and the Argonauts in the famous myth. Elsewhere in the sky, Puppis, Carina and Vela mark the remains of Jason's ship the Argo, and Ophiuchus, the Serpent Bearer, was once seen as Jason himself.

Triangulum, to the northwest of Aries, is a small V-shaped group of stars – recognized by the ancient Greeks largely because of its resemblance to the capital letter delta (Δ). Its stars are fairly faint, but it does contain one of the nearest galaxies to our own.

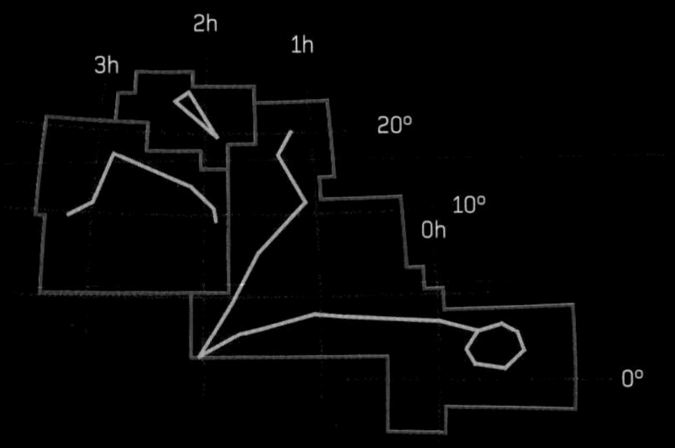

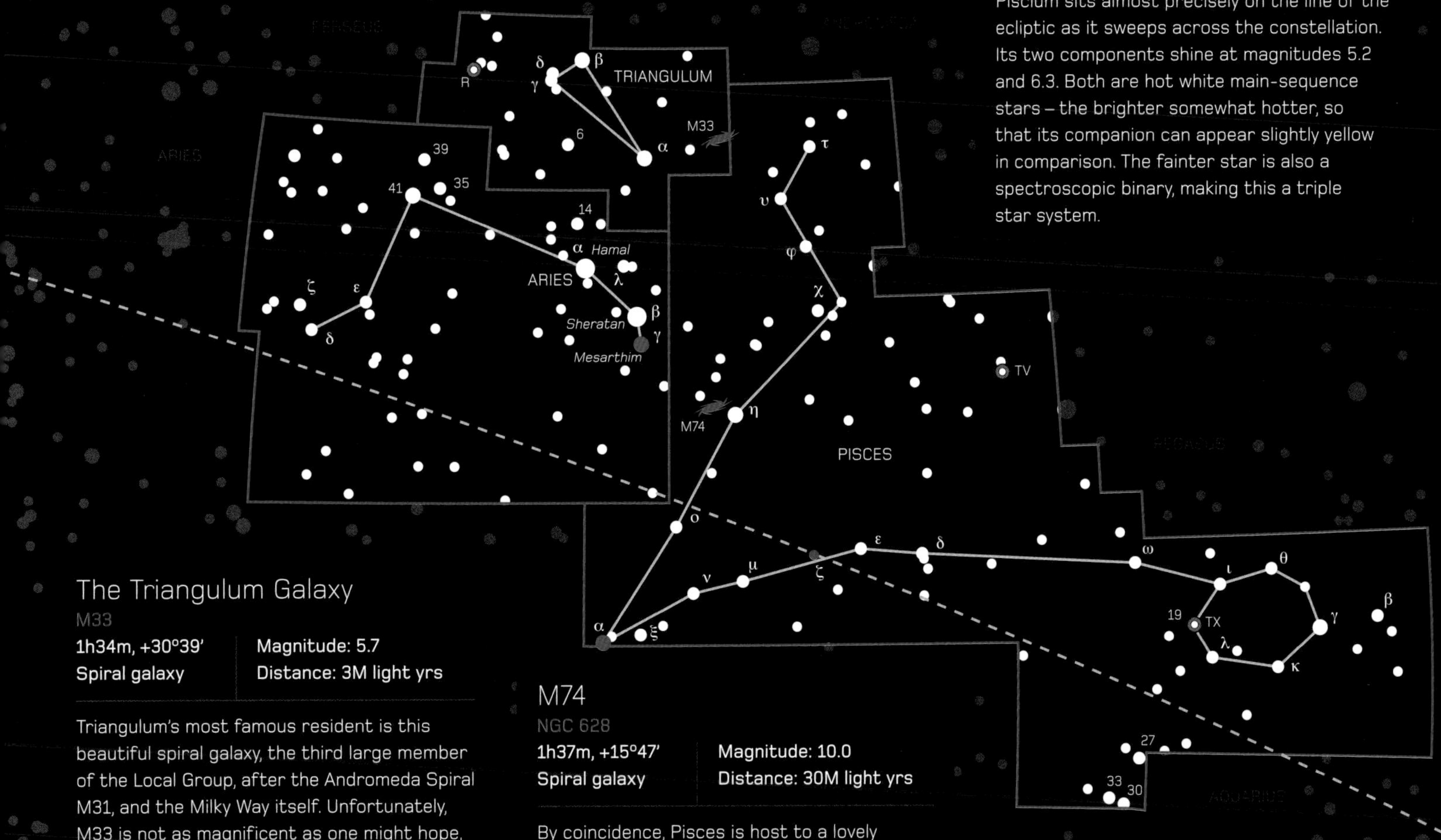

Zeta (ζ) Piscium

1h14m, +7°35′	Magnitude: 5.2
Multiple star	Distance: 148 light yrs

An easy double through any telescope, Zeta Piscium sits almost precisely on the line of the ecliptic as it sweeps across the constellation. Its two components shine at magnitudes 5.2 and 6.3. Both are hot white main-sequence stars – the brighter somewhat hotter, so that its companion can appear slightly yellow in comparison. The fainter star is also a spectroscopic binary, making this a triple star system.

Mesarthim

Gamma (γ) Arietis

1h54m, +19°18′	Magnitude: 3.9
Multiple star	Distance: 205 light yrs

This attractive star system consists of evenly matched white stars with magnitudes 4.6 and 4.7, easily divisible through the smallest telescope. These stellar twins are orbited by a more distant and fainter companion, an orange star of magnitude 9.6.

Mesarthim demonstrates an interesting anomaly of stellar temperatures. The dimmer star of the bright central pair (Gamma 1) is actually the more massive, hotter and (overall) more luminous of the two – it may even appear slightly bluer to some observers. Gamma 1 has an estimated mass of 2.8 Suns, while Gamma 2 weighs about 2.5 Suns, but because Gamma 1's surface is hotter, it emits more of its energy as invisible ultraviolet radiation. So while Gamma 1 pumps out about 56 times as much energy as the Sun, and Gamma 2 just 50 times as much, it is Gamma 2 that appears slightly brighter in visible light.

The Triangulum Galaxy

M33

1h34m, +30°39′	Magnitude: 5.7
Spiral galaxy	Distance: 3M light yrs

Triangulum's most famous resident is this beautiful spiral galaxy, the third large member of the Local Group, after the Andromeda Spiral M31, and the Milky Way itself. Unfortunately, M33 is not as magnificent as one might hope, partly because it is a distant third in terms of brightness, and also because it lies 'face on' to the Milky Way, so its light is spread out across an area of sky larger than the Full Moon. It is best seen on a dark night through binoculars or a small telescope on low power – only larger instruments will collect enough light to make higher magnifications rewarding.

The Triangulum Galaxy is a very different kind of spiral from either Andromeda or our own galaxy. Astronomers describe it as 'flocculent', which literally means 'clumpy'. Instead of being neatly aligned along well-defined spiral arms, its star-forming regions are spread out more randomly. Despite its lack of overall structure, medium-sized telescopes can pick out some impressive structures in M33. The most notable of these is the huge nebula NGC 604.

M74

NGC 628

1h37m, +15°47′	Magnitude: 10.0
Spiral galaxy	Distance: 30M light yrs

By coincidence, Pisces is host to a lovely spiral galaxy that can rival Triangulum's M33, even though it is almost ten times more distant. M74 is a so-called 'grand design' spiral, a face-on whirlpool of stars that displays a very well-defined structure. Small telescopes should be able to pick up the blurry glow of the galaxy's core in a dark sky and using low magnification, but medium-sized instruments are needed to show any detail.

M74 lies at the centre of its own small galaxy group, 30 million light years from Earth, and has been the subject for intensive study. Orbiting telescopes have imaged it at nearly all wavelengths, highlighting features that range from cool dust clouds and evenly distributed sunlike stars in the infrared, through brilliant clusters of hot young stars in visible and ultraviolet light, to violent black holes at X-ray wavelengths.

Alrescha

Alpha (α) Piscium

2h2m, +2°46′	Magnitude: 3.8
Multiple star	Distance: 140 light yrs

With a name that means 'the cord', Alrescha marks the tie between the constellation's two fishes. Despite its designation of Alpha, it is actually the second brightest star in Pisces, slightly fainter than Gamma. A moderate-sized telescope will split Alrescha in two, revealing a pair of white stars of magnitudes 4.2 and 5.2 in a 720-year orbit around each other. The components are thought to have masses of about 2.3 and 1.8 Suns respectively, and are both still on the 'main sequence', converting hydrogen to helium in their cores. The fainter component is probably a spectroscopic binary in its own right, and the brighter star may be too, but chemical peculiarities in the atmospheres of both stars make it hard to be certain in either case.

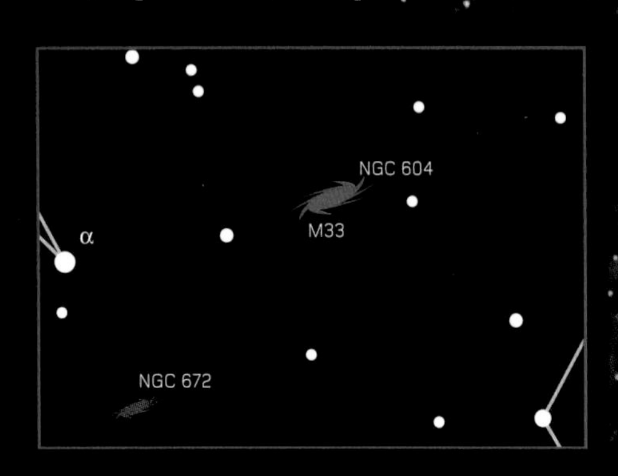

NGC 604

1h34m,+30°48′ Magnitude: 14.0
Diffuse nebula Distance: 3M light yrs

The huge starbirth region of NGC 604 is one of the largest emission nebulae in any known galaxy. More than 1,500 light years across, it contains at least 200 hot blue stars with the mass of 15 Suns or more, and countless smaller stars that escape our notice.

Ultraviolet radiation from these stellar giants causes the inner edges of the cavity to fluoresce, turning the entire nebula into a huge luminous grotto.

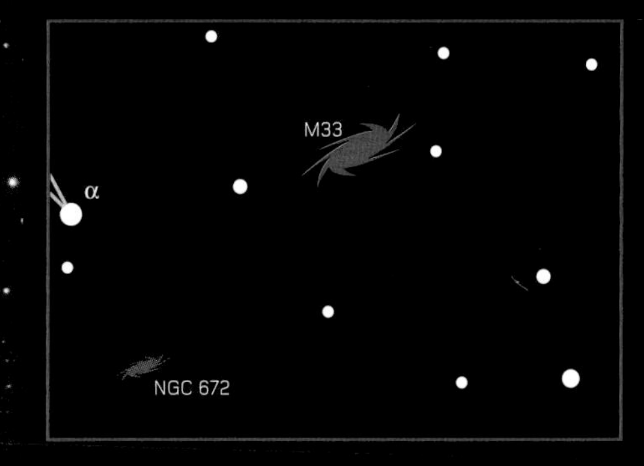

The Triangulum Galaxy

M33

1h34m, +30°39'
Spiral galaxy

Magnitude: 6.3
Distance: 3M light yrs

The loose 'flocculent' appearance of M33 indicates that its star formation is governed not by the spiral compression waves seen in other galaxies, but by more localized processes. Among the major influences are the shockwaves of supernovae – short-lived giant stars that, even in their death throes, trigger the next wave of starbirth.

Taurus

The Bull

One of the most obvious patterns in the sky, Taurus draws attention to itself through the bright red star Aldebaran, the V-shaped Hyades cluster beyond it, and the more compact but eyecatching Pleiades or Seven Sisters. The constellation has been seen as a bull almost universally – from the ancient Middle East to classical Greece, and even as far afield as pre-Columbian South America. The ancient Egyptians, for example, saw the constellation as an embodiment of the sacred bull Serapis, while in the classical world, the bull was Zeus, disguised in order to carry off the beautiful Europa, another of his infatuations. On another level, of course, the bull is simply a fearful beast charging at the mighty celestial hunter Orion. Only the Chinese had a different interpretation, seeing the central region of the constellation as a white tiger.

The Hyades and Pleiades, meanwhile, are two groups of nymphs or maidens in Greek myth – all daughters of the titan Atlas, though by different mothers. Other cultures saw the Hyades as a spoon, a net, or even a herd of camels, but strangely the association of the Pleiades with a group of maidens took root in many isolated cultures around the world, from Australia to North America, and even some Pacific Islands.

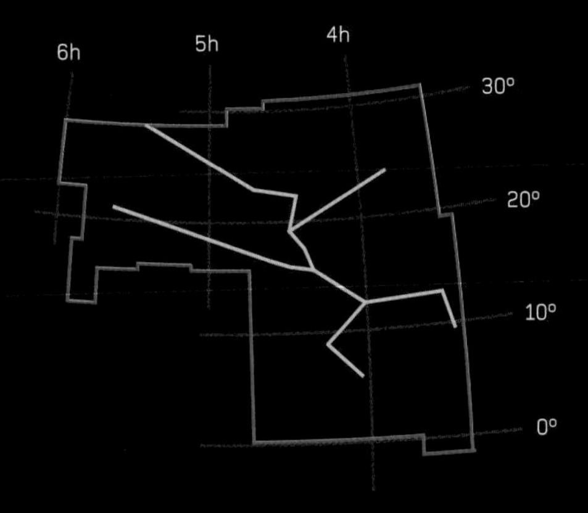

The Crab Nebula

M1

5h35m, +22°01′ | Magnitude: 8.4
Supernova remnant | Distance: 6,300 light yrs

Just to the northeast of Zeta Tauri, marking the tip of Taurus's lower horn, good binoculars or a small telescope will reveal a small fuzzy cloud of light. This is the Crab Nebula, the most famous supernova remnant in the sky, and an impressive sight through larger instruments or in long-exposure photographs.

Messier listed the nebula as object M1 in his famous catalogue, although it was first spotted a few decades earlier by British astronomer John Bevis, in 1734. It is the expanding remnant of a spectacular stellar explosion that was observed around the world in AD 1054, most notably by Chinese and Arab astronomers. The Crab supernova marked the death of a supergiant star far more massive than the Sun.

While most stars peter out in a planetary nebula after exhausting their core supplies of hydrogen and burning through the helium waste left behind, the most massive stars can continue to fuse the carbon and oxygen left behind by their helium phase, and then work through successively heavier elements. Each new element increases the density of the core still further, and generates less energy, but eventually, a turning point is reached. When the star tries to fuse atoms of iron, the process absorbs more energy than it releases. The core's power supply is abruptly cut off and, with no outward pressure of radiation to support it, the star collapses inwards. A shockwave rebounds off the core and rips the outer layers apart, compressing them so violently that they ignite in spontaneous fusion themselves.

A supernova can outshine an entire galaxy for several months, but eventually it fades to leave an expanding shell of gas – the supernova remnant, with the collapsed remnant of the core at the centre. In the case of the Crab, the remnant is a neutron star – a city-sized ball of subatomic particles with the weight of several Suns. The neutron star's intense magnetic field funnels its radiation into two narrow beams that spin around rapidly, turning it into a celestial lighthouse called a pulsar.

T Tauri

4h22m, +19°32′ | Magnitude: 9.6 (var)
Variable star | Distance: 575 light yrs

Just to the east of Epsilon Tauri at the upper tip of the Hyades' distinctive V lies an obscure star, typically of tenth magnitude, only visible through a telescope.

T Tauri varies unpredictably between magnitude 9.6 and as low as magnitude 14, and is one of the youngest stars known – barely a million years have passed since it began to form, and its core has not yet begun to fuse hydrogen into helium. Instead, the star shines because of heat generated as it collapses, and through the less demanding process of deuterium fusion, which can begin at lower temperatures and pressures. Intense stellar winds have cleared a gap in the cocoon of gas and dust allowing the star to be seen.

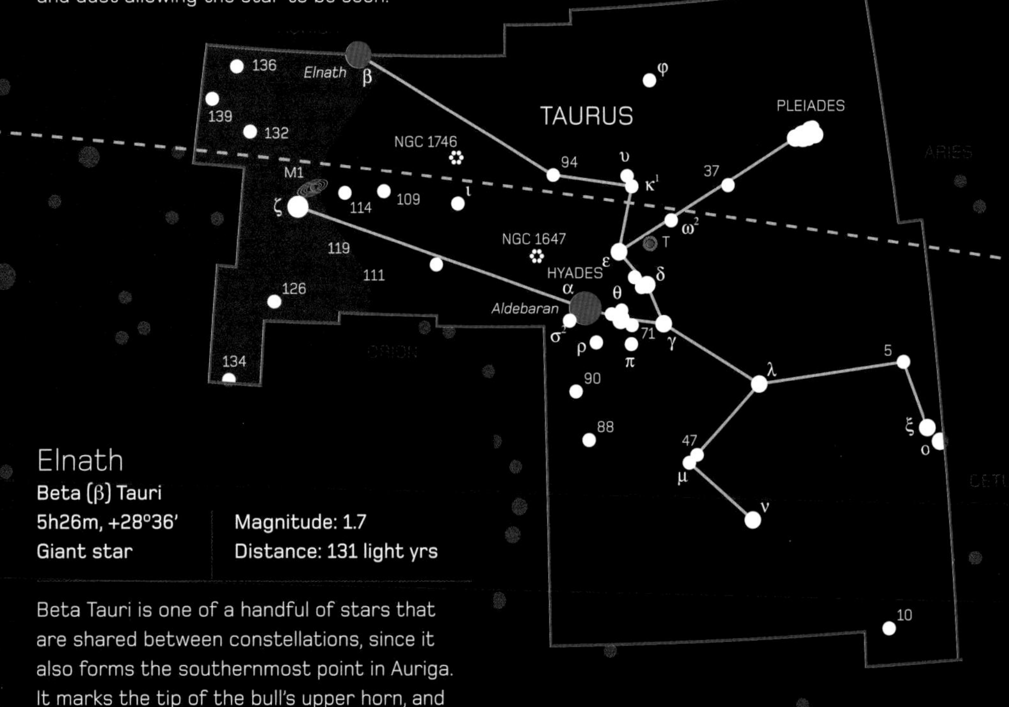

Elnath

Beta (β) Tauri

5h26m, +28°36′ | Magnitude: 1.7
Giant star | Distance: 131 light yrs

Beta Tauri is one of a handful of stars that are shared between constellations, since it also forms the southernmost point in Auriga. It marks the tip of the bull's upper horn, and its name means 'the butting one'.

Elnath is a blue-white giant at the end of life on the main sequence – it has exhausted the supply of hydrogen in its core, and fusion has begun to move out into an expanding shell, while the core slowly contracts and heats up. The shell-burning phase will see the star brighten and billow outwards, growing into a giant with a much cooler surface. Currently, Elnath generates about 700 times as much energy as the Sun, but more than half of this is released as ultraviolet radiation.

Aldebaran

Alpha (α) Tauri

4h36m, +16°31′ | Magnitude: 0.85 (var)
Variable star | Distance: 65 light yrs

Embedded in the midst of the Hyades, Aldebaran is the 13th brightest star in the sky. Its name means 'follower' (of the Pleiades), but it is not actually associated with either of Taurus's major star clusters, and is instead an independent orange giant closer to Earth. It orbits with a red dwarf of magnitude 13.5 that is normally lost in its glare.

Aldebaran currently has a diameter about half the size of Mercury's orbit, and its outer layers are pulsating without any obvious pattern, causing its brightness to vary by about 0.2 magnitudes.

The Hyades

Melotte 25

4h27m, +16° | Magnitude: -
Open star cluster | Distance: 150 light yrs

One of the closest open clusters to Earth, the Hyades are certainly the most impressive to the naked-eye observer. Only the Ursa Major group is closer, but it is so well scattered that it is difficult to notice its true nature. The Hyades, however, consist of about 200 stars (a dozen or more visible to the naked eye) scattered across several degrees of sky, and are a wonderful sight through binoculars. The brightest stars form a distinct V-shape that marks the head of Taurus, but the most obvious star of all, Aldebaran, is not a cluster member at all – it lies considerably closer to Earth. The brightest true member of the cluster, Theta 2 Tauri or Phaesyla, is a white giant of magnitude 3.4.

Astronomers estimate that the Hyades are a little under 800 million years old, and that they occupy a volume of space about 80 light years across (though the bright central stars are concentrated within about 10 light years of each other). Measurements of the motion of individual stars and the cluster as a whole suggest that it shares a common origin with the Beehive or Praesepe cluster, M44 in Cancer.

The Pleiades

M45

3h47m, +24° | Magnitude: 1.6
Open cluster | Distance: 440 light yrs

Perhaps the most famous star cluster in the sky, the Pleiades are visible to the naked eye as a hook-shaped cloud of stars. They have been recognized since ancient times, and were highly significant to several ancient cultures around the world.

Although the Pleiades are the 'Seven Sisters', people with average eyesight can usually pick out just six without optical aids. The cluster is relatively young – just 50 million years old – and so contains many stellar heavyweights that have not yet reached the end of their short main-sequence lives. One example is Pleione, also known as the variable star BU Tauri, which is a fast-spinning 'shell star' similar to Gamma Cassiopeiae. It varies in brightness unpredictably between magnitudes 4.8 and 5.5.

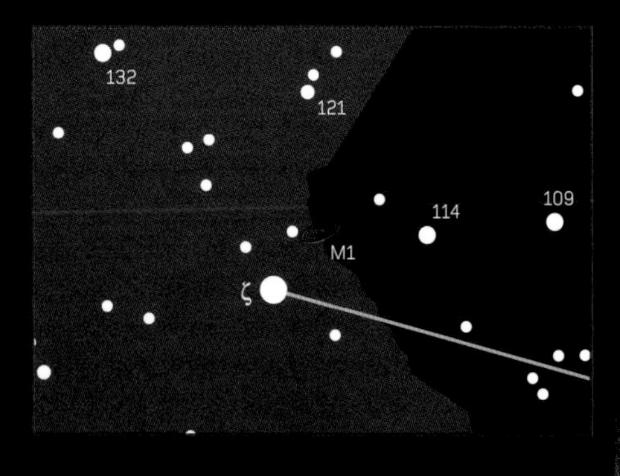

The Crab Nebula

M1

5h35m, +22°01′	Magnitude: 8.4
Supernova remnant	Distance: 6,300 light yrs

The shattered remains of a former giant star still glow brightly in northern Taurus, forming the sky's brightest supernova remnant. At its heart lies a rapidly spinning pulsar – a superdense star blasting beams of radiation across the heavens.

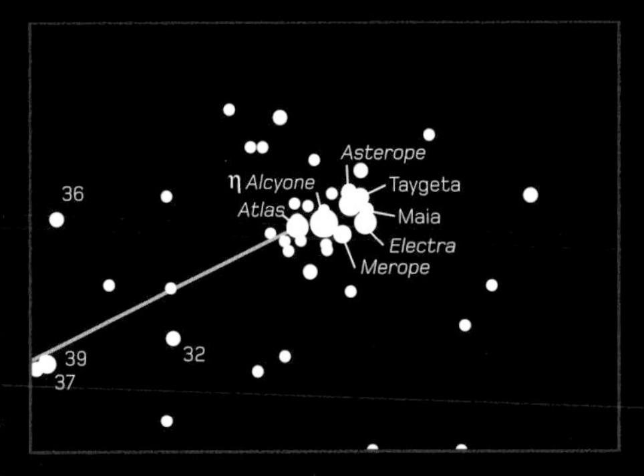

The Pleiades

M45

3h47m, +24°	Magnitude: 1.6
Open cluster	Distance: 440 light yrs

Binoculars or a small telescope transform this famous cluster into a field of brilliant blue and white stars – there are more than a hundred in total, of which the brightest is Alcyone, at magnitude 2.9. The stars are still surrounded by a faint nebula of gas left over from their formation, and traces of this can be glimpsed even with small instruments under clear, dark skies.

Gemini

The Twins

Easy to locate thanks to its twin bright stars Castor and Pollux, Gemini is a prominent constellation, found to the east of Taurus, especially near the start of the year when it lies opposite the Sun in the sky.

The proximity of these two stars drew the attention of ancient astronomers around the world, who almost always envisaged them as a pair of some sort. The Chinese saw them as the universal principles of Yin and Yang, the Egyptians as a pair of plants, the Arabs as two peacocks, the Phoenicians as a pair of kids or gazelles. The Romans sometimes associated them with Romulus and Remus, the founders of their nation, but the attribution that has stuck depicts them as Castor and Pollux, sometimes called the Dioscuri. Members of Jason's crew of Argonauts, Castor and Pollux were in fact 'half twins' – their mother was Queen Leda of Sparta, but Castor's father was her husband King Tyndareus, while Pollux was fathered by Zeus, who famously came to Leda in the form of a swan.

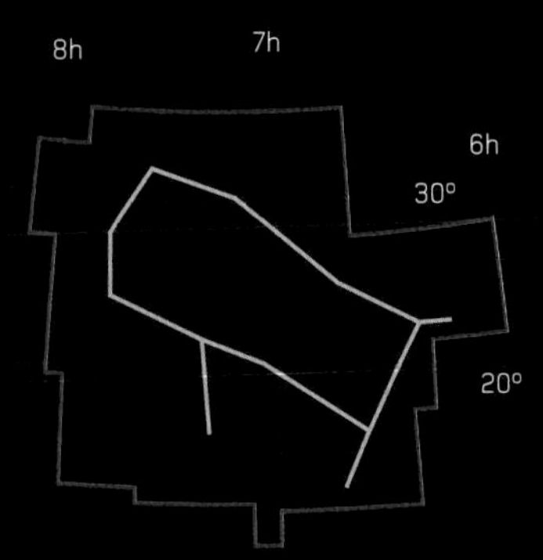

Mekbuda

Zeta (ζ) Geminorum

7h4m, +20°34′	Magnitude: 4.0 (var)
Variable star	Distance: 1,170 light yrs

This yellow supergiant, with a name that means 'the lion's paw', is a pulsating star that fluctuates in brightness every 10.2 days, in the same way as the famous variable star Delta Cephei. In fact, Zeta Geminorum is the brightest of all the 'Cepheid' variables, and its variations between magnitude 3.7 and 4.2 are easy to trace with the naked eye or binoculars. Optical aids will also show an apparent companion of magnitude 7.6. This is just a chance stellar alignment, but Zeta is also a spectroscopic binary with a real, though unseen, companion.

The longer a Cepheid's period, the greater the star's true luminosity, so Mekbuda must be considerably more brilliant than Delta Cephei itself (whose period is only 5.4 days). In fact, it is about 3,000 times as luminous as the Sun, while Delta Cephei is about two thirds as luminous.

Castor

Alpha (α) Geminorum

7h35m, +31°53′	Magnitude: 1.6
Multiple star	Distance: 52 light yrs

Fittingly for one of the 'heavenly twins', Castor is a famous multiple star. A small telescope should be able to split the bright blue-white star into a pair of magnitudes 1.9 and 2.9, which orbit each other in about 470 years. It will also reveal a third star in the system, a red dwarf of magnitude 9.3. Impressively, each of these stars is double in its own right. The bright Castor A and Castor B are spectroscopic binaries consisting of white stars slightly more massive than the Sun orbited by red dwarfs. The dimmer Castor C (also known as YY Geminorum) is an eclipsing binary pair of dwarfs, which drops in brightness to magnitude 9.8 when the stars pass in front of each other in a 19.5-hour cycle. Adding to the confusion, at least one of these dwarfs is a 'flare star', subject to violent stellar flares that can cause erratic changes in the system's brightness.

M35

NGC 2168

6h9m, +24°20′	Magnitude: 5.3
Open cluster	Distance: 2,800 light yrs

This impressive open star cluster packs several hundred stars into an area of sky about the size of the full Moon, with at least 100 members visible through a small telescope. Considering its distance of almost 3,000 light years, M35 must have an actual diameter of about 65 light years across, and it is thought to be about 100 million years old.

Observers using telescopes with wide fields and low magnifications to scan this beautiful cluster may spot a fuzzy ball of magnitude 8.6 to its southwest. This is NGC 2158, an incredibly old open cluster about 15,000 light years away.

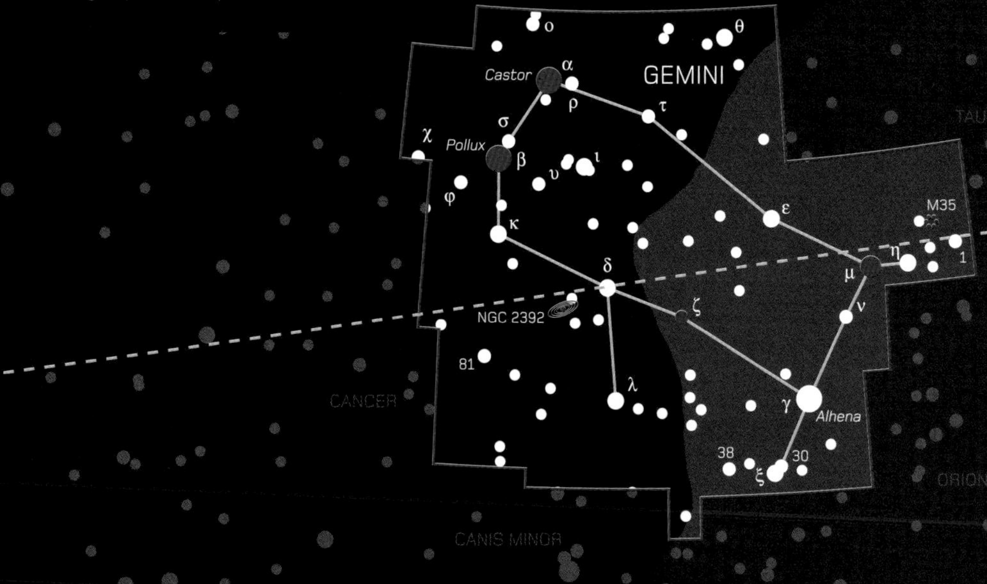

Eskimo Nebula

NGC 2392

7h29m, +20°55′	Magnitude: 9.1
Planetary nebula	Distance: 3,000 light yrs

This planetary nebula appears through a small telescope as a blue-green disc – it takes larger instruments to reveal the face-like structure formed by the lobes of gas in its centre, and the striated outer perimeter that gives the nebula its resemblance to a face peeking out of a furry hood. The nebula was discovered by William Herschel in 1787, and is formed from gas thrown off by its unusually bright, magnitude 10.5, central star.

Pollux

Beta (β) Geminorum

7h45m, +28°02′	Magnitude: 1.2
Extrasolar planetary system	Distance: 34 light yrs

In contrast to its sibling Castor, Pollux is a lone star – the brightest in Gemini by some margin. It is an orange giant 34 light years from Earth, and somewhat small for a star of its type. It has the mass of about 1.8 Suns, and is thought to be in the stage of its life where it shines largely by the fusion of helium in its core to form carbon and oxygen. As a result, it emits about 48 times the energy of the Sun, one third of that as infrared radiation.

Pollux is one of the brightest stars in the sky with a confirmed planetary system, although so far only one planet is known; 'Pollux b' has perhaps three times the mass of Jupiter, and orbits its star in a more-or-less circular orbit, slightly further out than Mars in our own solar system. Like most extrasolar planets discovered so far, Pollux b was identified from studies of the spectrum of light coming from its parent star. These revealed a periodic variation in the speed with which Pollux moves relative to Earth, caused as the planet's gravity tugs on it.

Tejat

Mu (μ) Geminorum

6h23m, +22°31′	Magnitude: 2.9 (var)
Variable star	Distance: 230 light yrs

This cool red giant is an irregular variable star, fluctuating in brightness between magnitude 2.75 and 3 as its outer layers pulsate in roughly 27 days. It is thought to be a star with the mass of three Suns, in the final stages of its life. Its core is already exhausted of secondary helium fuel, and is reduced to a gradually compressing planet-sized ball of carbon, oxygen and other relatively heavy elements. The star is now supporting itself through fusion of hydrogen and helium in expanding shells around the core, but although these cause its luminosity to increase greatly (including infrared radiation it pumps out about 1,500 times the energy of the Sun), they make the star unstable. Eventually, its oscillations will increase in intensity while its period will lengthen until it becomes a long-period variable similar to Mira in Cetus.

Cancer

The Crab

Cancer is the faintest of all the zodiac constellations – a rough triangle of stars best found by looking between Gemini's twin stars Castor and Pollux, and Leo's bright Regulus. The constellation has been seen as a crab since the earliest times – its true origin is unknown, but the ancient Greek astronomers explained this association with a tale that Cancer was a crab crushed underfoot by Hercules as he wrestled the serpent Draco. Other cultures saw it differently, however – to the Babylonians it was the Tortoise, and to the Egyptians a sacred Scarab Beetle.

One other intriguing association of the constellation is with darkness – it was known as the 'Dark Sign' into medieval times, and the name is usually explained by the constellation's faintness. However, it may have a deeper origin, since back in the very earliest days of astronomy, before written records, this was the constellation where the Sun lay during the northern winter solstice, when days were at their shortest.

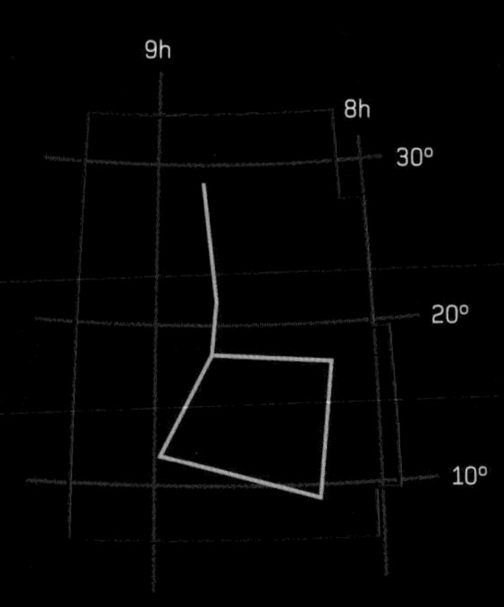

55 Cancri

8h53m, +28°20′
Extrasolar planetary system
Magnitude: 6.0
Distance: 41 light yrs

This binary star on the edge of naked-eye visibility is easily found with binoculars, although only the magnitude 6.0 yellow primary star is visible through most small telescopes, since its red dwarf companion shines at a dim magnitude 13.2.

The two stars orbit each other at roughly 1,000 times the Earth's distance from the Sun, and the primary is quite unusual – it is about 5 percent less massive than the Sun and 60 percent as luminous, but is 10 percent larger. This strange behaviour, contrary to the way most stars behave, may be due to 55 Cancri's high concentration of metals.

However, this star's main claim to fame is its planetary system – so far it has four known planets, designated 55 Cancri b through e. Three of these worlds orbit closer to the star than Mercury's orbit around the Sun, while the fourth, with a mass of perhaps four Jupiters, rings the system at roughly Jupiter's distance from the Sun. Other, smaller planets could well exist in the large gap between the known planets, and they might orbit in the 'habitable zone' where liquid water can exist on the surface of a planet. Such small planets are currently beyond the limits of detection, but if they are found, 55 Cancri would become a prime target in the search for extraterrestrial life.

Beehive Cluster

M44
8h40m, +19°59′
Open cluster
Magnitude: 3.7
Distance: 580 light yrs

One of the sky's finest open star clusters, the Beehive (also known as Praesepe, the manger), is easily visible to the naked eye on a dark night. The cloud of roughly 200 stars covers about 1.5 degrees of sky – about three times the diameter of the full Moon. It has been known from ancient times, and was first recorded in the third century BC, but Italian astronomer Galileo Galilei, famous for his discoveries with an early telescope, was the first to resolve it into individual stars.

M44's brightest star is Epsilon Cancri, a white main-sequence star of magnitude 6.3, lying on the near edge of the cluster at about 550 light years from Earth.

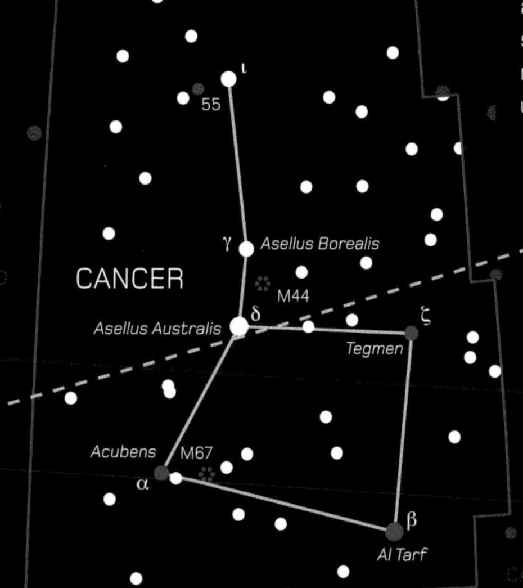

Acubens

Alpha (α) Cancri
8h58+11°51′
Multiple star
Magnitude: 4.3
Distance: 174 light yrs

Alpha Cancri is only the fourth brightest star in Cancer, but it is still interesting in its own right. Its name means 'the claw', and a moderate-sized telescope will reveal that it has a faint companion of magnitude 11.9, in an orbit that must last for several thousand years. However, each of these stars is in fact a double in its own right. Since the star lies close to the ecliptic, it is occasionally blocked out or 'occulted' by the Moon, and the 'double blinks' caused as the stars of Acubens disappear and reappear during these events confirms the evidence from analysing their spectra.

The primary pair is formed by identical stars of magnitude 5.1, orbiting each other in about six years. At first glance these appear to be normal white main-sequence stars with about twice the mass of the Sun, shining with about the same luminosity as Sirius, but their spectral lines show the chemical signature of metals such as zinc and strontium, that are, unusually, collecting on the surface of each star.

M67

NGC 2682
8h50m, +11°49′
Open cluster
Magnitude: 6.1
Distance: 2,700 light yrs

Discovered by German astronomer Johann Koehler in 1779, M67 is a remarkable open cluster. Normally, these groups of stars are born in huge clouds of gas and dust, but drift apart relatively quickly over tens of millions of years. M67, in contrast, has held together for an incredible period of time – at around four billion years old, it is barely younger than the solar system. This makes the cluster an ideal place to test our theories of stellar evolution, since all its stars have the same age, and all lie at roughly the same distance from us (so their apparent magnitudes mirror their true luminosities).

Most open clusters disintegrate due to a combination of external gravity from encounters with other stars, the outward blast caused by the detonation of the most massive stars in the cluster's youth, and the kicks of acceleration produced when two stars in a cluster have a close encounter with one another. So far, no one has produced a convincing reason why M67 has resisted these pressures to evolve, so it seems the cluster is simply a bizarre but lucky fluke.

Al Tarf

Beta (β) Cancri
8h17m, 9°11′
Giant star
Magnitude: 3.5
Distance: 290 light yrs

The constellation's brightest star, marking one of the Crab's legs, is Al Tarf ('the end'). It is a typical orange giant with about three times the mass of the Sun, and about 270 times as luminous in visible light, though it pumps out even more energy as invisible infrared radiation. Al Tarf has a faint, very wide companion of magnitude 12.9, that appears to lie at the same distance and share the same motion through space. It is almost certainly a true binary system, but the red dwarf companion must take around 80,000 years or more to complete its orbit.

Tegmen

Zeta (ζ) Cancri
8h12m+17°38′
Multiple star
Magnitude: 4.7
Distance: 83 light yrs

Another interesting multiple star, Zeta Cancri contains at least four, and perhaps more, stars, with an overall magnitude of 4.7 as seen from Earth. A small telescope will separate the two main components, which appear as yellow stars of magnitudes 5.1 and 6.2, orbiting each other in 1,000 years or more. Moderate telescopes will resolve the brighter of these two stars into almost identical stars (magnitudes 5.6 and 6.0) locked in a 60-year orbit that sees them at their widest separation in 2020.

Professional telescopes have successfully split the fainter of the two main components to reveal that it is orbited by a magnitude 10 red dwarf, and there is some evidence that this faint companion might be a binary in its own right.

Leo & Leo Minor

The Lion and the Lesser Lion

Leo is one of the handful of constellations that truly resembles the creature it supposedly represents – in this case, a resting lion. In the classical world, the pattern was associated with the Nemaean lion, fought by the demigod Hercules as one of his twelve famous tasks. Nearly every other culture also recognized the lion in these stars – the only notable exception were the Chinese, who saw the same figure as a celestial horse.

Leo is, of course, a zodiac constellation, and the Sun, Moon and planets all pass through it regularly. It is best seen around March, when the Earth's annual orbit puts it on the opposite side of the sky from the Sun. Some distance from the Milky Way, its chief attractions are some interesting stars, and a number of distant galaxies.

Leo Minor, in contrast, bears no resemblance to anything – it is a late addition to the sky, slotted into a convenient gap by Polish astronomer Johannes Hevelius and first published in his atlas of 1687.

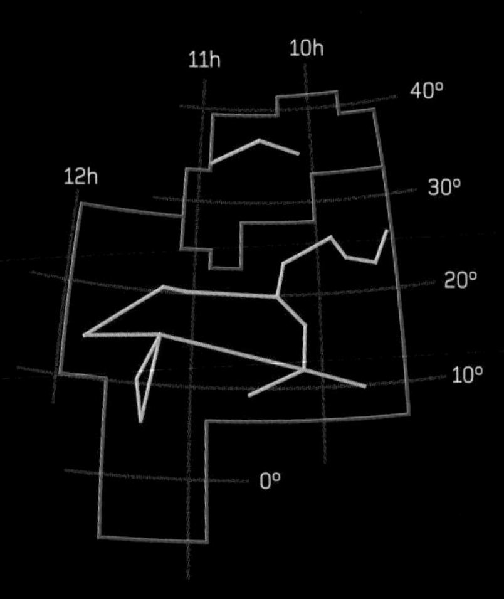

Regulus

Alpha (α) Leonis

10h8m, +11°58′	Magnitude: 1.35
Multiple star	Distance: 77 light yrs

The name of Leo's brightest star means 'the little king' in Latin, but this name far pre-dates the astronomers of classical antiquity. Even in ancient Mesopotamia, Regulus was regarded as one of four 'royal stars' (the others were Fomalhaut, Antares and Aldebaran).

The star itself is white, and about 140 times as luminous as the Sun. It is relatively young, and spins very rapidly (rotating once every 15 hours), which gives it a distinctly ovoid shape that has been detected by analysing its light. The primary star is orbited at a great distance by a pair of fainter companions, the brighter of which, at magnitude 8.1, can be seen with a small telescope.

Because Regulus lies almost on the ecliptic, it is often seen in spectacular conjunctions with the planets, and is also frequently occulted by the Moon. Such events are a fascinating sight for any astronomer.

Algieba

Gamma (γ) Leonis

10h20m, +19°50′	Magnitude: 2.0
Binary star	Distance: 126 light yrs

This attractive double star, with a name that means 'the forehead' (despite its location on what is usually seen as the lion's neck) is a pair of orbiting giants that circle each other in about 600 years. Small telescopes will reveal that the stars are orange with magnitude 2.6, and yellow with magnitude 3.5. However, the proximity of the two stars exaggerates the colour difference, and some observers see the hotter, yellow star as white or even green. Each star has a mass about twice that of the Sun, but astronomers are uncertain of their precise evolutionary states, so they cannot be sure which star is actually the more massive. One thing that is certain is the difference in their luminosities – in visible light the orange and yellow stars are about 150 and 50 times as brilliant as the Sun respectively. Because the cooler orange star emits more of its total energy as infrared, it is actually about 180 times as luminous as the Sun in total.

Praecipua

46 Leonis Minoris

10h53m, +34°13′	Magnitude: 3.8
Giant star	Distance: 98 light yrs

Leo Minor's brightest star has a name that means 'the Chief' – a useful reminder that it is the constellation's brightest star despite the fact it only has a 'Flamsteed number' designation. Praecipua is a fairly typical orange giant 98 light years from Earth and about 22 times as luminous as the Sun in visible light. However, this is only two-thirds of the star's overall energy output.

R Leonis

9h48m, +11°26′	Magnitude: 7.5 (var)
Variable star	Distance: 390 light yrs

One of the most easily observed variable stars in the sky, R Leonis is a long-period variable – a slowly pulsating red giant similar to the famous Mira (Omicron Ceti). At maximum it can be spotted with the naked eye, usually hovering around magnitude 5.8, but sometimes reaching a peak of magnitude 5.0 or even brighter. At minimum it sinks to magnitude 10 or 11. Binoculars are ideal for tracking it through the majority of its 312-day cycle.

M95

NGC 3351

10h44m, +11°42′	Magnitude: 9.7
Barred spiral galaxy	Distance: 38M light yrs

Leo contains two small groups of galaxies – the M65/M66 spiral pair near the lion's hindquarters, and M95, M96 and M105 beneath its belly. The central group is best located with a small telescope by scanning north of the naked-eye star 57 Leonis. M95 and M96 appear side by side as circular or oval patches of light, and a more powerful instrument will reveal their structure. M95 is a barred spiral with a small, concentrated nucleus, and almost-circular spiral arms connected to it by a long faint bar. M96 is a normal spiral with a larger and more diffuse nucleus. A little way to the north, another faint smudge of light marks the location of M105. In this case, though, higher-powered instruments do not reveal extra detail – M105 is an elliptical galaxy, nothing but a huge ball of stars.

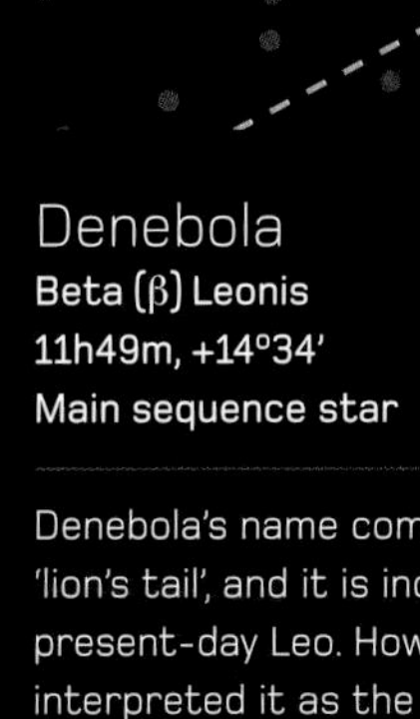

Denebola

Beta (β) Leonis

11h49m, +14°34′	Magnitude: 2.1
Main sequence star	Distance: 36 light yrs

Denebola's name comes from the Arabic for 'lion's tail', and it is indeed the hindmost star in present-day Leo. However, most artists have interpreted it as the creature's rump, often with the tail itself extending to a tuft in Coma Berenices. This relatively nearby star is yellow-white in colour and about 15 times as luminous as the Sun, suggesting it has just over twice the Sun's mass. It is thought to be about 400 million years old, and infrared studies have revealed that, like the bright stars Vega and Fomalhaut, it is surrounded by a cool disc of potentially planet-forming dust. Curiously, although this star is resolutely on the 'main sequence' of stellar evolution, it is slightly variable (by fractions of a magnitude in a period of a few hours). The pattern of variations matches so-called 'Delta Scuti' stars, but such stars are usually more evolved 'subgiants' that have exhausted their core supplies of hydrogen and begun to swell into giants.

NGC 4414

12h26m, +31°13′
Spiral galaxy

Magnitude: 11.0
Distance: 62M light yrs

This somewhat fluffy spiral is a 'flocculent' galaxy similar to the Triangulum Galaxy M33 in our own Local Group. Here the major regions of star birth and death are not confined to tightly regimented spiral arms, but spread out in clumps where the life and death of one generation of stars triggers the genesis of the next.

Black Eye Galaxy

M64

12h57m, +21°41′ | Magnitude: 8.5
Spiral galaxy | Distance: 19M light yrs

M64's thick foreground dust lane gives it an obvious resemblance to a bruised eye. It is a remnant of this cannibal galaxy's last meal – a smaller galaxy whose only trace is a halo of gas and dust surrounding the larger spiral. The galactic collision, perhaps a billion years ago, still makes itself known through the swathe of rich, star-forming regions that shine through the dust.

Virgo

The Maiden

This constellation has been almost universally seen as representing a celestial maiden. For the Greeks and Romans she was usually Persephone, daughter of Ceres the harvest goddess. The pattern's brightest star, Spica, was an ear of wheat held in the maiden's hand. Virgo's association with the harvest, and Spica's with corn, is found around the world, but the classical world had an alternative way of seeing the figure, as a goddess of justice holding scales in her hand.

As a zodiac constellation, Virgo often plays host to the Sun, Moon and planets. Although most of its stars are undistinguished, Spica stands out in a comparatively empty region of sky. Observers in the northern (and much of the southern) hemisphere can find it easily by extending the curve of stars in the handle of the Plough, or Big Dipper, into an arc that passes first through Arcturus, and then through Spica. But Virgo's real treasures are telescopic – it is home to the Virgo Cluster, the nearest major galaxy cluster to Earth.

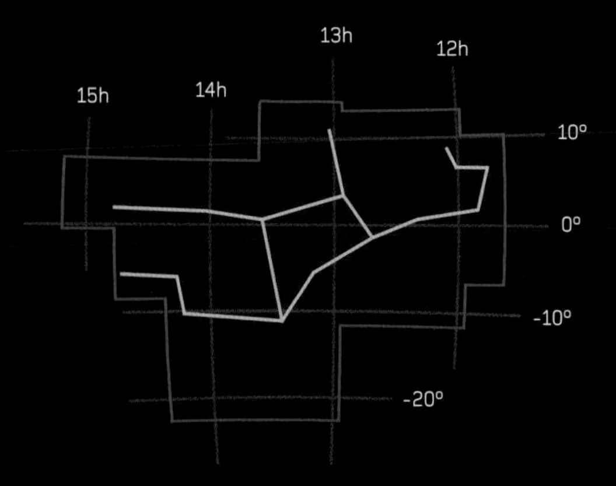

Spica

Alpha (α) Virginis

13h25m, -11°10'	Magnitude: 1.0
Multiple star	Distance: 260 light yrs

Virgo's brightest star, Spica, is also one of the brightest stars in the sky, and a useful marker for locating the constellation as a whole. In Latin, the name means 'ear of wheat', indicative of Virgo's association with the harvest.

The star itself is a binary, consisting of two very similar blue main-sequence stars orbiting each other in just 4 days. The brighter of the two stars is a pulsating variable similar to Beta Cephei, while the entire system is also variable because the stars distort each others' shapes – as the system spins around its centre of gravity, we see different amounts of each star's surface.

Spica's components both have searing surfaces of around 20,000°C. As a result, most of the energy they pump out escapes as ultraviolet radiation rather than visible light. So while optically Spica has the luminosity of 2,100 Suns, it is actually radiating about 13,000 times as much energy as our star.

Because Spica lies close to the ecliptic, it is frequently seen close to the planets, and occasionally the Moon can pass in front of it in an event called an occultation. By studying the way that Spica's light falls away during these events, astronomers have found evidence for three more faint stars in the Spica system.

Porrima

Gamma (γ) Virginis

12h42m, -1°27'	Magnitude: 2.9

M49

NGC 4472

12h30m, +8°00'	Magnitude: 8.4
Elliptical galaxy	Distance: 60M light yrs

This prominent member of the Virgo galaxy cluster was first spotted in 1771 by French comet-hunter Charles Messier, who added it to his catalogue of potentially confusing objects. A glance through a small telescope shows why – this elliptical, fuzzy 'star' looks distinctly comet-like. In reality, though, it is a huge elliptical ball of stars, 160,000 light years long. M49 is one of several 'giant ellipticals' that form the gravitational ballast of the Virgo Cluster – but it is dwarfed by its neighbour M87.

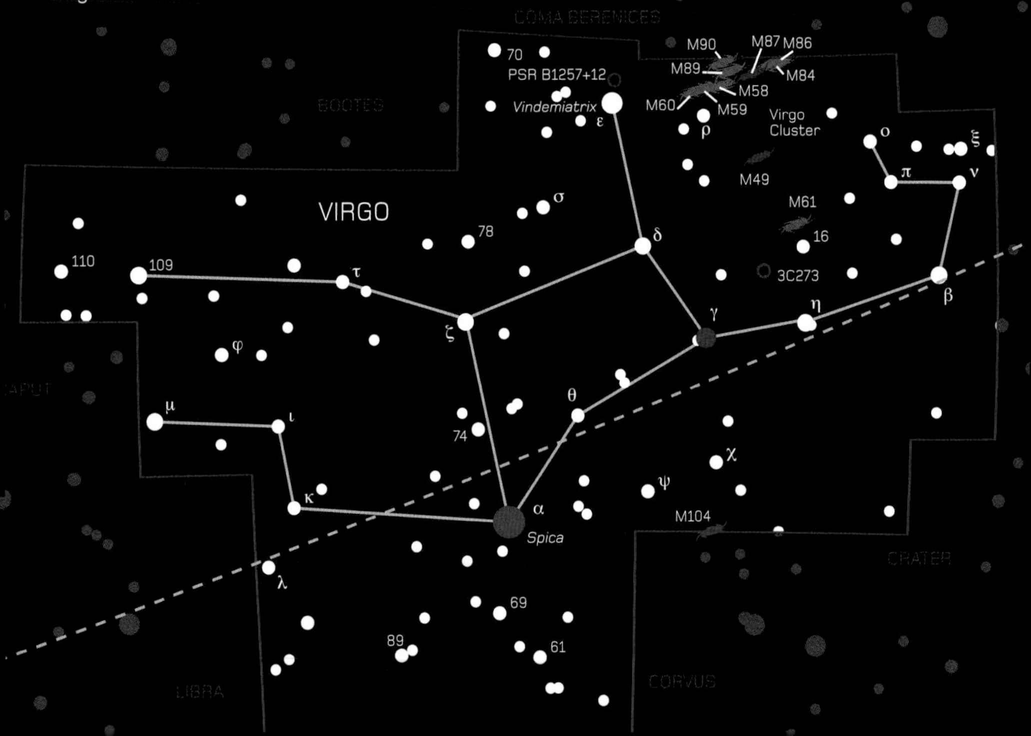

M87

NGC 4486

12h31m, +12°24'	Magnitude: 8.6
Elliptical galaxy	Distance: 60M light yrs

This huge spherical ball of stars appears as a fuzzy rounded sphere through amateur telescopes, not dissimilar to some globular clusters. But the reality is rather different – M87 is one of the largest known galaxies, containing perhaps a million million stars in the central region (about 160,000 light years across), and with many more in a halo that stretches out to a diameter of about 500,000 light years in total. It is the central galaxy of the Virgo Cluster, and its gravity affects all the other galaxies around it – billions of years from now it may swallow them all up to become even more massive.

It is also an active galaxy: it coincides with a source of radio waves called Virgo A, and long-exposure photographs reveal a jet of material shooting out from its centre. This all suggests that the supermassive black hole at the centre of M87 (with a mass of perhaps 2 billion Suns) is currently being fed with gas and dust. Hubble Space Telescope images have even shown a disc of material spinning around this central region – probably the vestiges of the last galaxy cannibalized by this cosmic monster.

PSR B1257+12

13h0m, +12°40'	Magnitude: -
Extrasolar planetary system	Distance: 980 light yrs

The planetary system around this remote stellar remnant was the first to be discovered, thanks mainly to its bizarre nature.

Serpens & Ophiuchus

The Serpent and the Serpent-Bearer

The large but comparatively barren constellation of Ophiuchus, the Serpent Bearer, is often called the thirteenth sign of the zodiac, since the ecliptic, the plane of the solar system extended into space, passes through it in between Sagittarius and Scorpius; hence the planets, Sun and Moon can all sometimes be found here. Ophiuchus is usually depicted wrestling a snake that is depicted in the sky's only two-part constellation: Serpens Caput marks the snake's head, and Serpens Cauda its tail. The figure has been seen as a representation of various figures from ancient mythology, including Cadmus wrestling a dragon, and Jason in pursuit of the Golden Fleece of Aries.

Today the group is given a more benevolent interpretation as Asclepius, the god of medicine, who carries with him a staff with a snake entwined around it (a symbol still used by the medical profession to this day). The constellations are best found by looking to the north of brilliant Antares in Scorpius.

M5

NGC 5904

15h19m, +02°05′	Magnitude: 5.6
Globular cluster	Distance: 24,500 light yrs

The brightest of several globular clusters in this region of the sky (others include M10 and M12 in Ophiuchus), M5 is just on the border of naked-eye visibility, and an impressive sight through binoculars or a small telescope. The outer reaches of this 170-light-year ball are easily resolved into stars, but the central region is an ill-defined blur to even the largest instruments. The cluster is particularly rich in variable stars of the RR Lyrae or 'cluster variable' type, allowing a very accurate measurement of its distance.

M16

NGC 6611

18h19m, -13°47′	Magnitude: 6.4
Open cluster	Distance: 7,000 light yrs

This impressive open star cluster is one of the youngest in the sky, still embedded in the famous Eagle Nebula from which its stars are forming. Astronomers know that the cluster only formed around 5 million years ago because it contains very hot, massive blue stars of type O, which have very short lifespans – if the cluster had been around much longer, these stars would have disappeared in supernova explosions.

Stars are still emerging from the nearby pillars of dust and gas in the nebula, and the entire region has been subjected to intense study with the Hubble Space Telescope and other instruments providing many new insights into the processes of starbirth. The cluster itself was discovered by Swiss astronomer Philippe Loys de Chéseaux in 1745, but reports never reached Charles Messier, who found it independently in 1764, and was the first to note the nebulosity around it. Although spectacular in long-exposure photographs, the nebula is too faint to observe directly through any but the largest amateur telescopes – though it adds tell-tale blur to the stars of M16 when they are viewed through binoculars.

Rasalhague

Alpha (α) Ophiuchi

17h35m, +12°34′	Magnitude: 2.1
Main sequence star?	Distance: 47 light yrs

Alpha Ophiuchi's Arabic name means simply 'the head of the serpent bearer'. It is a slightly variable white star, 30 times more luminous than our Sun, and there is some evidence that it has just begun to evolve off the main sequence of stellar evolution, towards giant status. Rasalhague is also a spectroscopic binary.

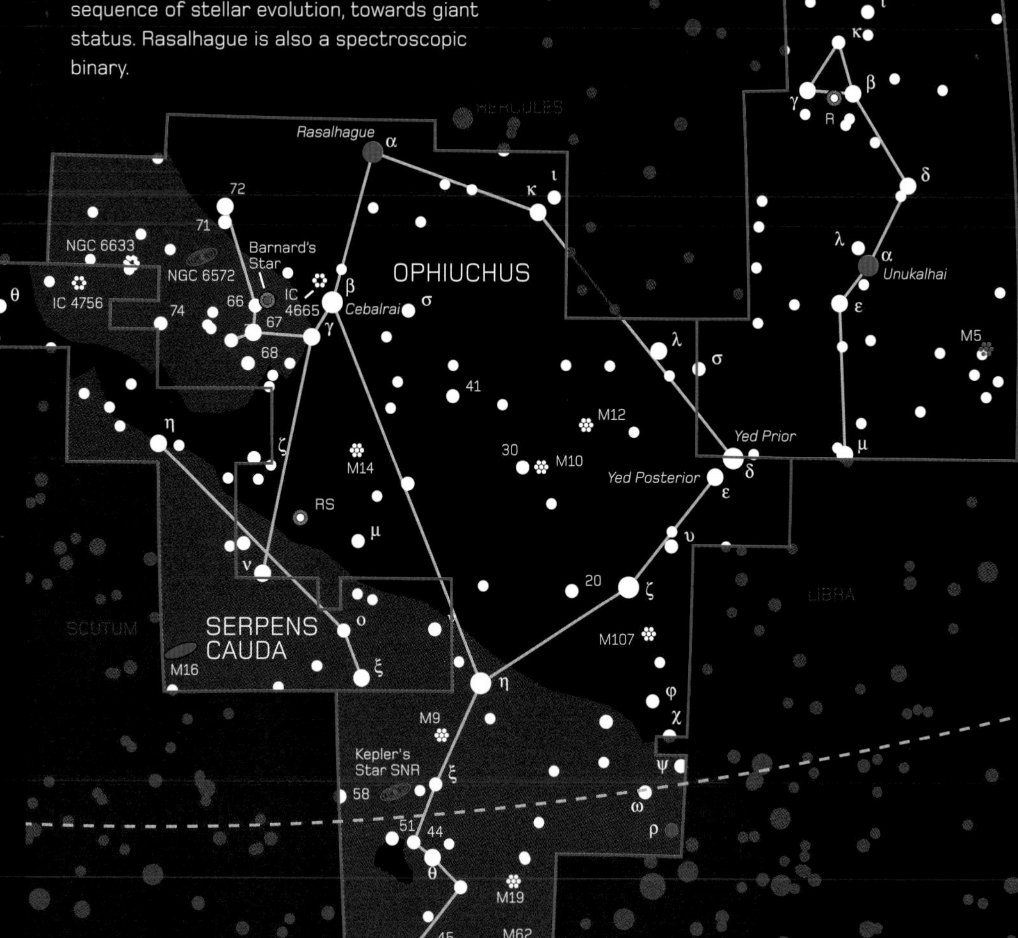

Unukalhai

Alpha (α) Serpentis

15h44m,+06°25′	Magnitude: 2.6
Giant star	Distance: 73 light yrs

This orange giant star has an Arabic name meaning 'neck of the serpent'. Small telescopes will show that it has an apparent companion of magnitude 11.8, but this is thought to be a mere chance alignment. Unukalhai itself is a fairly typical orange giant, with a surface temperature of about 4,000°C.

Rho (ρ) Ophiuchi

16h26min, - 23°27′	Magnitude: 4.6
Multiple star	Distance: 395 light yrs

This comparatively faint star is a beautiful multiple system, still embedded in the nebula from which it has only recently begun to shine. Small telescopes reveal a magnitude 5.0 central star with a magnitude 5.9 companion, flanked by fainter stars of magnitudes 6.7 and 7.3, best seen at low magnification or through binoculars. The entire system is embedded in rich clouds of star-forming material, which probably absorb much of the stars' light on its way to us – the stars' true luminosity is estimated at around 1,000 times that of the Sun, so if its light reached us unobscured, it the system would be of around magnitude 2.5. Interstellar absorption affects all stars to some extent, and accounting for it is always a challenge when estimating the true brightness of stars.

Barnard's Star

V2500 Ophiuchi

17h58m, +-4°42m	Magnitude: 9.54
Dwarf star	Distance: 5.9 light yrs

One of the best known stars in the sky despite its faintness, Barnard's Star's claims to fame are that it is one of our closest neighbours in space, and the star with the fastest 'proper motion' across the sky – the combination of its own movement through space with that of our Solar System ensures that it drifts across the sky by half a degree (a full Moon's width) every 180 years.

Barnard's Star is less than 6 light years away, but shines with just 0.35% of the Sun's luminosity, and was only discovered by US astronomer E.E. Barnard in 1916. This faintness is not unusual – it merely indicates that the star is a red dwarf, with a mass perhaps one fifth of the Sun's.

This star has existed for billions of years longer than our Sun – perhaps since the birth of the Milky Way itself – and persists for so long because the nuclear reactions in its core proceed at a fraction of the speed of those in more massive stars. Yet despite this, Barnard's Star and red dwarfs like it can produce surprising and unpredictable activity in the form of huge stellar flares and dark starspots on their surface.

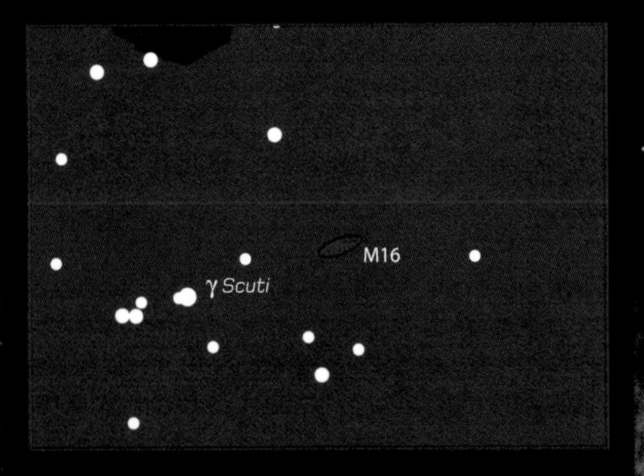

Eagle Nebula

M16

18h19m,-13°47'
Open cluster and emission nebula

Magnitude: 6.4
Distance: 7,000 light yrs

As fierce radiation from young, hot stars born in the Eagle Nebula blasts away at the stellar nurseries of gas and dust, it has created the finger-like projections famous as the Pillars of Creation. Their surrounding cocoon is illuminated by the same ultraviolet radiation, creating a glowing cavern in the heavens.

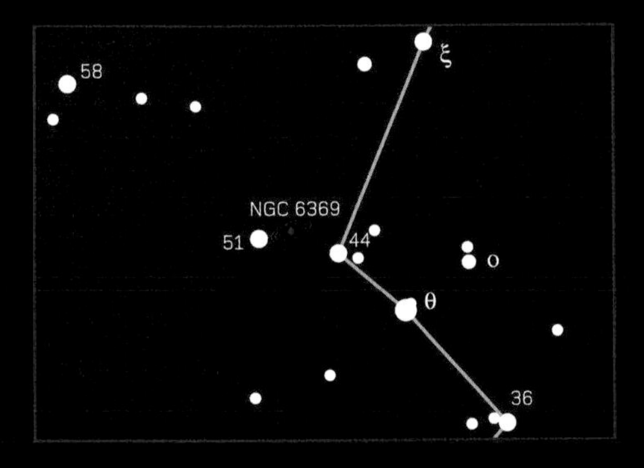

Little Ghost Nebula

NGC 6369

17h29′,-23°46′
Planetary nebula

Magnitude: 13
Distance: 3,500 light yrs

The ragged rings of the Little Ghost Nebula offer a glimpse of our Sun's future. As ultraviolet radiation floods out from the hot but faint white dwarf at its centre, it interacts with the shells of gas previously shrugged off by the dying star, causing them to glow. Oxygen atoms in the expanding nebula glow blue, hydrogen glows green, and nitrogen emits red light.

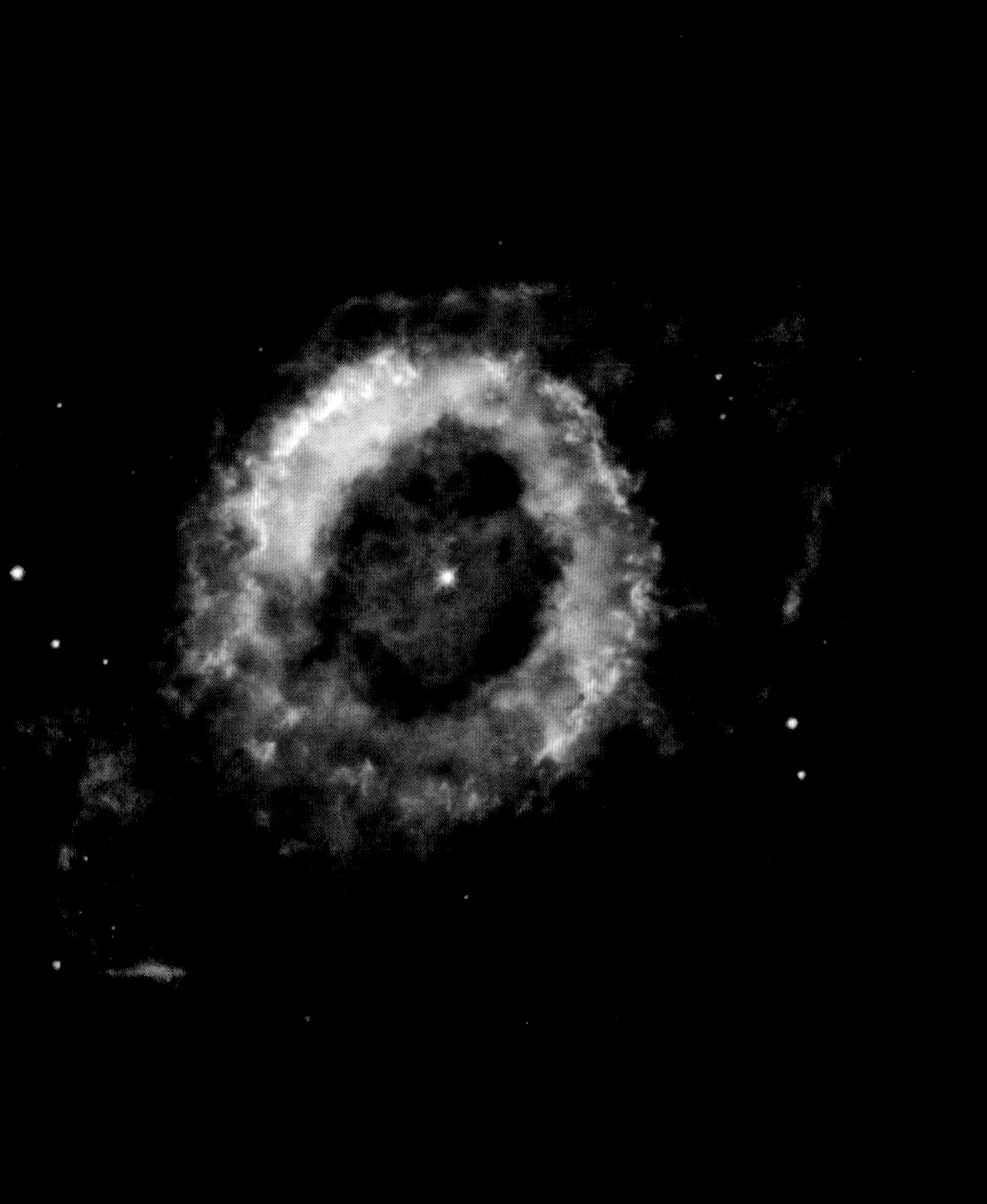

Orion

The Hunter

One of the best known, and certainly most beautiful, constellations in the entire night sky, Orion straddles the celestial equator, so that its heavenly treasures are visible to observers anywhere on Earth. Representing a hunter from Greek and Roman myth, bright stars mark his shoulders and knees, while fainter ones trace the position of the hunter's head, his raised shield, and his club. The centrepiece of the constellation is a distinctive chain of three fairly bright stars, marking the hunter's belt, and the fainter, shorter chain, including the Great Orion Nebula M42, that shows the location of his sword.

Orion is the centre of a complex scene involving many of the surrounding constellations. Canis Minor and Major, directly behind him in his precession around the sky, represent his hunting dogs, and Taurus an attacking bull that he is fending off, while the hare Lepus cowers unseen at his feet. Orion is also connected mythologically with Scorpius, a zodiac constellation on the opposite side of the sky.

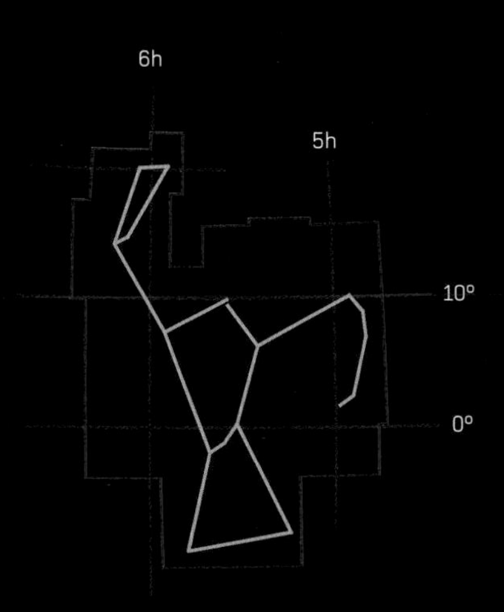

Rigel

Beta (β) Orionis

5h15m, -08°12′	Magnitude: 0.1
Main sequence star	Distance: 780 light yrs

Despite its designation, Beta Orionis, or Rigel, is the constellation's brightest star. In striking contrast to Betelgeuse, it is a brilliant blue-white colour. Some have speculated that the mislabelling of the two stars suggests Betelgeuse actually outshone Rigel in the historic past. Brilliant blue-white 'supergiants' such as Rigel are some of the most massive stars known – Rigel has an estimated mass of 17 Suns, a surface temperature of 11,000°C, and a luminosity 66,000 times that of our own Sun. Rigel has a companion star (an 'optical double' – the two are not directly related) that is theoretically within reach of binoculars at magnitude 6.8, but is hard to find when lost in the brighter star's glare.

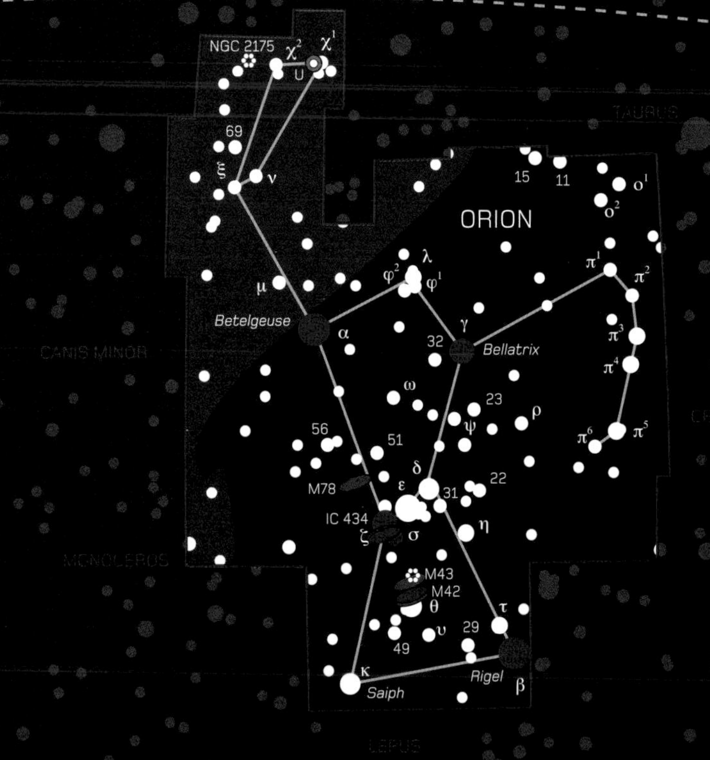

Great Nebula

M42

5h35m, -05°27′	Magnitude: 4.0
Emission nebula	Distance: 1,500 light yrs

Orion's centrepiece is the Great Nebula, otherwise known as M42. Visible to the naked eye as a smudge of light in the short chain of stars representing the hunter's sword, the nebula is in fact a huge stellar nursery, one of the brightest gas clouds in the entire sky. Long-exposure photographs or large telescopes reveal glowing tendrils of gas stretching from its bright core constrained into a flower-like form by darker lanes of obscuring dust around them.

Seen from Earth, M42 itself has a diameter twice that of a full Moon – in reality, it is more than 15 light years across, and 1500 light years away. Several large blobs of gas extend above and below it, and the whole sword region is the centre of a broader complex of star-forming clouds that stretches into neighbouring constellations and has a diameter of several hundred light years.

At the centre of the nebula lies a multiple star (Theta Orionis) called the Trapezium. Binoculars should show it as a double, while telescopes will reveal that it is in fact a 'double double' consisting of two closely-bound pairs of young stars in orbit around one another.

Sigma (σ) Orionis

5h39m, -02°36′	Magnitude: 3.8
Multiple star	Distance: 1,150 light yrs

The beautiful multiple star Sigma Orionis appears as a normal blue-white star to the naked eye, hanging just below Orion's belt. Binoculars will reveal that it has a fainter companion of magnitude 6.6, and small telescopes will show a second, closer companion of equal magnitude, and perhaps a fourth, fainter member of the group at magnitude 9 on the opposite side of the primary star. The most obvious of the companions is itself a close eclipsing binary (though its variations are too small to be easily detected), and an unrelated triple star, Struve 761, lies nearby.

Alnitak

Zeta (ζ) Orionis

5h41m, -01°57′	Magnitude: 1.8
Multiple star	Distance: 820 light yrs

The easternmost star of Orion's belt (its name means 'the girdle"), Alnitak is a close binary star, with a magnitude 4.0 companion that can only be separated through a small telescope. A third member of the group is much fainter at magnitude 9.5, and can only be seen through larger instruments. Running south of Alnitak is a long faint strand of nebulosity – the famous but elusive Horsehead Nebula.

Bellatrix

Gamma (γ) Orionis

5h25m, +06°21′	Magnitude: 1.6
Main sequence star	Distance: 240 light yrs

The blue-white Bellatrix makes a striking contrast to Betelgeuse, which marks the hunter's other shoulder. It is technically a blue giant, 8 times the mass of the Sun and 4,000 times as luminous – considerably fainter than Rigel even though it is one-third of the distance, at roughly 350 light years.

Betelgeuse

Alpha (α) Orionis

5h55m, +07°24′	Magnitude: 0.3 – 1.2 (var)
Giant star	Distance: 430 light yrs

Alpha Orionis is the distinctive red supergiant Betelgeuse. Despite its 'alpha' designation, and its rank as the tenth brightest star in the sky, it is not the brightest star in Orion – that honour goes to Rigel. Betelgeuse is a fascinating star, however – so large that it is unstable and fluctuates in both size (between 300 and 400 times the diameter of the Sun) and brightness (between magnitudes 1.3 and 0.0). The huge size of Betelgeuse also allowed the Hubble Space Telescope to perform a notable first – directly imaging the surface of a distant star.

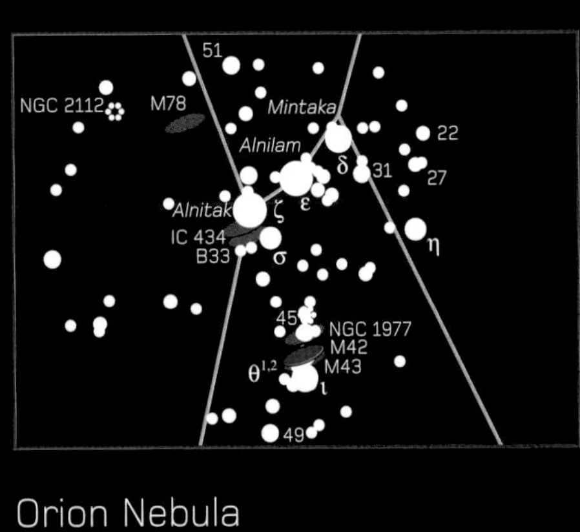

Orion Nebula

M42

5h35m, -5°27'	**Magnitude: 4.0**
Emission nebula	**Distance: 1,600 light yrs**

The brightest and most impressive nebula in the entire sky, Messier 42 is a luminous gleam on Orion's sword. It is illuminated from within by fierce radiation from a tightly bound quartet of hot, young stars, known as the Trapezium.

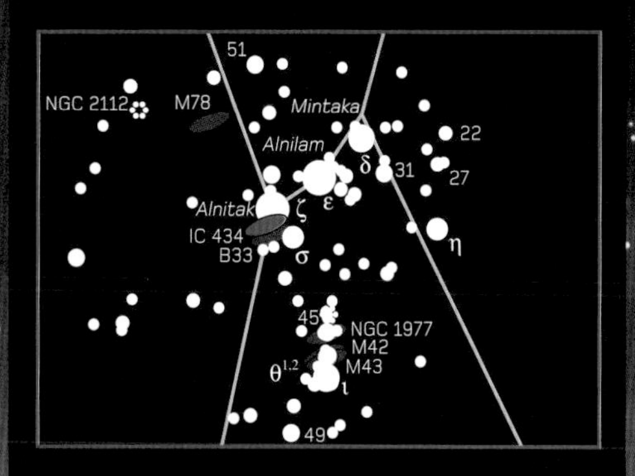

Horsehead Nebula

Barnard 33/IC 434

5h41m, -02°28′	**Magnitude: –**
Dark nebula	**Diameter: 1,500 light yrs**

Perhaps the most famous denizen of the Orion Molecular Cloud, the renowned celestial chesspiece is actually loop of dark obscuring dust roughly 3.5 light years across. It is visible in stark silhouette against shimmering curtains of more distant hydrogen, glowing with energy they receive from nearby Sigma Orionis.

Monoceros & Canis Minor

The Unicorn and the Lesser Dog

Directly to the east of Orion lie the large, relatively faint constellation of Monoceros and the compact Canis Minor, marked by its single bright star Procyon. The W-shape of the Unicorn bears no real resemblance to any animal, but it lies across a rich stretch of the Milky Way and contains some fascinating objects. The Lesser Dog is barely a pattern at all – simply a line between two stars, but is easily found thanks to Procyon.

Canis Minor is an ancient constellation, and it was widely associated with a dog throughout late antiquity, although its name, definitively grouping it with Canis Major as Orion's hunting dogs, did not emerge until the first century BC.

The origins of Monoceros are something of a mystery. It is widely supposed to have been invented by Dutch theologian Petrus Plancius in 1613 (inspired by a reference to a unicorn – perhaps really a rhinoceros – in the Bible). However, several early historians of astronomy claimed to have found it in charts of Arabic and Persian origin from medieval times.

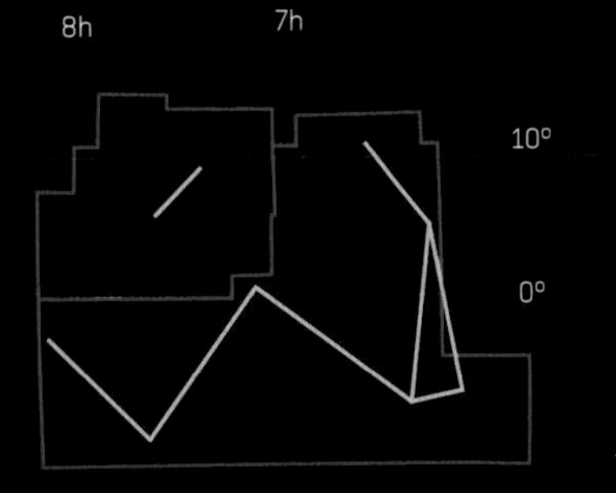

Procyon

Alpha (α) Canis Minoris

7h39', +05°13'	Magnitude: 0.3
Binary star	Distance: 11.4 light yrs

The brightest star in Canis Minor, Procyon's name comes from the Greek for 'before the dog', on account of its position in the sky, rising before Sirius in Canis Major. In fact, Procyon and Sirius have a lot of similarities – they are both moderate-sized white stars that rank among the brightest in Earth's skies simply because they are in our cosmic back yard – Procyon is the fourteenth closest star to Earth.

With a mass of 1.4 solar masses, Procyon is almost seven times more luminous than the Sun, but it is slightly larger than might be expected if it shone purely by burning hydrogen in its core. Instead, it is on its way to becoming a giant, swelling as the hydrogen fusion process moves out from the largely exhausted core, and into a surrounding spherical shell.

The other major similarity to Sirius is that Procyon, too, has a white dwarf companion star. The existence of Procyon B was suggested in 1844, from the evidence of wobbles in the primary star's motion, but it was not directly observed until fifty years later. Procyon B shines at magnitude 10.8 as seen from Earth, and has the mass of 1.4 Suns – close to the upper mass limit for a white dwarf (larger stellar cores collapse into neutron stars in the aftermath of supernova explosions). The pair orbit each other in 41 years, and the dwarf's faintness and proximity to a brilliant neighbour mean it can only be seen with a fairly large telescope.

Beta (β) Monocerotis

6h29m, -7°02'	Magnitude: 3.9
Multiple star	Distance: 690 light yrs

This beautiful triple star is visible through even the smallest telescope. Its stars are all blue-white, and form a close pair of magnitudes 5.4 and 5.6, orbited by a slightly more remote but brighter star of magnitude 4.6. All three are stellar heavyweights with masses between 6 and 7 times that of the Sun. They were born a little over 30 million years ago, and are squandering their fuel at a tremendous rate – the brightest star pumps out more than 3,000 times the energy of the Sun, the fainter pair about half that. Despite this, all three stars still have enough hydrogen in their cores to shine steadily for several million more years.

M50

NGC 2323

7h3m, -8°20'	Magnitude: 5.9
Open cluster	Distance: 3,200 light yrs

Easily located with binoculars as a patch of light half the size of the full Moon, this substantial cluster of about 200 stars includes 80 or so visible through a small telescope. It was probably discovered by Italian astronomer Gian Domenico Cassini before 1711.

The cluster is relatively colourful – its brightest individual star is blue-white and shines at magnitude 9.8, but it is rivalled by a contrasting red giant and several yellow giants, formed as the more massive stars in the cluster near the end of their lives and evolve away from the 'main sequence'. Measuring the most massive stars that have not yet gone down this path gives an estimated age for the cluster of 78 million years.

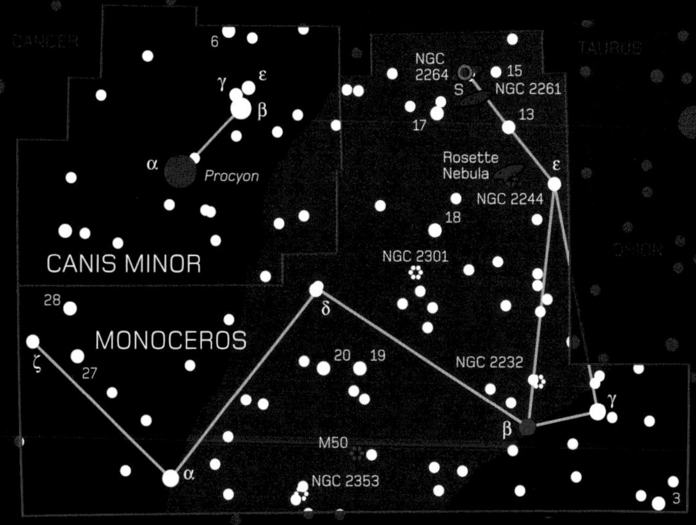

Hubble's Variable Nebula

NGC 2261

6h39m, +8°44'	Magnitude: 9.5 (var)
Reflection nebula	Distance: 2,500 light yrs

This small and unusual nebula surrounds the variable star R Monocerotis, which varies erratically between magnitude 9.5 and 12. This newborn star is thought to be at the same distance as the Christmas Tree Cluster NGC 2264. Thick dust lanes still surround it, and as they move across its face, they cast shadows out onto the surrounding gas. The gas cloud itself varies in brightness and changes its shape quite rapidly, but it can only be seen with quite powerful instruments.

The Christmas Tree Cluster

NGC 2264

6h41m, +9°53'	Magnitude: 3.9
Open cluster	Distance: 2,400 light yrs

This cluster of 40 or so stars, visible with the naked eye or easily through binoculars, is associated with a complex nebula that only becomes visible through larger telescopes. Fierce radiation from its young stars is exciting the surrounding gas clouds and causing them to glow, silhouetting an enormous pillar of dust – a dark nebula called the 'Cone Nebula'.

NGC 2264's most interesting star is S Monocerotis, which appears to the naked eye as a blue-white star of magnitude 4.7. In reality, this star is a binary system containing two monstrous blue stars, with masses around 30 and 20 times the Sun. The pair orbit each other in about 25 years, and are about 100,000 and 50,000 times as luminous as the Sun (although most of their radiation is emitted in the ultraviolet.

Rosette Nebula

NGC 2244

6h32m, +4°52'	Magnitude: 4.8
Emission nebula	Distance: 5,500 light yrs

The small star cluster NGC 2244 can be spotted with the naked eye in dark skies, or easily picked up with binoculars, but the nebula that gave birth to it is far more challenging. The Rosette Nebula has four NGC numbers to itself – one for each of the brighter clumps of gas that are visible through large amateur telescopes. Only long-exposure photographs will reveal the full extent of this faint, extensive nebula, but the eagle-eyed can sometimes spot a slightly lighter patch of sky through good binoculars on a dark, moonless night.

The Rosette itself is a huge cloud of gas and dust, perhaps 130 light years across. Star formation is still continuing, and is probably triggered in places where the gas is compressed by fierce stellar winds from previous generations of stars.

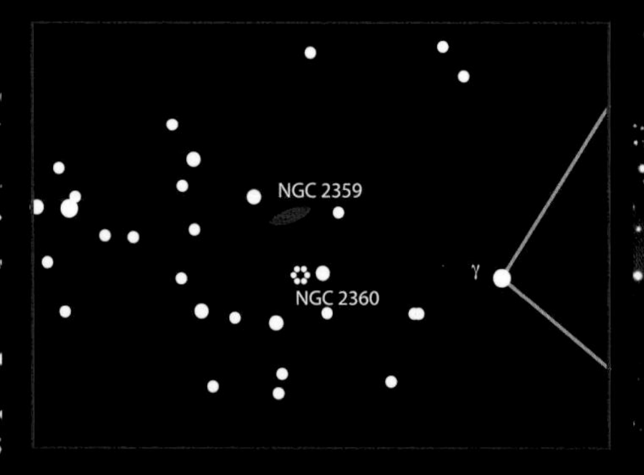

Thor's Helmet

NGC 2359

7h19m, -13°14'	Magnitude: 11.5
Emission nebula	Distance: 13,000 light yrs

This distinctive celestial bubble, with a diameter of about 30 light years, is formed from material blown away on the stellar winds of one of a massive Wolf-Rayet star – a stellar giant so fearsome that its radiation is driving away its own outer layers to expose hotter layers inside. As the shells of expelled material billow out into the surrounding space, they encounter other clouds of gas that sculpt the helmet's 'wings'.

NGC 2207/IC 2163

6h16m, -21°22′ | Magnitude: 12.0
Colliding galaxies | Distance: 114M light yrs

This spectacular image captures the aftermath of a grazing collision between two spiral galaxies in Canis Major. 40 million years ago, IC 2163 (on the right) made its closest approach to the larger NGC 2207.

Gravitational tides raised between the two galaxies have severely distorted the smaller one. Now IC 2163 lies behind its partner, and its light helps to silhouette dust lanes in the foreground galaxy.

Centaurus

The Centaur

One of two centaurs that occupy prominent positions in the sky (the other is Sagittarius, the archer), Centaurus fills a large area of the heavens between Hydra and Crux. Its brightest stars, Rigil Kentaurus and Hadar, both lie in its southeastern corner, but the constellation's central region is marked by a rough pentagon of quite bright stars, with a beautiful centrepiece in the form of the Omega Centauri globular cluster.

This group of stars were not always recognized as a centaur – it was occasionally given the name of another hybrid – the Minotaur. The astronomers of ancient Babylon, meanwhile, saw it as a bull with no human characteristics. However, by classical times it was established as a Centaur, and usually associated with the wise and scholarly figure of Chiron, mythical scholar to both Hercules, and Achilles (of Trojan War fame). According to the legend, Chiron was accidentally shot by Hercules with a poisoned arrow, but was raised to immortality in the sky by Zeus. The constellation is closely associated with the neighbouring stars of Lupus.

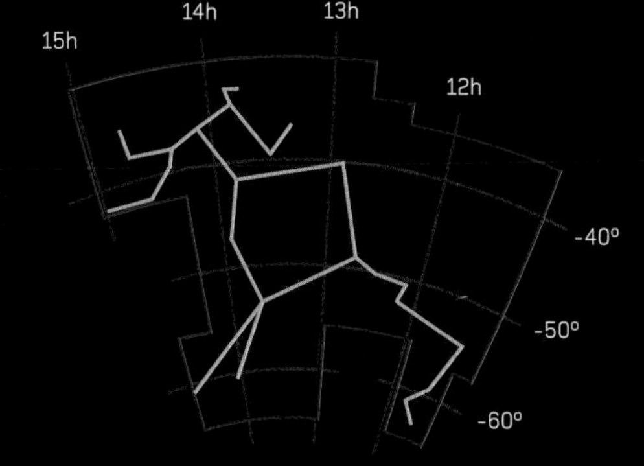

Rigil Kentaurus

Alpha (α) Centauri

14h40m, -60°50′	Magnitude: -0.1
Multiple star	Distance: 4.36 light yrs

The brightest star in Centaurus, and third brightest in the entire sky, also marks the closest star system to Earth. The Alpha Centauri system's proper name means 'foot of the Centaur', but it is sometimes abbreviated to 'Rigil Kent'. A small telescope will easily reveal that Alpha is in fact a binary system, with a yellow primary of magnitude 0.0, and an orange companion of magnitude 1.4. The primary star, Rigil Kent A, is the same colour as our Sun – its surface temperature is the same to within 10°C, but it is about 10 percent more massive, 25 percent larger, and 50 percent more luminous. Rigil Kent B, in contrast, is slightly less massive than the Sun, smaller, and less luminous. The two stars take 80 years to circle each other.

The weight of evidence suggests that the Alpha Centauri system is somewhat older than the Sun – perhaps 8 billion years old. Rigil Kent A may be on the verge of exhausting its core supply of hydrogen and swelling to become a giant, in which case it will overtake Canopus and Sirius to become the brightest star of all.

But there is a third member of this system that claims the honour of being the closest star to the Sun. Tiny Proxima Centauri is a red dwarf, typical of countless stars in the sky. It shines at a faint magnitude 11 – just detectable in small telescopes if you know where to look – and is two degrees away from its brighter siblings in the sky. Nevertheless, it is gravitationally bound to them, in an orbit that lasts for hundreds of thousands of years. It has a mass just one-fifth that of the Sun, and produces just 0.2 percent of the Sun's energy, but like many faint stars of its kind, it is a 'flare star', with a fairly strong magnetic field that produces unpredictable activity on its surface.

Hadar

Beta (β) Centauri

14h4m, -60°22′	Magnitude: 0.6
Multiple star	Distance: 525 light yrs

To the naked eye, Hadar appears as a single blue-white star, but medium-sized telescopes and high magnifications should be able to separate a magnitude 4 companion (Hadar B) from the glare of the primary star. Hadar A is itself a spectroscopic binary, consisting of twin blue stars, each with the mass of 15 Suns, in a 320-day orbit around each other, while Hadar B has five times the mass of the Sun.

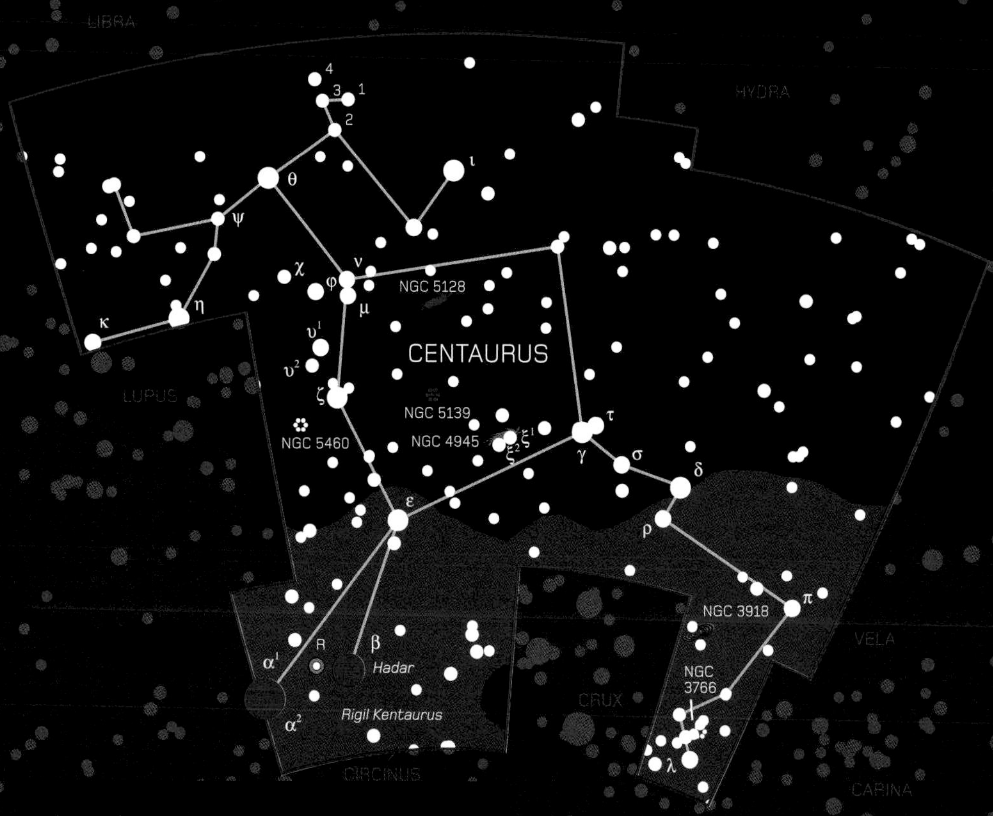

The Blue Planetary

NGC 3918

11h50m, -57°11′	Magnitude: 8.1
Planetary nebula	Distance: 2,600 light yrs

This small planetary nebula appears as a blue disk of light in a moderate-sized telescope. More detailed studies have revealed a faint spherical outer shell, about 0.3 light years across, surrounding an elongated oval of hotter gas. This expanding envelope is thought to be driven outwards by hot stellar winds from the magnitude 11 central star.

Centaurus A

NGC 5128

13h26m, -43°01′	Magnitude: 7.0
Active galaxy	Distance: 15M light yrs

The bright and unusual galaxy NGC 5128 can be seen through binoculars on a dark night, and the smallest telescopes reveal its generally elliptical structure. Larger instruments and long exposures are needed to show its most unusual feature, however, as a broad, dark band of dust bisects the spherical ball of stars.

NGC 5128 coincides with a source of radio signals known as Centaurus A – a pair of vast lobes of gas extending to either side of the galaxy. For this reason, the galaxy itself is often known simply as Centaurus A. Astronomers classify it as an active galaxy, with a supermassive black hole at its centre that is feeding voraciously and producing the radio jets as a side effect. In this case, the dust lane offers some evidence for the cause of activity – it's likely that NGC 5128 has recently swallowed up a smaller spiral galaxy, and is still feasting on the remains.

Omega (ϖ) Centauri

NGC 5139

13h27m, -47°29′	Magnitude: 3.7
Globular cluster	Distance: 16,000 light yrs

The most spectacular globular cluster of them all, Omega Centauri is among a handful of star clusters that were classified as stars before their true nature was realized. It is not only the largest globular cluster in orbit around the Milky Way, but one of the closest to Earth.

The cluster contains about 10 million stars, in an area about 150 light years across. At its centre, the average distance between stars is measured in light days rather than light years. Small telescopes and even binoculars will resolve some of the individual stars around the cluster's outer edges.

Omega is a monster of a cluster, about 10 times heavier than any of the others that orbit the Milky Way. It is also unusual because its stars did not all form at the same time – instead, they seem to have been created in waves over about 2 billion years, with the last burst of star formation happening about 10 billion years ago.

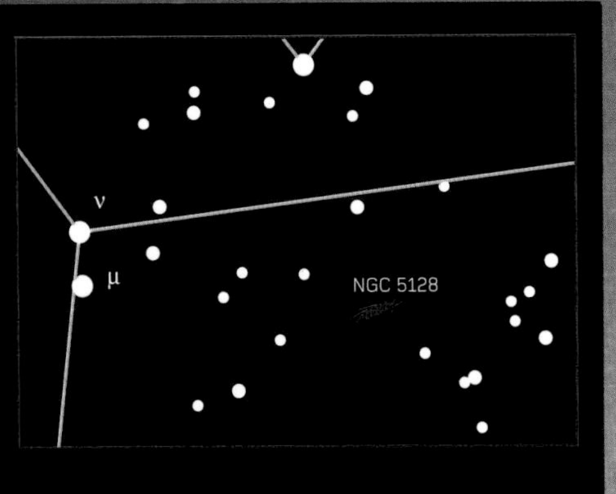

Centaurus A

NGC 5128

13h26m, -43°01′
Active galaxy

Magnitude: 7.0
Distance: 15M light yrs

Nearby cannibal galaxy NGC 5128 is still clearly scarred by the struggle with its last victim – a dark gash across the centre of this elliptical ball of stars is all that remains of the spiral arms from a galaxy that has now been absorbed. The influx of new material has awoken the supermassive black hole in the galaxy's heart, and as dust, gas and stars are pulled to their doom, they emit streams of particles and violent radiations.

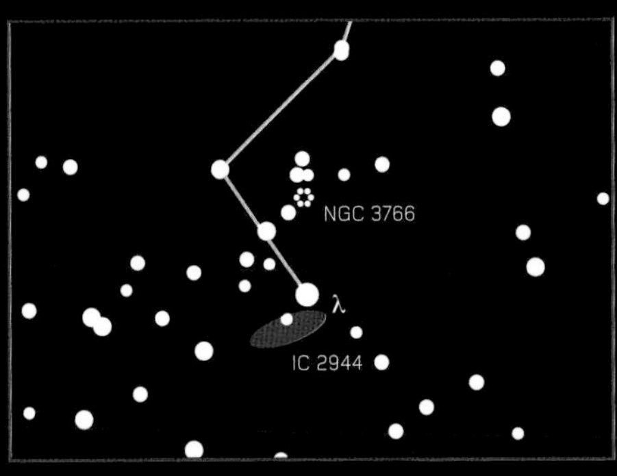

Thackeray's Globules

IC 2944

12h41m, -40°59′	Magnitude: 12.3
Diffuse nebula	Distance: 108M light yrs

Deep within a glowing cloud of gas known as the Running Chicken Nebula, but properly catalogued as IC 2944, lie dark knots of gas where stars are gestating. Typical of a class of objects called Bok Globules, the largest, Australia-shaped mass in Thackeray's Globules is thought to contain as much as 15 Suns' worth of material.

Lupus

The Wolf

Embedded in the Milky Way between Scorpius and Centaurus, this constellation is a somewhat confusing mass of middle-ranking stars, best located along the line from Antares to Hadar or Beta Centauri. Its most interesting features are a range of clusters and double stars.

Today Lupus is the celestial Wolf, but while the constellation itself is an ancient one, its identification did not come until medieval times. In the classical world, these stars were often simply Bestia or Fera, both names for an unspecified beast. Other astronomers linked them with neighbouring Centaurus, depicting the beast as an animal impaled on a lance by the centaur Chiron, with names like Bestia Centauri or Victima Centauri. The association with a wolf may come from a mistranslation of Al Fahd, a title used by the Arabs who saw this constellation as a leopard.

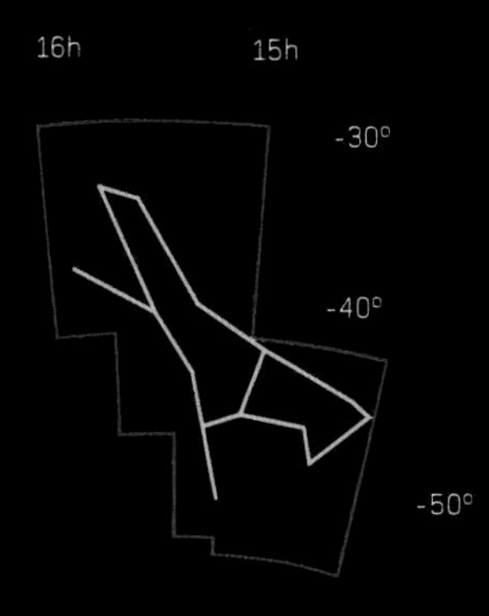

Kakkab

Alpha (α) Lupi

14h42m, -47°23′	Magnitude: 2.3
Giant star	Distance: 550 light yrs

The brightest star in Lupus is this stellar giant – a hot blue star with a surface temperature of more than 21,000°C. Kakkab has an estimated luminosity of 20,000 Suns, but typically for such a hot star, it pours out most of its energy as invisible ultraviolet radiation – only about one sixth of it is released in visible light. Like many other stars in Lupus and the surrounding skies, Kakkab is part of the same 'OB association' – a group of bright, hot stars that all formed in the same period (around 20 million years ago) as an open cluster, and have since scattered themselves across the sky. The stars of this group, the 'Scorpius-Centaurus Association' appear particularly widely scattered because they are the closest such group to Earth.

NGC 5822

15h4m, -54°24′	Magnitude: 6.5
Open cluster	Distance: 6,000 light yrs

This large open cluster forms a dense knot of light against the background Milky Way starclouds. It is easily spotted through binoculars or a small telescope in a dark sky, and contains about 120 or more stars scattered across an area of sky slightly larger than the full Moon.

IC 4406

14h22m, -44°09′	Magnitude: 10.5
Planetary nebula	Distance: 2,000 light yrs

This faint and small planetary nebula is a target for amateurs with larger telescopes using higher magnifications, but detailed photographs reveal a curious object that appears to be rectangular in shape. The reality is that this nebula is cylindrical, but we are seeing it side-on. From the top, it would appear circular, like many other ring nebulae. IC 4406 has been sculpted into its unusual shape because the dying star at the centre is shedding its outer layers in expanding 'doughnut' rings around its equator.

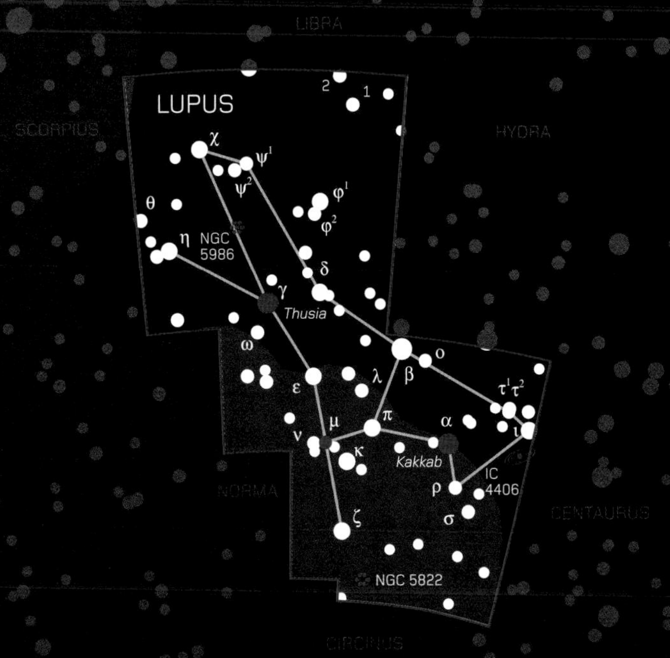

Thusia

Gamma (γ) Lupi

15h35m, -41°10′	Magnitude: 2.5
Binary star	Distance: 570 light yrs

A fairly large telescope is usually needed to separate the two stars of this tight binary system. The primary star, magnitude 2.8, is orbited by a companion of magnitude 3.7 in about 147 years. Both stars are hot and blue – the less massive companion is still firmly in the 'main sequence' phase of its life, shining by the fusion of hydrogen in its core, The more massive primary star, meanwhile, is on its way to becoming a giant – it has burnt through its main fuel supply and is now burning hydrogen in a shell around the core in order to keep shining.

NGC 5986

15h46m, -37°47′	Magnitude: 7.1
Globular cluster	Distance: 34,000 light yrs

The brightest of several globular clusters in Lupus, NGC 5986 is easy to spot through binoculars as a fuzzy blob of light about one third the diameter of a full Moon. Larger instruments will reveal an attractive cluster with its density increasing smoothly towards the centre. It was discovered by Scottish astronomer James Dunlop in 1826, while he was working in Australia at an observatory built by Sir Thomas Brisbane.

Mu (μ) Lupi

15h19m, -47°53′	Magnitude: 4.3
Multiple star	Distance: 290 light yrs

This attractive multiple star appears through a small telescope as a magnitude 4.3 blue-white star with a well-separated companion of magnitude 7.2. Medium-sized telescopes, however, should reveal that the brighter star is itself a double, made up of near-twin stars with magnitudes 5.1 and 5.2. In visible light, each of these stars shines roughly 70 times brighter than the Sun, implying a mass of about three Suns each.

Scorpius

The Scorpion

One of the oldest constellations, Scorpius has been seen in more or less its current form since the earliest astronomical records. It does indeed bear some resemblance to a scorpion, although some civilizations saw it as a hybrid of man and scorpion. Today the creature's body is truncated at its head – its extended pincers have been appropriated by neighbouring Libra, which was once known as Chelae Scorpionis, the Scorpion's Claws.

In classical mythology, this Scorpion had two roles. In one story, it was the beast that frightened the horses of the sun god's chariot and sent his reckless son, Phaeton, crashing to his doom. In another variation, it was a monster sent by a jealous god (either Apollo or the goddess Hera) to kill the hunter Orion. When it had done its worst, the god relented and placed Orion and Scorpius on opposite sides of the sky.

The constellation is easy to find from its bright northern star Antares, and its distinctive flanking stars Sigma and Tau Scorpionis. Crossed by a dense band of the southern Milky Way, it is rich in fascinating objects.

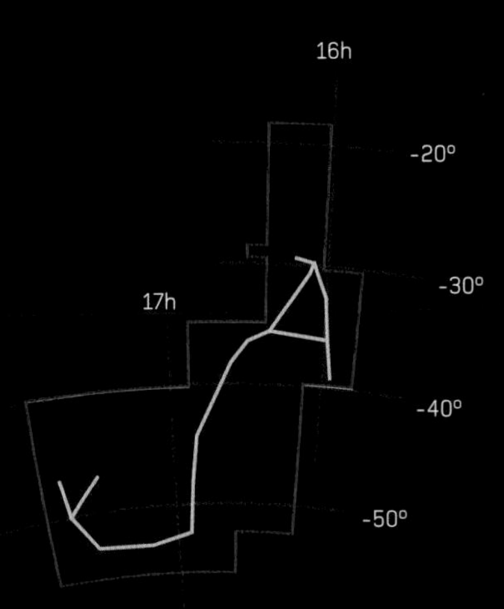

Antares

Alpha (α) Scorpii

16h29m, -26°26′	**Magnitude: 1.0 (var)**
Variable star	**Distance: 600 light yrs**

With a name that means 'rival of Mars', Antares is an appropriately brilliant and red star, unmistakable in evening skies during northern summer. This red supergiant is extremely luminous – about 10,000 times as brilliant as the Sun in visible light, but pumping out six or seven times that energy in the infrared. To be so brilliant and yet so red (with a surface temperature of just 3,300°C) Antares must be truly monstrous – with a diameter three-quarters the size of Jupiter's orbit around the Sun, it would easily engulf all the inner planets in our Solar System.

Antares has the mass of about 16 Suns, but is so huge that it is one of the least dense stars known. As a result, its gravity is so weak at its outer surface that it is constantly losing large amounts of material from its outer layers, blown away by its fierce radiation. It is now cocooned within a nebula of its own making, alongside its blue companion star.

This secondary star, Antares B, shines at magnitude 5.5 but is usually swamped by the light of its neighbour, which it orbits once every couple of thousand years. As a result, the pair can only be separated with medium-sized telescopes – and the existence of Antares B was not even suspected until the primary star was blocked out by the moon during an occultation in 1919.

Xi (ξ) Scorpii

16h4m, -11°22′	**Magnitude: 4.2**
Multiple star	**Distance: 93 light yrs**

Xi Scorpii is another impressive multiple star, but this is one with a difference. Small telescopes will split it into two pairs. One consists of white stars of magnitudes 4.2 and 7.6, the other of orange stars with magnitudes 7.4 and 8.1. The two groups must be separated by at least 8,000 astronomical units – roughly 40 times the diameter of Pluto's orbit round the Sun – and so will take tens of thousands of years to whirl around each other.

The white pair of stars are on their way to becoming giants – they have recently exhausted the hydrogen in their cores and begun to expand. The orange pair, meanwhile, are stellar lightweights slightly less massive than the Sun, and still firmly on the 'main sequence'. A larger telescope will reveal that the system's brightest star is itself a close double, consisting of near-twin stars of magnitudes 4.8 and 5.1, orbiting each other every 46 years.

Acrab

Beta (β) Scorpii

16h5m, -19°48′	**Magnitude: 2.3**
Multiple star	**Distance: 530 light yrs**

Small telescopes will easily separate the two bright components of this fascinating multiple star, whose name means simply 'scorpion', revealing a pair of hot blue stars with magnitudes 2.6 and 4.9. Much larger telescopes can reveal another, 10th-magnitude companion in orbit around the brighter star, and analysis of the light from the system reveals that both bright elements are also spectroscopic binaries, inseparable with any current telescope. All the stars in the system are hot and blue with the exception of the 10th-magnitude companion star. As a result they pump out most of their energy as ultraviolet radiation. The two components of the brighter pair at least are so massive (weighing the same as ten Suns each) that they will end their lives in supernova explosions. There are even suggestions of more elements to be discovered.

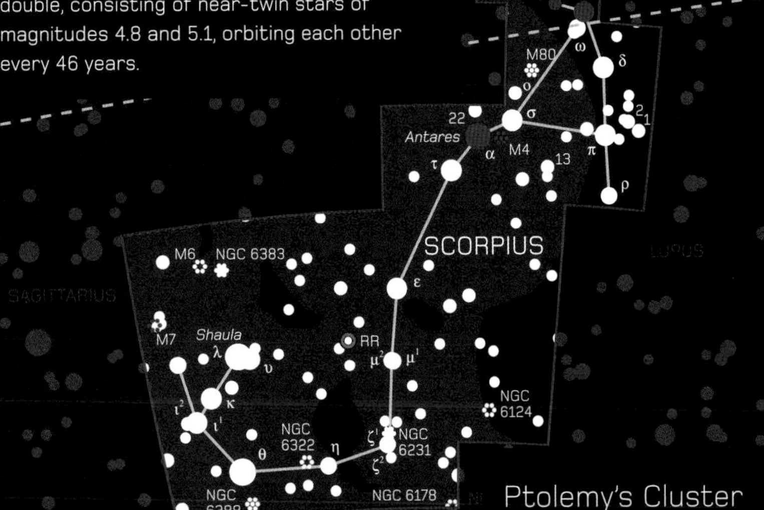

M4

NGC 6121

16h24m, -26°32′	**Magnitude: 5.6**
Globular cluster	**Distance: 7,200 light yrs**

The brightest globular cluster in Scorpius, M4 is best found through binoculars, since it has a comparatively loose structure which ensures its light is spread evenly across a wide area of sky (about the size of a full Moon). This is the closest globular cluster to Earth, and would be far more impressive if not for the obscuring clouds of gas and dust that its light has to pass through on its way to us. Small telescopes can resolve its outlying stars, and it has an estimated diameter of about 75 light years.

Ptolemy's Cluster

M7

17h54m, -34°49′	**Magnitude: 3.3**
Open cluster	**Distance: 800 light yrs**

This bright and dense open cluster is also known as the 'Scorpion's Tail'. For naked-eye observers, it forms a distinct knot of light in the Milky Way. Binoculars or a telescope with low power and a wide field reveal stunning vistas of 6th-magnitude stars against a field of fainter ones. In total, there are about 80 stars within range of binoculars or a small telescope, and the brightest form a rough cross-shape. Covering more than a degree of sky at a distance of 800 light years, the cluster has a diameter of about 20 light years.

Jabbah

Nu (ν) Scorpii

16h12m, -19°28′	**Magnitude: 3.9**
Multiple star	**Distance: 440 light yrs**

This stunning multiple star rivals the famous 'double double' Epsilon Lyrae as the best example of its type in the sky. The smallest telescopes will easily separate it into a pair of blue-white stars, magnitudes 4.0 and 6.3, but a moderate-sized instrument will split each of these two components in turn, revealing a complex system of four hot blue stars waltzing around one another. The brighter eastern couplet, Nu Scorpii A and B, shine at magnitudes 4.4 and 6.9 respectively, while the fainter western pair, C and D, have magnitudes 6.5 and 7.9. The two pairs orbit one another in something like 100,000 years, and the system is further complicated because Nu Scorpii A is a spectroscopic binary, meaning that Jabbah (whose name indicates the 'forehead' of the Scorpion) is a quintuple system.

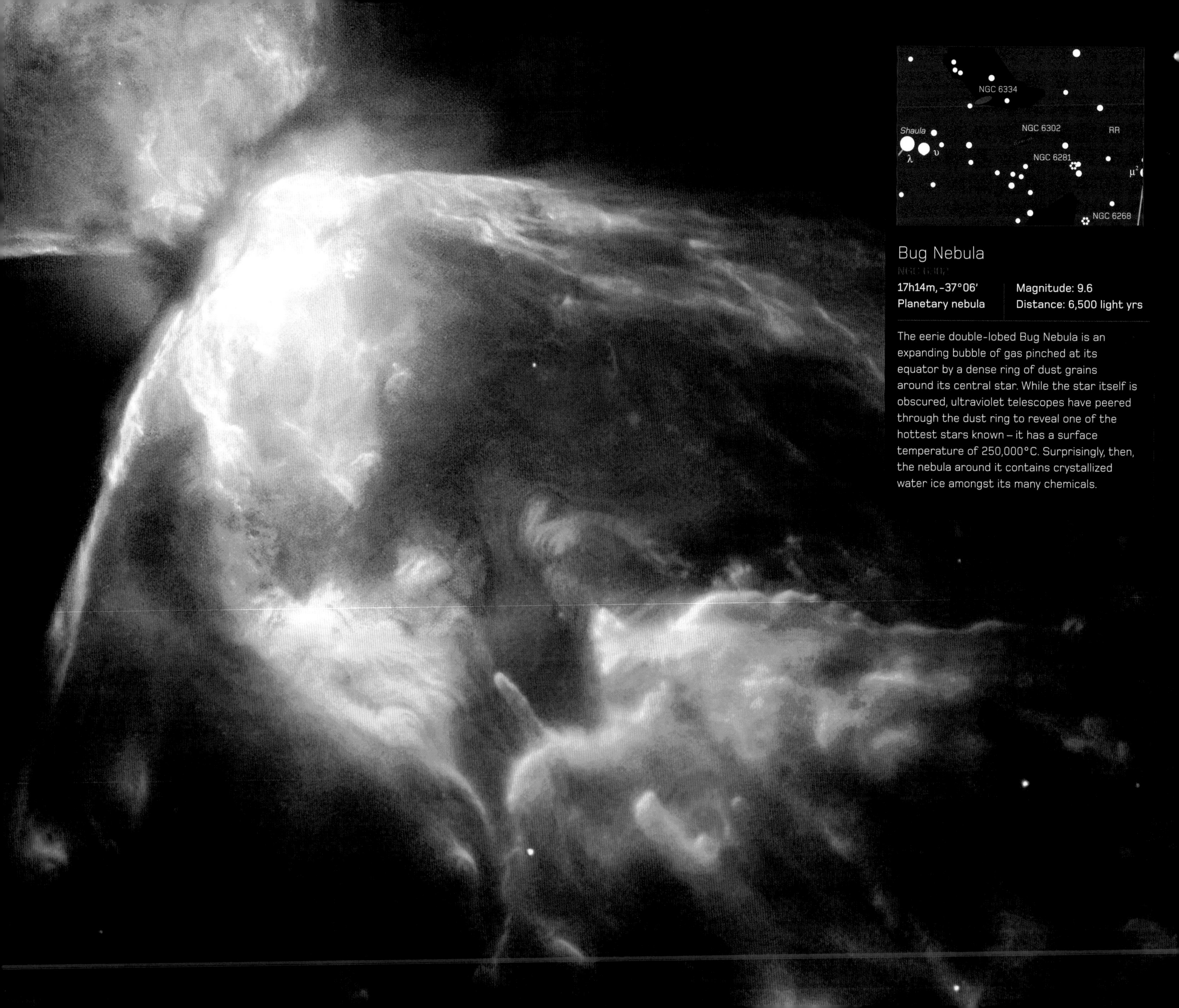

Bug Nebula

NGC 6302

17h14m, -37°06′
Planetary nebula

Magnitude: 9.6
Distance: 6,500 light yrs

The eerie double-lobed Bug Nebula is an expanding bubble of gas pinched at its equator by a dense ring of dust grains around its central star. While the star itself is obscured, ultraviolet telescopes have peered through the dust ring to reveal one of the hottest stars known – it has a surface temperature of 250,000°C. Surprisingly, then, the nebula around it contains crystallized water ice amongst its many chemicals.

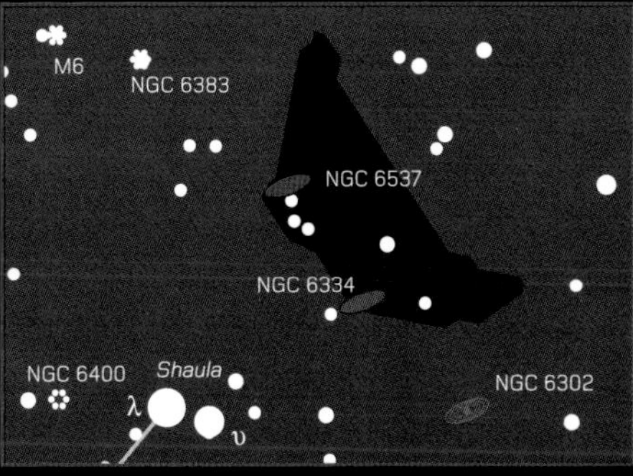

Red Spider Nebula

NGC 6537

18h5m, -19°51′	Magnitude: 13
Planetary nebula	Distance: 1,900 light yrs

The heart of the Red Spider Nebula conceals a star even hotter than that in the Bug Nebula, with a surface temperature of half a million °C. Fierce stellar winds blasting from the star's surface have sculpted this nebula into a unique form, creating waves of gas tens of billions of kilometres long.

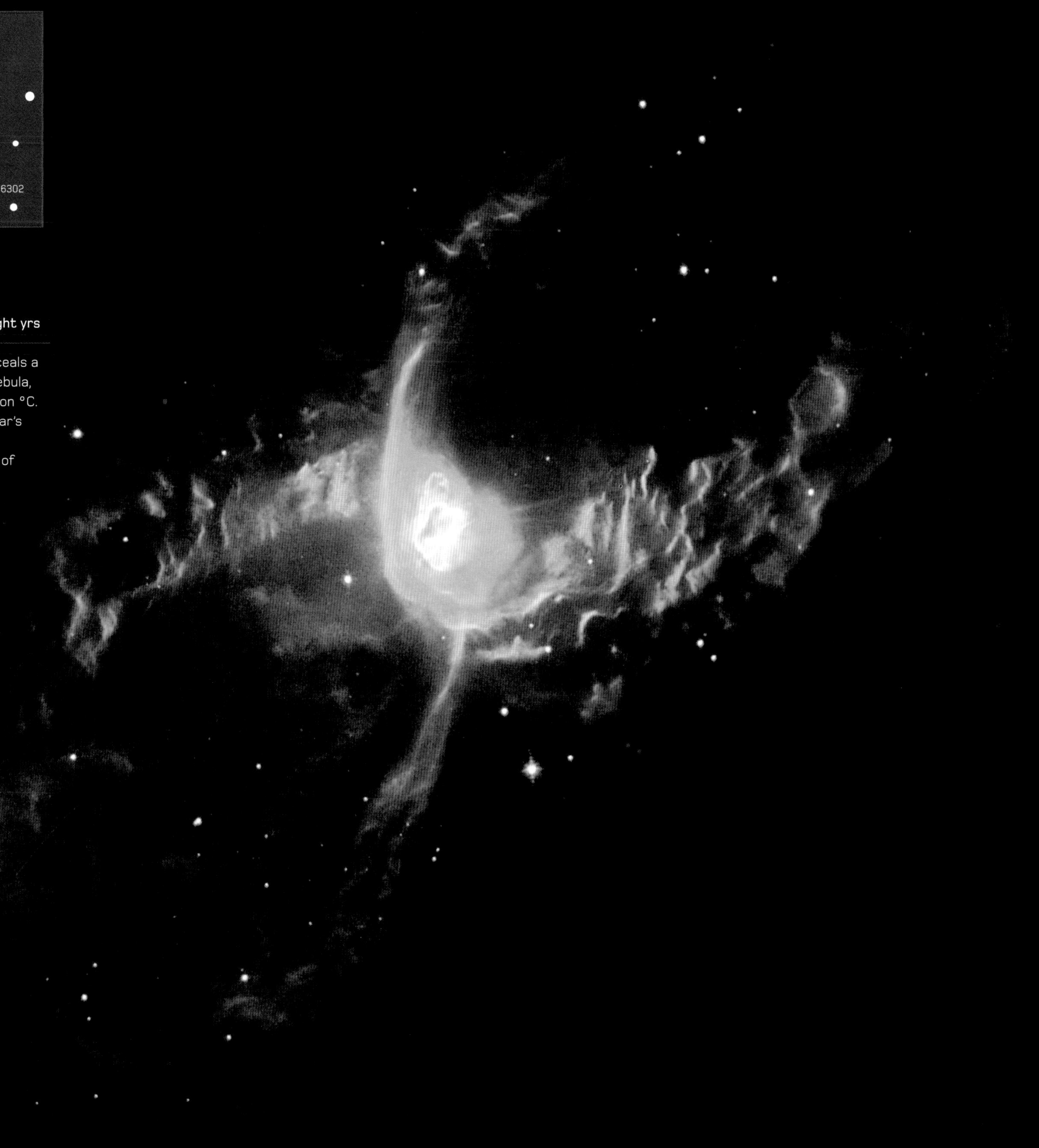

Sagittarius

The Archer

The constellation of the Archer, Sagittarius occupies one of the richest regions of the sky since its stars happen to lie in the direction of the centre of our Milky Way galaxy. In myth and astrology, the stars were usually seen as a centaur (half man, half horse) armed with a bow and arrow. While Centaurus was identified with the wise and scholarly Chiron, Sagittarius was a more primal beast, a warrior centaur of the type depicted in many Greek sculptures. Older Middle Eastern myths saw the constellation as a Warrior God or King, mounted on horseback, while in India the constellation was simply the Horse, and in China it was the Tiger.

However, most modern observers are more likely to see it as a distinctive teapot-shape of moderately bright stars. The starfields that lie behind it are filled with deep-sky wonders, including dense star clusters and glowing nebulae.

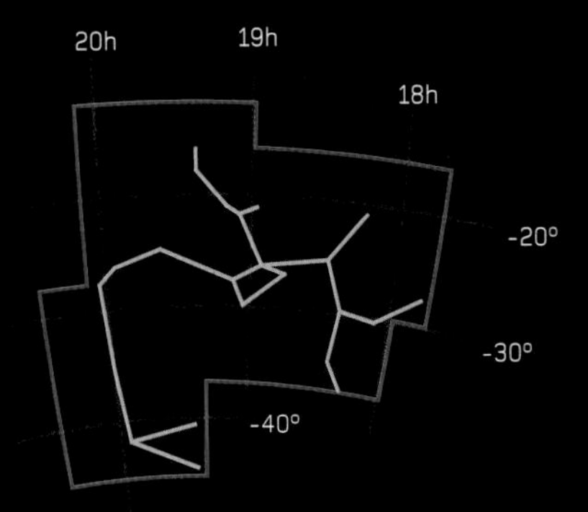

Sgr A*

17h46m, -29°
Supermassive black hole

Magnitude: -
Distance: 26,000 light yrs

Hidden behind the densest star clouds of the Milky Way, a sleeping monster lurks at the very centre of our galaxy. It reveals itself only through radio signals from an object called Sgr A* (pronounced Sagittarius A-star), and its effect on the stars around it.

The object is a supermassive black hole, with the same mass as perhaps four million Suns. It grew out of collapsing gas clouds when the Milky Way itself was coalescing, and once feasted on everything within its gravitational reach. Now, though, it is slumbering – in order to survive in the galactic centre, the stars and major gas clouds have to keep in strictly circular orbits at a safe distance, and it is only the slow, steady drift of gas into the black hole's grasp that causes it to gently glow at radio wavelengths.

The presence of a giant black hole at the galaxy's centre was suspected for many years – the region around it is a strange and violent place, crowded with stars, violent stellar remnants, and mysterious clouds of radiation, many of which would be best explained by the influence of such a monster. But it was only when astronomers pierced the clouds of dust and gas around the centre that they realized the stars there were moving so fast that they must be orbiting an incredibly massive object occupying a volume of space just half the diameter of Earth's orbit around the Sun. Now it seems that most if not all galaxies have such black holes at their centre, and they may be an integral part of the process by which galaxies form.

RY Sagittarii

19h17m, -33°31′
Variable star

Magnitude: 6.5 (var)
Distance: 9,000 light yrs

This bizarre star is the southern hemisphere's brightest equivalent to R Coronae Borealis. It shines fairly steadily at around magnitude 6.5 for much of the time, but occasionally undergoes catastrophic drops in brightness to around magnitude 13. The star is thought to be an elderly red giant surrounded by a shell of carbon-rich vapour. Occasionally this carbon condenses into dark clouds in a runaway reaction that blocks much of the star's light until the clouds are blown away.

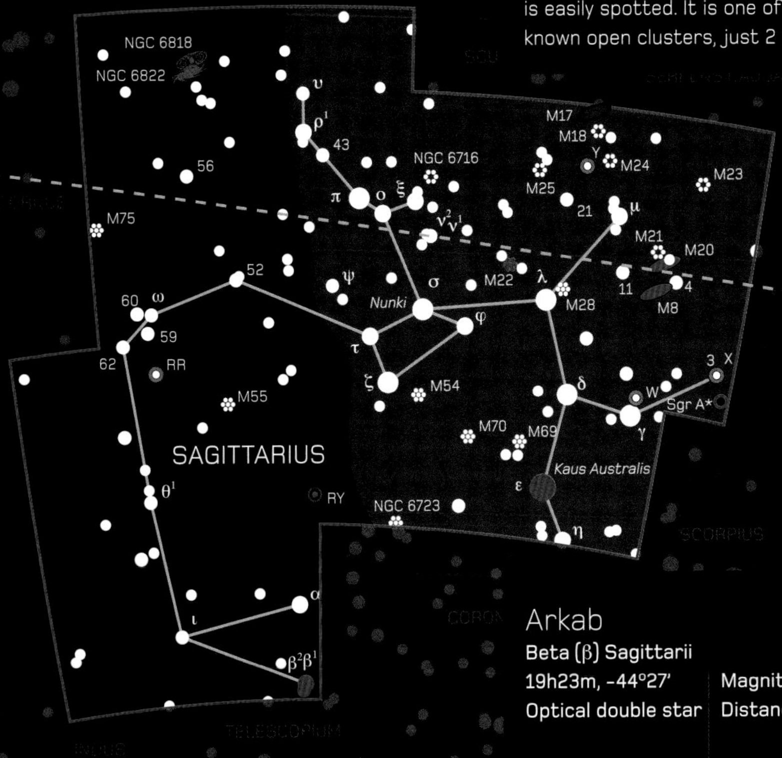

Lagoon Nebula

M8

18h4m, -24°23′
Emission nebula

Magnitude: 6.0
Distance: 5,000 light yrs

This bright emission nebula is one of several scattered through the Milky Way in north-western Sagittarius. It is barely visible to the naked eye in dark skies as a hazy patch in the Milky Way, and binoculars will reveal its brighter nucleus and a broad expanse of gas extending to cover 1.5 degrees of sky.

On the eastern side of the nebula, and just in front of it from our point of view, lies a naked-eye open cluster, containing about 100 stars. NGC 6530 shines at magnitude 4.6, and is easily spotted. It is one of the youngest known open clusters, just 2 million years old.

Arkab

Beta (β) Sagittarii

19h23m, -44°27′
Optical double star

Magnitude: 4.0, 4.3
Distance: 380 light yrs
140 light yrs

Arkab, at the southern end of Sagittarius, is a naked-eye or binocular double consisting of two star systems at different distances. The northerly Arkab Prior is the brighter star at magnitude 4.0. Southerly Arkab Posterior is slightly fainter at magnitude 4.4.

Arkab Posterior is the closer of the two stars – a white giant 140 light years away. Arkab Prior, meanwhile, is the more distant and more luminous star, 380 light years away. A small telescope will show that it is a physical double too, with a white companion of magnitude 7.1.

Kaus Australis

Epsilon (ε) Sagittarii

18h24m, -34°23′
Binary star

Magnitude: 1.8
Distance: 145 light yrs

The brightest star in Sagittarius, Kaus Australis, is one of the sky's enigmas. Its basic properties are well understood – it is a bright 'white giant', a heavyweight star with perhaps four times the Sun's mass, that has exhausted its core fuel supply and is now swelling, on the path to becoming an orange or red giant.

The mystery lies in its spectrum, which reveals a chemical composition with very few heavy elements. In theory, the interstellar clouds where stars are born should have become steadily enriched in these elements (so-called 'metals') through the history of our galaxy, so any stars born as recently as Kaus Australis should contain plenty of metals. One possible explanation is that somehow the star formed from a particularly metal-poor nebula. Another is that it is concealing its true nature behind a shell of metal-deficient gas, and it is the light from this, rather than the star itself, that gives rise to the strange spectrum.

M22

NGC 6656

18h36m, -23°54′
Globular cluster

Magnitude: 5.1
Distance: 10,400 light yrs

This globular cluster was the first object of its kind to be discovered, probably by German astronomer Johann Abraham Ihle in 1665. Visible to the naked eye as a Moon-sized patch of light, binoculars reveal its somewhat elliptical shape, while small telescopes will resolve individual stars around its edges. M22 is one of the closest globular clusters to Earth, and the finest visible to most observers in the northern hemisphere, which has led to its being intensely observed.

One particularly intriguing study made use of the fact that M22 lies in front of the more distant star clouds of the Milky Way's centre. As light from these stars passes through the cluster on its way to us, it can be slightly deflected by the gravity of objects within M22 – a phenomenon known as gravitational lensing. Slight changes to the light of the more distant stars have revealed that the cluster contains large numbers of relatively small objects – either large planets or small 'brown dwarf' stars.

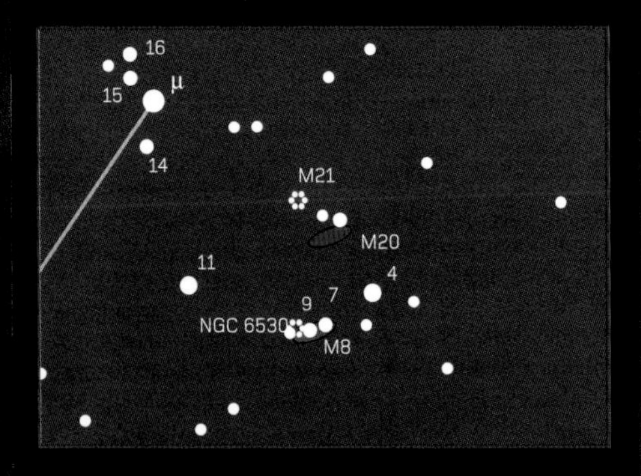

Lagoon Nebula

M8

18h4m, -24°23′
Emission nebula

Magnitude: 6.0
Distance: 5,200 light yrs

Deep inside the Lagoon Nebula, stellar winds whip gas and dust into interstellar tornadoes, starkly silhouetted against the background radiation of the nebula's more distant gas. The tornadoes may form in the same way as their earthly counterparts, twisted by the temperature variations between their hot outer surface and cold interior.

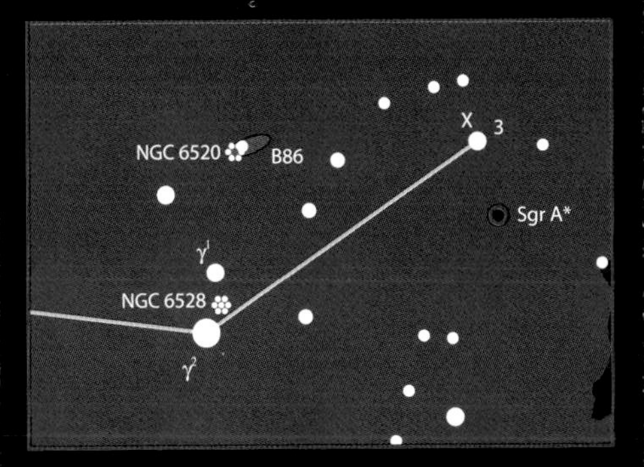

Galactic Centre

Sgr A*

17h46m, -29°0'

Supermassive black hole

Magnitude: -

Distance: 26,000 light yrs

Infrared cameras pierce the obscuring veil of dust, gas, and intervening stars to unveil the central regions of the Milky Way. In a broiling central cauldron of rapidly spinning nebulae, some of the galaxy's largest stars are born and die in thrall to the unseen central black hole.

Omega Nebula

M17

18h21m, -16°11'
Emission nebula

Magnitude: 6.0
Distance: 5,000 light yrs

Also known as the Swan Nebula, M17 is the northernmost in a chain of bright nebulae that cross the Milky Way in Sagittarius. Here the radiation from new stars blows clouds of hydrogen, helium and other elements such as sulphur into enormous billowing waves.

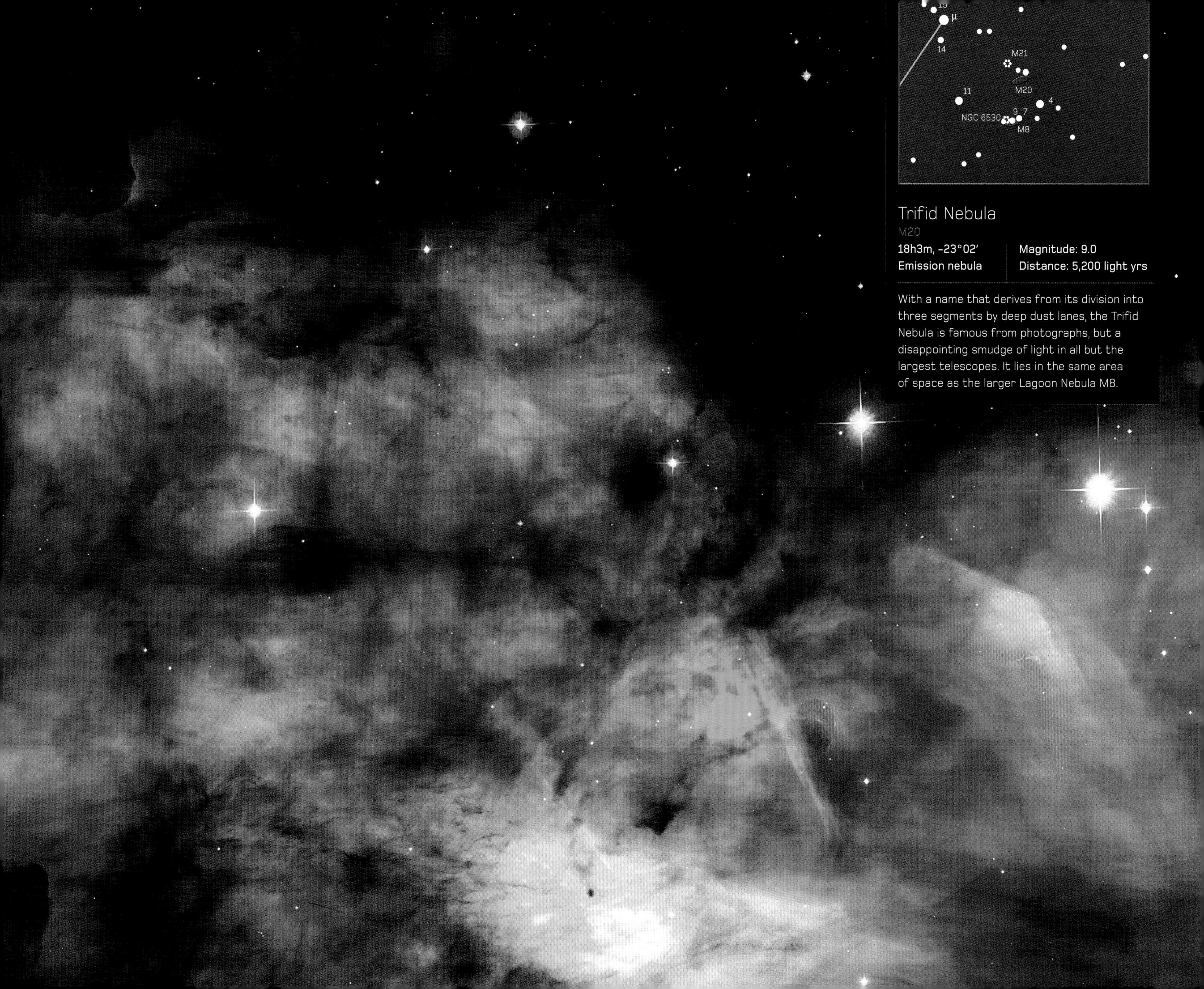

Trifid Nebula

M20

18h3m, –23°02′	Magnitude: 9.0
Emission nebula	Distance: 5,200 light yrs

With a name that derives from its division into three segments by deep dust lanes, the Trifid Nebula is famous from photographs, but a disappointing smudge of light in all but the largest telescopes. It lies in the same area of space as the larger Lagoon Nebula M8.

Capricorn

The Sea Goat

One of the strangest creatures in the sky is surely Capricornus, an animal with the head and body of a goat, but the tail of a fish. Its pattern is indistinct, and best found by looking for the close pairing of Alpha and Beta Capricorni east of Sagittarius or northwest of Fomalhaut.

The ancient Greeks saw Capricornus as a representation of the goat-headed god Pan, who turned himself into a fish when pursued by the monster Typhon. However, the constellation pre-dates the Greeks by a thousand years or more, and the figure of the 'goat-fish' was known to the Assyrians and Babylonians. Other cultures also saw some kind of grazing animal here – a normal goat, an ibex, or in the case of the Chinese, an ox.

2,500 years ago, when the astrological zodiac became 'fixed', the Sun reached its southernmost point in the sky in this constellation, and as a result the line on the Earth's surface at 23.5 degrees south, where the Sun is overhead at this extremity, is known as the Tropic of Capricorn. However, the relentless effects of precession, the 25,800-year 'wobble' of Earth's poles, has moved the Sun's present-day southernmost point into neighbouring Sagittarius.

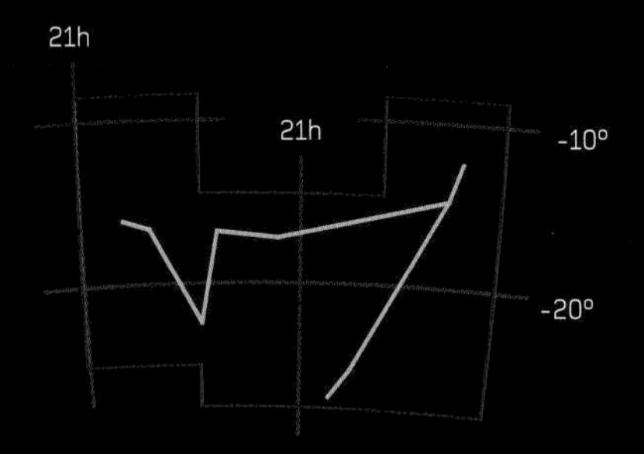

Deneb Algedi

Delta (δ) Capricorni

21h47m, -16°08'	**Magnitude: 2.9 (var)**
Multiple star	**Distance: 38 light yrs**

In comparison with some of the stars in Capricornus, the constellation's brightest star, Deneb Algedi (meaning 'the goat's tail') is relatively simple, although it still has its share of mysteries. The star is an eclipsing binary system, dropping in brightness by just 0.2 magnitudes once every 24.5 hours. The size of the magnitude dip and the lack of an obvious 'secondary minimum' when the eclipse is reversed suggests that almost all the system's light comes from the white primary, roughly eight times as luminous as the Sun, and the eclipsing companion must be a very dim dwarf star. The primary's spectrum, meanwhile, raises other questions – it is unusually rich in some elements but unusually poor in others, and as a result astronomers can't even agree on what stage the star has reached in its evolution.

NGC 6907

20h25m–24°49'	**Magnitude: 11.3**
Spiral galaxy	**Distance: 140M light yrs**

This faint but beautiful barred spiral galaxy had an intriguing tale to tell. Moderate-sized telescopes or long-exposure photographs reveal an elongated nucleus at the centre of two swirling spiral arms, one of which has a distinctly bright 'knot' within it. For more than a century, this was accepted as a feature of the galaxy itself – perhaps a region of intense star formation. But in 2004, astronomers established that it was the centre of another galaxy in its own right – a 'lenticular' system like a spiral stripped of its arms. What's more, this is no chance alignment – studies of the light from the two galaxies reveals that they are moving away from us at the same speed, and therefore (because the Universe is expanding) they lie at the same distance from us. It's now accepted that this is an act of galactic cannibalism 'caught in the act' – NGC 6907 is in the process of swallowing up its smaller companion.

M30

21h40m, -23°11'	**Magnitude: 7.2**
Globular cluster	**Distance: 26,000 light yrs**

Discovered by Charles Messier in 1764, and first resolved into stars 20 years later by William Herschel, M30 is a huge ball of stars about 90 light years across, and easily spotted through small telescopes. A slightly larger instrument will reveal individual stars, and show that M30 is somewhat unusual for a globular cluster – it is one of a handful that are said to have 'collapsed' cores. This means that most of its stars are concentrated at its very centre in a space perhaps just 20 light years wide, leaving the outer regions relatively empty – through a telescope they appear to be composed of loose chains of stars around a fuzzy, unresolved core.

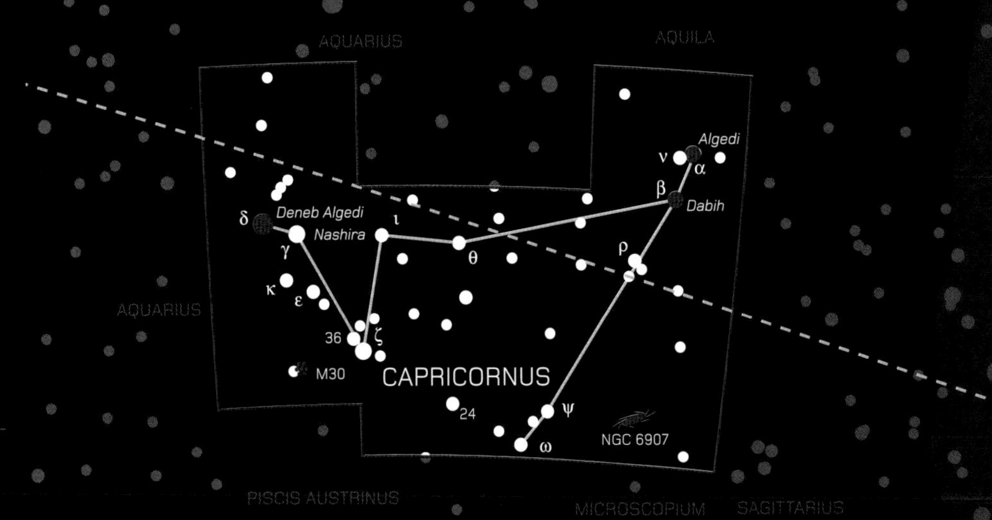

Dabih

Beta (β) Capricorni

20h21m, -14°47'	**Magnitude: 3.1**
Multiple star	**Distance: 330 light yrs**

The second brightest star in Capricornus, Dabih is also one of the most complex groups in the sky. Binoculars will easily separate it into two components – an orange star of magnitude 3.1 and a blue-white companion of magnitude 6.1. Even though these stars are widely separated, they are at the same distance and are thought to be bound by gravity, orbiting each other in about a million years. What's more, each of these stars is a multiple system in its own right.

The brighter star, Beta 1 Capricorni, is an orange giant. Very large telescopes can split Beta 1 from its close companion, a blue main-sequence star of magnitude 7.2, so close that the system takes just 3.8 years to complete an orbit. Evidence from the spectra of each star indicates that both are again binaries in their own right – the blue star certainly has a companion that orbits it in just 9 days.

The fainter Beta 2 Capricorni is a binary that is within the range of moderate telescopes, and there are two other faint stars in the location that may or may not be associated with the system.

Algiedi

Alpha (α) Capricorni

20h18m, -12°31'	**Magnitude: 4.2, 3.6**
Optical double/ multiple stars	**Distance: 635 light yrs 108 light yrs**

The two bright stars of Alpha Capricorni can be separated with the naked eye, or easily with binoculars. These naked-eye stars are a chance alignment in the sky, but a small telescope will reveal that, coincidentally, each is a binary system in its own right – the brighter and more easterly alpha 2 has a companion of magnitude 11 (which is itself a double), while the fainter alpha 1's partner shines at magnitude 9. The bright stars of each system are both yellow and have similar magnitudes, but they are at vastly different distances, indicating that alpha 1 is actually far more luminous in reality. It is thought to be a yellow supergiant with five times the mass of the Sun, while alpha 1 is a yellow giant with about twice the Sun's mass.

Piscis Austrinus & Microscopium

The Southern Fish and the Microscope

The constellation of Piscis Austrinus is an ancient one, despite being deep in the sky's southern hemisphere. Classical astronomers probably found it hard to ignore because of its single bright star, the brilliant Fomalhaut. The rest of the constellation is a scattering of much less impressive stars, however.

Mythologically, this fish is often depicted as the parent of the fishes in Pisces, which lie some way to its northeast. It is usually imagined swimming in (or drinking from) the water pouring from the jug of neighbouring Aquarius.

Microscopium, in contrast, is a relatively recent invention, devised by Nicolas de Lacaille in the 1750s. Like most of Lacaille's constellations, it represents a piece of scientific equipment, and like most, it is small, faint, and almost entirely obscure.

Fomalhaut

Alpha (α) Piscis Austrini

22h58m, -29°37'	**Magnitude: 1.2**
Extrasolar planetary system	**Distance: 25 light yrs**

The brightest star by far in Piscis Austrinus, Fomalhaut's name means 'mouth of the fish' in Arabic. It is a fairly typical white 'main sequence' star, shining by turning hydrogen to helium in its core. At a distance of 25 light years, it must be about 16 times as luminous as the Sun, which suggests it weighs a little over twice as much.

Fomalhaut is a fairly young star – perhaps 300 million years old – and when infrared telescopes were first trained on it during the 1980s, they revealed that it was surrounded by a vast disc of cool material about five times the size of our Solar System. Later studies have shown that the disc is like a ring doughnut, with a hole in the middle roughly the diameter of our own planetary system – and there have even been hints of distortions in the ring that might be caused by the gravity of large planets.

Beta (β) Piscis Austrini

22h32m, -32°21'	**Magnitude: 4.3**
Multiple star	**Distance: 148 light yrs**

Small telescopes will reveal that this white star is double, with a well-separated companion star of magnitude 7.7. At a distance of almost 150 light years, the primary is a main-sequence star with more than 2.5 times the mass of the Sun, and 35 times its luminosity.

Gamma (γ) Piscis Austrini

22h53m, -32°53'	**Magnitude: 4.5**
Binary star	**Distance: 220 light yrs**

This double star is more challenging than Beta Piscis Austrini, mainly because the two components are closer together, so light from the brighter primary tends to swamp its magnitude 8.0 companion. Nevertheless, a medium-sized telescope should reveal the two stars – a hot white giant 60 times as luminous as the Sun in visible light, and a normal yellow-white main sequence star just 2.5 times brighter than the Sun.

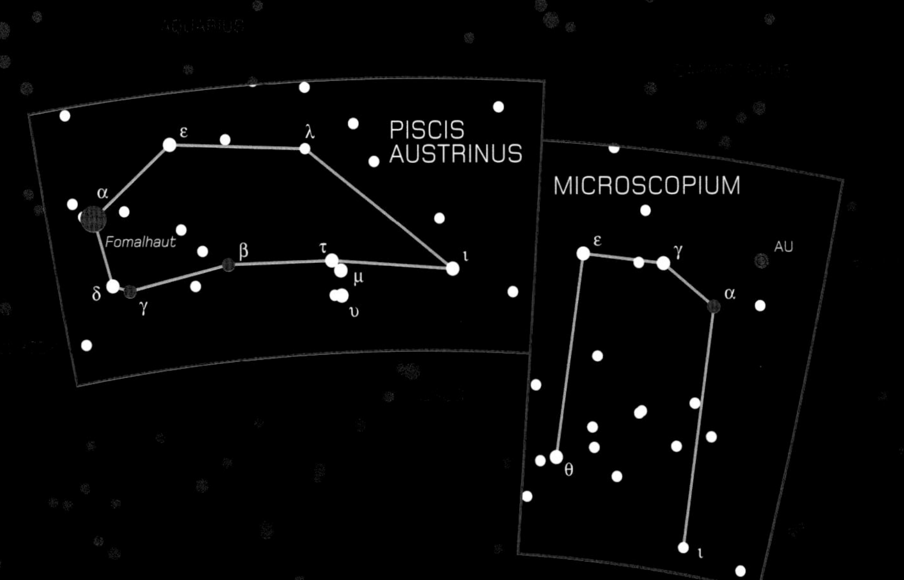

Alpha (α) Microscopii

20h50m, -33°47'	**Magnitude: 4.9**
Multiple star	**Distance: 380 light yrs**

One of the few objects of interest in the barren constellation of Microscopium, Alpha is not actually the brightest star – it is 0.2 magnitudes fainter than either Gamma or Epsilon. A small telescope will reveal that this yellow giant star, roughly 130 times as luminous as the Sun, has a dwarf companion of magnitude 10.6.

AU Microscopii

20h45m, -31°20'	**Magnitude: 8.6 (var)**
Extrasolar planetary system	**Distance: 32 light yrs**

Located to the northwest of Alpha Microscopii are the faint stars AU and AT. AT is a binary system of dim 10th magnitude red dwarfs, 32 light years away, and orbited at a distance of more than a light year by a more interesting star, AU Microscopii.

AU is a dwarf itself – it has roughly half the Sun's mass, two thirds its diameter, and only 3 percent of its luminosity. However, it is also a 'flare star' – a type of young dwarf star that has an extremely active magnetic field and can occasionally produce stellar flares that cause it to brighten considerably.

According to current estimates, AU might only have begun to shine about 10 million years ago – the blink of an eye in astronomical terms. It is still surrounded by a disc of debris left behind by its formation, and the presence of 'gaps' in this disc suggests that planets may be forming around the young star.

Eridanus

The River

Winding across the sky from a source near the celestial equator to an estuary in the far south of the sky, Eridanus is one of the longest constellations in the sky, and the sixth largest. Its path is hard to trace, and scattered with fairly faint stars, but at least its beginning and end are clearly marked. The second star in its long chain, magnitude 2.8 Cursa (the footstool) lies close to brilliant Rigel in Orion, while the southernmost star, Achernar, is one of the brightest stars in its own right.

The name Eridanus comes from the mythical river into which Phaeton, the unruly son of the sun god Helios, crashed the chariot of the Sun (supposedly after his horses took fright at the monstrous Scorpius). However, it has also been associated with many real rivers, including the Po in Italy, and the Nile of Egypt. Although its origins are ancient, the constellation was only extended to Achernar when European voyagers saw the brilliant star that lay beyond the river's end. Previously it ended at Theta Eridani or Acamar – and the names of both this star and Achernar come from the same Arabic phrase for 'river's end'.

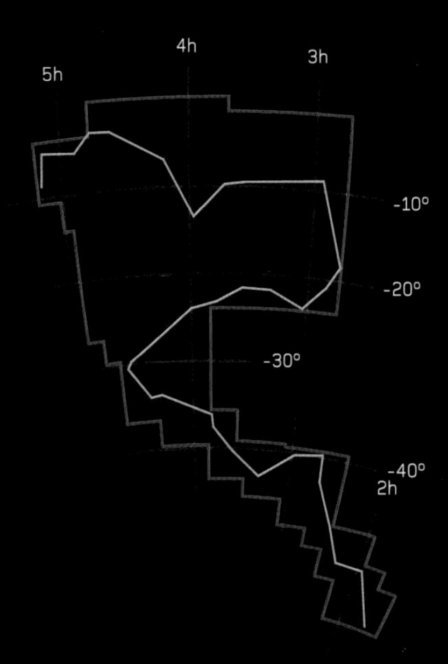

Keid

Omicron (ο) 2 Eridani

4h15m, -07°39′	Magnitude: 9.5
Multiple star	Distance: 16.5 light yrs

Omicron Eridani is a pair of closely-spaced stars of magnitudes 4.0 and 4.4, which can usually be separated with the naked eye. Omicron 1, the brighter northwestern star, is a distant 125 light years away, but Omicron 2 Eridani (also known as 40 Eridani) is on our cosmic doorstep, just 16 light years from Earth. A small telescope will reveal that our near-neighbour is a multiple star, with a magnitude 9.5 companion.

40 Eridani B, as it is often known, is the most easily seen white dwarf in Earth's skies – the burnt-out but still fiercely glowing core of a star similar to the Sun. It is all that remains after the original star expanded to a red giant and shrugged off its outer layers in a planetary nebula. Although Sirius B and Procyon B are both even closer to Earth, they are far harder to observe because their light is lost in the glare of their bright companions. Careful observers may spot that 40 Eridani B is a double star in its own right – the white dwarf is accompanied by a faint red dwarf of magnitude 11.2.

NGC 1300

2h20m, -19°25′	Magnitude: 11.4
Barred spiral galaxy	Distance: 70M light yrs

This remote galaxy is a challenging object for amateurs, requiring a medium-sized telescope, but it is still fascinating. Large telescopes reveal it as a spectacular barred spiral, with a central bar of stars extending perhaps 80,000 light years across the nucleus, linking onto a pair of swirling spiral arms at each end. The galaxy's core is intriguing in its own right – Hubble Space Telescope photos have revealed a secondary spiral structure of dust lanes embedded deep inside it – the main bar seems to extend from either end.

Although astronomers have a good idea of how galaxy spiral arms form with the passage of pressure waves through a region of gas and dust, they are less clear on the origins of barred spirals. The fact that some galaxies seem to have this 'spiral within a spiral' structure is even more intriguing.

Cleopatra's Eye

NGC 1535

4h14m, -12°44′	Magnitude: 9.6
Planetary Nebula	Distance: 5,800 light yrs

Despite its low overall magnitude, this planetary nebula is condensed into a very small area, so it has high contrast against the background sky, and small telescopes will reveal it as a fuzzy blue-green 'star'. Larger instruments will show the pale bubble of the nebula's structure, and perhaps the magnitude 12.2 central star.

Achernar

Alpha (α) Eridani

1h38m -57°14′	Magnitude: 0.5
Main sequence star	Distance: 143 light yrs

The brightest star in Eridanus lies at its extreme southern end, and since it barely rises from Mediterranean latitudes, it was not part of the original constellation in classical times. Achernar (whose name means 'river's end') is the hottest of the sky's really bright stars. It has a mass more than six times the Sun, and generates as much energy as several thousand Suns. As with many truly massive stars, Achernar rotates at incredibly high speeds which cause its equator to bulge out relative to the poles by as much as 50 percent. The varying density this creates in different parts of the star causes its surface temperature to range between 14,000°C around the equator, and 19,000°C near the poles.

Epsilon (ε) Eridani

3h33m, -09°28′	Magnitude: 3.4
Extrasolar planetary system	Distance: 10.5 light yrs

Epsilon Eridani is another of the Sun's near neighbours, barely 10 light years from Earth. It is a yellow dwarf star with about 80 percent of the Sun's mass, radiating about one-third of its energy, and is (at present) the closest star known to have a solar system. In the 1990s, astronomers discovered a dust ring in orbit around it, and there is at least one planet – a giant the size of Jupiter or bigger, in an eccentric orbit that varies between 2.4 and 5.8 astronomical units from its star. The fact that the dust ring seems to have been cleared close to the star raises the intriguing possibility that there might be other, more Earthlike, planets still to be discovered in this region.

Lepus, Caelum & Columba

The Hare, the Chisel and the Dove

Just south of the bright stars of Orion lie three constellations of varying age and obscurity. Lepus is one of the original 48 constellations recorded by the Greek-Egyptian astronomer Ptolemy in the second century AD. It represents a hare fleeing the approach of Orion's hunting dogs. A distinctive bow-tie shape makes it easy to spot, and Lepus would probably be better known if it were not overshadowed by its bright northern neighbour.

Columba was invented in the late 16th century, and has a chequered history. It was probably named by Dutch navigators Pieter Dirkszoon Keyser and Frederick de Houtman, and may have been intended to represent the bird that flew ahead of the Argo on its dangerous passage through the crashing rocks Scylla and Charibdis. Dutch astronomer and theologian Petrus Plancius later turned the bird into the dove sent out from Noah's Ark to prove that the Flood had receded – all part of his scheme to inject some Christian imagery into the pagan heavens.

Caelum, the Chisel, is one of the sky's also-rans – merely a line between two stars in an otherwise barren area of sky. Unsurprisingly it was devised in the 1750s by that inveterate constellation-maker, Nicolas de Lacaille.

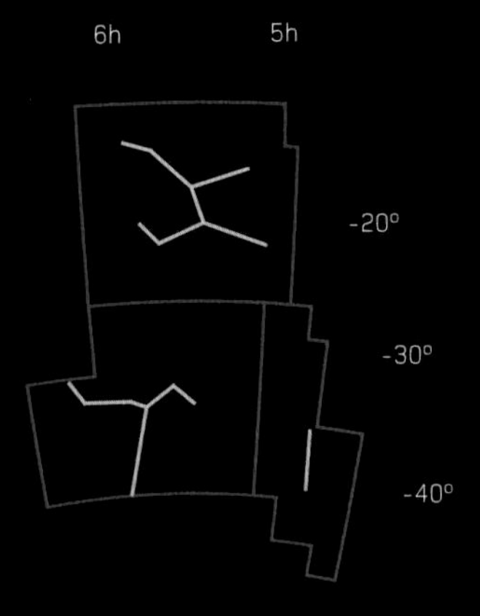

Arneb

Alpha (α) Leporis

5h33m, -17°49′	Magnitude: 2.6
Multiple star	Distance: 1,300 light yrs

The brightest star in Lepus has a name that means 'the hare' in Arabic. It is one of the sky's rare white supergiants – a star about 1,300 light years distant, but still brilliant enough to outshine its neighbours in Lepus, which are much closer to Earth. It is thought to be about 13,000 times as luminous as the Sun, so to appear white (implying a hot surface temperature of 6,700°C), it must be comparatively small for such a bright star – perhaps about the diameter of Mercury's orbit in the solar system.

Because the size of a star depends on the balancing act between the outward blast of radiation and the inward pull of its own gravity, a luminous supergiant such as Arneb must also be a stellar heavyweight, and estimates suggest Arneb has the mass of about eight Suns – putting it right on the borderline between stars that end their lives in planetary nebulae, and those that detonate in supernova explosions. It's not clear how far Arneb has gone along its evolutionary path – it may be swelling to become an even larger and brighter red giant, or it could be entering the relatively calm period during which it burns helium in its core.

M79

5h25m, -24°33′	Magnitude: 7.7
Globular cluster	Distance: 42,000 light yrs

This globular cluster is dimly visible through binoculars, but easily identified with a small telescope. Its location in the sky marks it out as unusual, since most globulars lie towards the centre of the Milky Way in Sagittarius – M79 is on the opposite side of the sky, and we are closer to the centre than it is.

What's more, the cluster is moving in the 'wrong' direction, retreating from us and the galactic centre at around 200 kilometres a second. This gives weight to a theory that M79 is an immigrant cluster, originating in a small dwarf galaxy and flung into its unusual orbit around the Milky Way when our galaxy consumed its parent.

Gamma (γ) Caeli

5h4m, -35°29′	Magnitude: 4.5
Binary star	Distance: 185 light yrs

This orange giant is the second brightest star in a dim constellation, shining only at magnitude 4.5. Medium-sized telescopes will reveal that it has a companion star of magnitude 8.1.

Hind's Crimson Star

R Leporis

5h0m, -14°48′	Magnitude: 7.7 (var)
Main sequence star	Distance: 820 light yrs

One of the reddest stars in the sky, R Leporis gets its distinctive colour (and name) because it is one of the coolest stars in the sky – a red giant whose light is tinted by the large amounts of carbon in its outer atmosphere. It is an unmistakable sight through binoculars, but sometimes requires a small telescope as it varies in brightness between magnitude 5.5 and 11.7. R Leporis is a long-period variable similar to the famous Mira in Cetus – it changes its brilliance continuously as it pulsates over a period of 430 days.

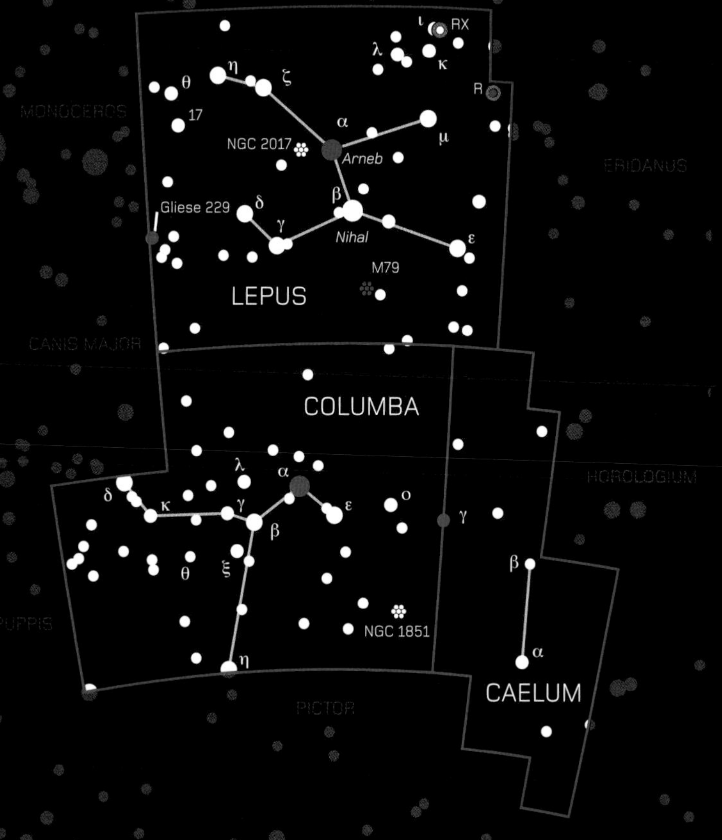

Phact

Alpha (α) Columbae

5h40m, -34°04′	Magnitude: 2.6
Variable star	Distance: 530 light yrs

Columba's brightest star has a name that comes from the Arabic for 'collared dove'. It is a brilliant blue-white star with more than four times the mass of the Sun, that has exhausted the hydrogen supply in its core and begun to evolve towards a true giant. It is also a rapidly rotating 'Be' star similar to Gamma Cassiopeiae – as it expands into a giant, it flings off material from around its equator, encasing it in semi-opaque rings that cause its brightness to vary slightly.

Gliese 229

6h10m, -21°52′	Magnitude: 8.1
Dwarf star system	Distance: 19 light yrs

An average red dwarf star, with just a third of the Sun's mass and two percent of its luminosity, Gliese 229 is only within the reach of small telescopes because of its relative proximity to the Solar System. Its fame, however, rests on its almost undetectable companion, Gliese 229B.

This 'brown dwarf' star was the first of its kind to be imaged directly, using the power of the Hubble Space Telescope to confirm the existence of what had, until 1995, been a hypothetical class of 'failed stars'. 229B is about the same size as Jupiter, but packs in between 25 and 65 times that planet's mass, creating a low-mass 'star' that is too small to begin the nuclear fusion of normal hydrogen in its core, but instead shines dimly thanks to gravitational contraction and the less demanding fusion of small amounts of deuterium (a heavier form of hydrogen that is more susceptible to nuclear reactions).

The dividing line between dim red dwarfs, brown dwarfs, and large Jupiter-class planets is very blurred, and they are usually defined by how they form rather than their size.

Brown dwarfs are created in the same way as stars, and space is thought to be scattered with lone dwarfs too dim to be detected. It is only in binary systems such as Gliese 229 that their gravitational tug on a brighter star gives away their existence.

Puppis & Pyxis

The Stern and the Compass

The large and bright constellation of Puppis is the northernmost of three groupings that in ancient times formed a single huge pattern known as Argo Navis. Since split into Puppis the Stern, Carina the Keel and Vela the Sails, the constellations together represent the famous mythical ship Argo, on which Jason and the Argonauts set sail in search of the Golden Fleece.

Puppis is aligned along a bright swathe of the Milky Way running south from Canis Major, and is host to a number of bright star clusters and other interesting objects. Its neighbour Pyxis is one of the sky's least convincing constellations – little more than three faint stars in a row. Nicolas de Lacaille named it after its supposed resemblance to a magnetic compass, so although the Argonauts would never have had such a device, the constellation is at least appropriately placed near the sky's great ship.

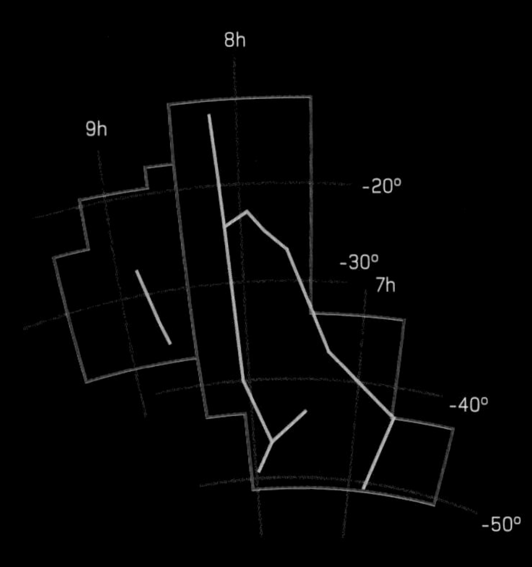

Naos

Zeta (ζ) Puppis
8h4m, -40° | **Magnitude: 2.2**
Giant star | **Distance: 1,400 light yrs**

The brightest star in Puppis is also one of the most extreme stars in the entire sky. Naos (whose name simply means 'ship') is among the hottest stars known, with a surface temperature of about 42,000°C. This gives it an obvious blue tint, and causes it to emit 97 percent of its energy in the ultraviolet. But Naos is such a huge and luminous star that the small amount of its energy emitted as visible light still makes it noticeable across 1,400 light years. At visible wavelengths, Naos is more than 21,000 times more luminous than the Sun, but overall it generates perhaps 750,000 times the energy of our star. To put it another way, if Naos emitted all its energy in visible light, it would be the brightest star in the sky, outshining Sirius at magnitude -1.6, despite being 160 times more distant.

In order to pump out so much energy, Naos must be a truly giant star, and indeed it is one of the most massive stars known, thought to weigh as much as 60 Suns. It has squandered the vast supply of hydrogen at its core in order to shine so brightly, and evolved into a blue supergiant that is now burning its way through the helium ash left behind by its brief 'main sequence' lifetime. It is doomed to end its life as a spectacular supernova in the next few million years.

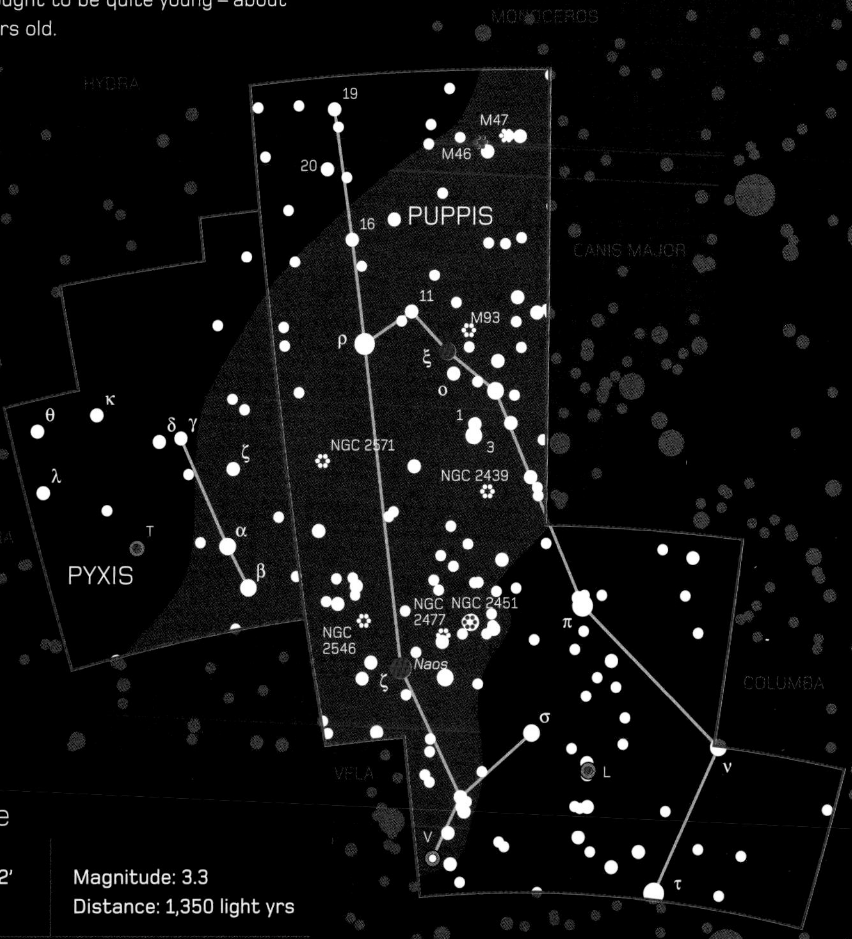

T Pyxidis

9h5m, -32°23′ | **Magnitude: 13.8 (var)**
Recurrent nova | **Distance: 6,000 light yrs**

This faint star, normally beyond the range of small amateur telescopes, is a recurrent nova, prone to occasional outbursts that bring it within range of binoculars or even (just) to naked-eye visibility. It is a binary system in which a dense white dwarf is pulling material away from a bloated giant star. Captured gas piles up to form a thick 'atmosphere' around the white dwarf, until it reaches such high temperatures and pressures that it ignites and burns away in a burst of nuclear fusion. The explosions happen every couple of decades, but are not entirely predictable – the most recent outbursts were in 1890, 1902, 1920, 1944, 1966, and 1996.

NGC 2451

7h45m, -37°58′ | **Magnitude: 2.8**
Open cluster | **Distance: 850 light yrs**

This cluster, far smaller and looser than M46, contains just 40 stars, but is obvious to the naked eye because of its proximity – just 850 light years away. Binoculars or a small telescope will reveal a beautiful star field dominated by the orange giant c Puppis. The cluster is thought to be quite young – about 36 million years old.

Asmidiske

Xi (ξ) Puppis
7h49m, -24°52′ | **Magnitude: 3.3**
Multiple star | **Distance: 1,350 light yrs**

Just slightly closer to Earth than Naos, Xi Puppis is a yellow supergiant, with a colour and surface temperature slightly cooler than the Sun, but far more luminous – pumping out 8,300 times more energy than the Sun. It is thought to have the mass of about eight Suns.

A moderate-sized telescope will reveal a faint but well-separated yellow companion of magnitude 13 that is genuinely similar to the Sun – it must take about 26,000 years to circle the primary. Studies of Asmidiske's light suggest that it also has a much closer, unseen companion, orbiting in about a year.

L Puppis

7h44m, -28°57′ | **Magnitude: 4.9, 4.4 (var)**
Optical double star | **Distance: 181 light yrs, 198 light yrs**

The two stars that mark L Puppis make an easily distinguished pair through binoculars, and even naked-eye observers can sometimes tell that L is double. The stars lie at different distances from Earth and are a chance alignment, but the nature of L2 Puppis gives the system a unique appeal.

A small telescope will reveal an attractive pairing of the blue-white L1 to the south, and the red L2 to the north. L1 shines at a steady magnitude 4.9, but L2 is a semi-regular variable, varying in brightness more-or-less predictably over a 141-day cycle. At its brightest, L2 reaches magnitude 2.6, making it easily the brighter of the two stars. At its faintest it disappears beyond naked-eye visibility to magnitude 6.2. L2 is a dying red giant star, probably on its way to becoming a long-period variable like Mira in Cetus. It is one of the sky's brightest variable stars, and all the more spectacular for having a close companion in the sky.

M46

NGC 2437
7h42m, -14°49′ | **Magnitude: 6.0**
Open cluster | **Distance: 5,400 light yrs**

This expansive open cluster, just beyond naked-eye visibility but easily found in binoculars, contains about 150 stars visible through a small telescope. It is thought to be about 300 million years old, but a slightly larger telescope may detect what seems to be a probem with this picture – a glowing planetary nebula, NGC 2438, on its northern edge. Any star with the comparatively low mass required to form a planetary nebula should have a lifespan of far longer than 300 million years.

Fortunately, analysis of light from the cluster stars and the nebula reveals an explanation. While the cluster is moving away at an average speed of 41 km/s, NGC 2438 is retreating much more quickly, at 77 km/s. This is strong evidence that the two objects are not physically related, and the planetary nebula just happens to lie along our line of sight to M46.

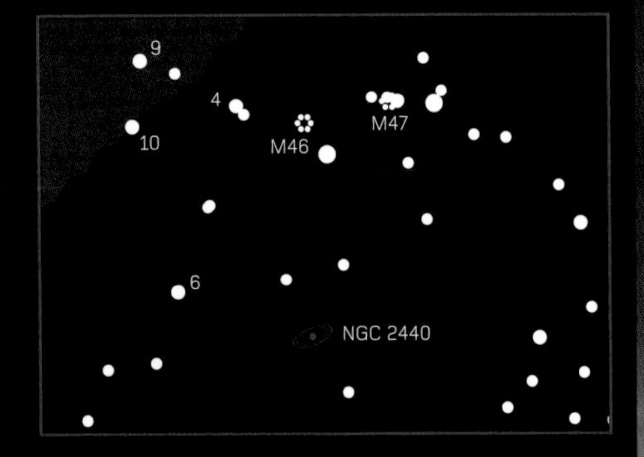

NGC 2440

7h42m, -18°12′
Planetary nebula

Magnitude: 11.0
Distance: 3,600 light yrs

At the heart of this faint planetary nebula lies one of the hottest white dwarf stars known – it has a scorching surface temperature of 200,000°C. The fierce heat comes from the gravitational collapse of a stellar core that has already burnt out and cast off its outer layers.

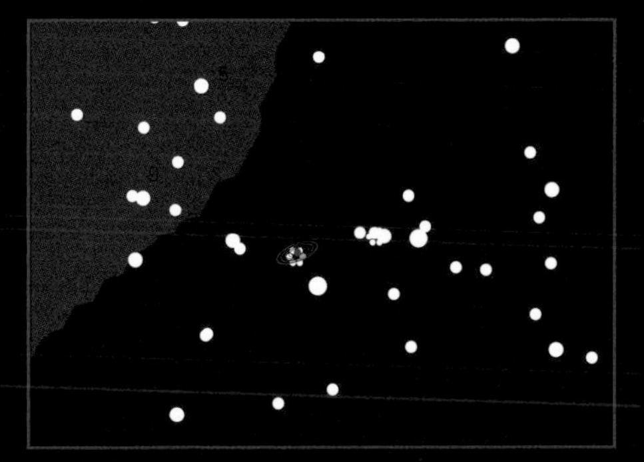

Rotten Egg/Calabash Nebula

OH 231.8+4.2

7h42m, -14°43′	Magnitude: 9.5
Planetary nebula	Distance: 5,000 light yrs

The Calabash Nebula gets its common nickname from the presence of sulphurous compounds in its expanding gases, which on Earth cause noxious smells. Jets of this yellow gas, expanding at 15 million kph, are ploughing into static interstellar gas clouds and causing them to emit cold blue light.

Vela

The Sail

A rough octagon of bright stars marks the location of Argo's sail – one of three surviving fragments of the once-great constellation of Argo Navis. The decision to rend the mighty ship asunder was made by the French astronomer Nicolas de Lacaille in the 1750s, and was soon widely adopted. One of the results was the curious distribution of the Greek Bayer letters in this area of the southern sky. Supposedly to avoid confusion, the stars kept their original Greek letters, even though they were divided among the constellations. The Alpha and Beta stars went to Carina, leaving Gamma Velorum as Vela's brightest star, and Zeta Puppis as the brightest star in the Stern.

Vela itself covers a large swathe of sky including a stretch of the southern Milky Way. Its most impressive features are two glowing nebulae that cover much of the constellation – supernova remnants of vastly different ages.

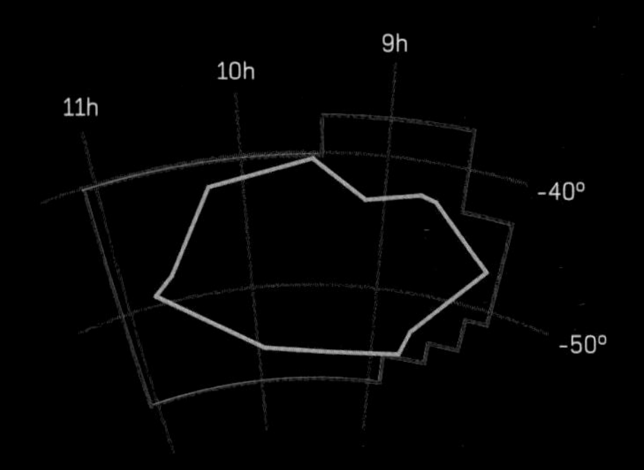

Regor

Gamma (γ) Velorum

8h10m, -47°20′	**Magnitude: 1.8**
Multiple star	**Distance: 1,200 light yrs**

The brightest star in Vela, Regor's name is surprisingly modern – it was invented as a joke by astronaut Gus Grissom in honour of his colleague Roger B. Chaffee, but when both men were killed in a tragic fire at the beginning of the Apollo programme, the name lived on.

Binoculars or a small telescope will show the beginnings of Regor's complex multiplicity, revealing a wide companion star of magnitude 4.3. Two other stars around the primary are also part of the system. One shines at magnitude 8.5, while the other, at magnitude 9.4, is a binary in its own right.

But it is the bright primary star that is most fascinating – studies of its spectrum reveal that it is a binary with a difference. Most of its light comes from a hot blue star with a surface temperature of 32,000°C, pumping out perhaps 180,000 times the energy of the Sun, and with 30 times its mass. Orbiting with it is a star with just ten times the mass of the Sun, but with an incredibly hot surface of around 60,000°C, and an atmosphere rich in carbon.

This Wolf-Rayet star began life as a monster with 40 times the Sun's mass, so hot that it generated a ferocious stellar wind which has gradually blasted the star's outer layers away into space. With each lost layer of hydrogen, a new and even hotter surface was exposed, and the winds intensified. Over a few million years, the star's intensity has cost it three quarters of its mass, until now the deep layers of its interior, enriched with products of the helium fusion that keeps it shining, are exposed to space.

Omicron (ο) Velorum Cluster

IC 2391

8h40m, -53°04′	**Magnitude: 2.5**
Open cluster	**Distance: 580 light yrs**

This bright open cluster contains about 50 stars, centred on blue-white magnitude 3.2 Omicron Velorum, also known as Xestus (the Greek god of ocean currents). This is a high-mass, blue-white star that has exhausted its core supply of hydrogen fuel and just begun to swell towards giant status. It varies by about 0.1 magnitudes in 67 hours.

The cluster as a whole is thought to be some 30 million years old, and was known to Persian astronomer Al Sufi as early as 964. More recently, it has been the subject of an intense search for faint brown dwarf stars, which fade as they age and so will be at their brightest in clusters like this. Twenty four probable brown dwarfs have been found, indicating that these 'failed stars' may be almost as numerous as proper stars.

NGC 3201

10h18m, -46°25′	**Magnitude: 6.8**
Globular cluster	**Distance: 15,000 light yrs**

This impressive globular cluster, discovered by Scottish astronomer James Dunlop during his visit to Australia in 1826, is among the nearest to the Sun, and fairly easy to pick up with binoculars. Medium-sized telescopes are needed to pick out individual stars, and will reveal that the cluster has an unusually sparse centre, with stars resolvable nearly all the way to the core.

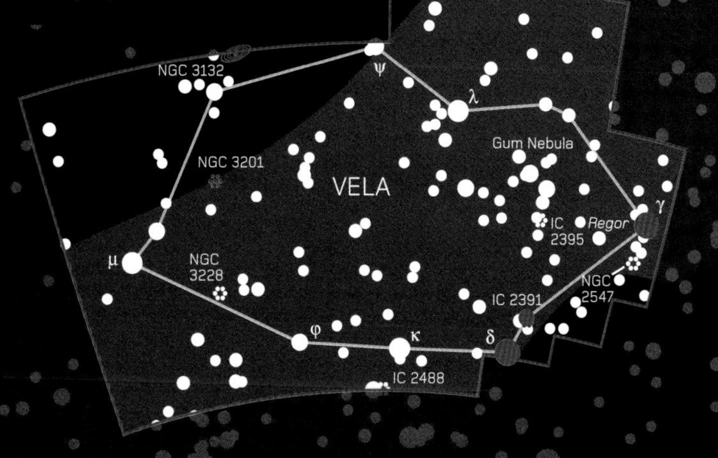

Delta (δ) Velorum

8h45m, -54°43′	**Magnitude: 2.0**
Multiple star	**Distance: 80 light yrs**

Vela's second star is a complex multiple system – a moderate sized telescope will reveal all of its elements, but the naked eye can detect its most striking feature. The primary star is an eclipsing binary, whose brightness drops by a noticeable 0.4 magnitudes for a few hours every 45 days when its fainter element, a white star with the mass of two Suns, partially blocks out light from a brighter star that is 30 percent more massive and six times more luminous.

This close pair, inseparable through the largest telescope, is accompanied by a yellow, sunlike star of magnitude 5.1 with a highly elliptical orbit. This companion orbits the central pair in 142 years, and is currently retreating from a close approach in 2000, when it could only be separated through a fairly large telescope. As the gap between the stars widens, they make a good test of telescope optics. Any telescope currently large enough to separate these stars should also pick up the faint light of another binary pair – red dwarfs of magnitude 11 and 13.5 that orbit one another in a couple of centuries, and the entire system in tens of thousands of years.

The Gum Nebula

Gum 12

8h35m, -45°	**Magnitude: -**
Supernova remnant	**Distance: 1,300 light yrs**

Much of Vela is covered with a layer of faint nebulosity, forming a background to the more prominent Vela Supernova Remnant (SNR). It was only in the 1950s that Australian astronomer Colin Gum realised that the various patches of softly glowing gas might all be part of one huge structure.

Today, astronomers believe that this structure, now known as the Gum Nebula, is a faded and distended supernova remnant in its own right. But in contrast to the brighter Vela SNR, which was formed by an explosion around 10,000 years ago, the blast that created the Gum Nebula happened about a million years ago.

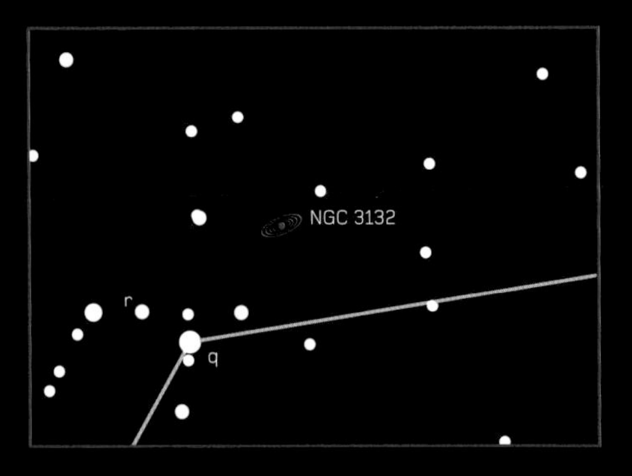

Eight-burst Nebula

NGC 3132

10h7m, -40°26′	Magnitude: 9.9
Planetary nebula	Distance: 2,000 light yrs

Bringing to mind a limpid tropical pool, the expanding shell of Vela's beautiful Eight-burst Nebula is illuminated and excited not by the bright, obvious central star, but by the faint companion. However, the presence of the brighter star is not a mere line-of-sight coincidence – the two form a true binary pair.

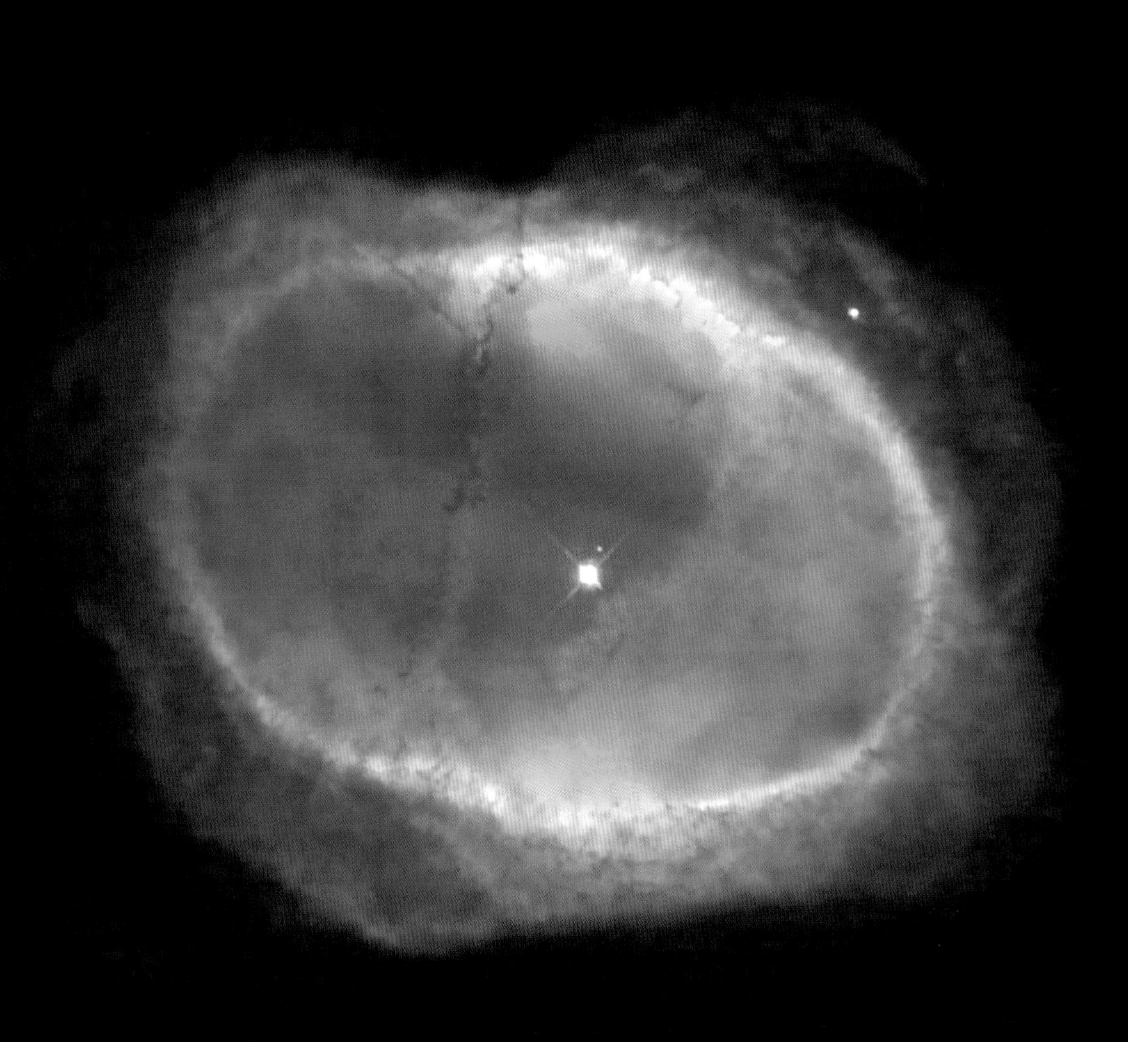

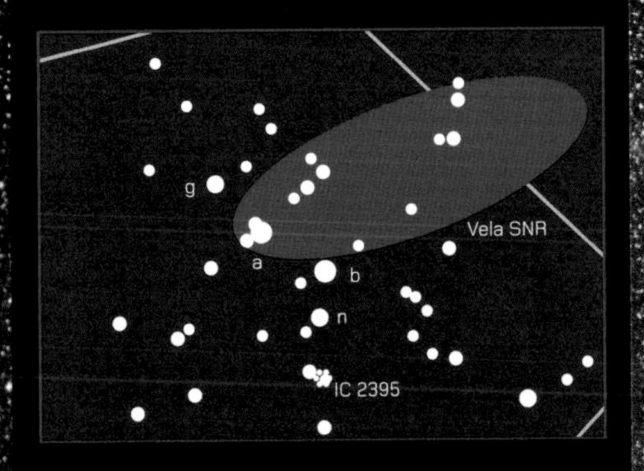

Vela Supernova Remnant

8h35m, -45°10′
Supernova remnant

Magnitude: 12
Diameter: 800 light yrs

Scattered across a large area of northwestern Vela lie the still-glowing ashes of a star that died in a spectacular supernova about 10,000 years ago. At the heart of this supernova remnant are the mortal remains of their progenitor – a rapidly spinning neutron star just 20 kilometres across.

Carina

The Keel

The most prominent part of the ancient Argo Navis constellation survives today as Carina, the Keel. It is unmistakable thanks to the presence of Canopus, the second brightest star in the entire sky, as well as five of the next ten brightest stars in Argo. Most of its interesting objects are clustered in the northeastern corner, around the huge Carina Nebula. Canopus is rather isolated to the northwest.

One obvious and occasionally confusing pattern in the constellation is the so-called 'False Cross' – a slightly larger lookalike of the Southern Cross, formed from the bright stars Iota and Epsilon Carinae, with Kappa and Delta Velorum. The False Cross lies a little to the west of the true cross, Crux, and its long axis is almost exactly parallel.

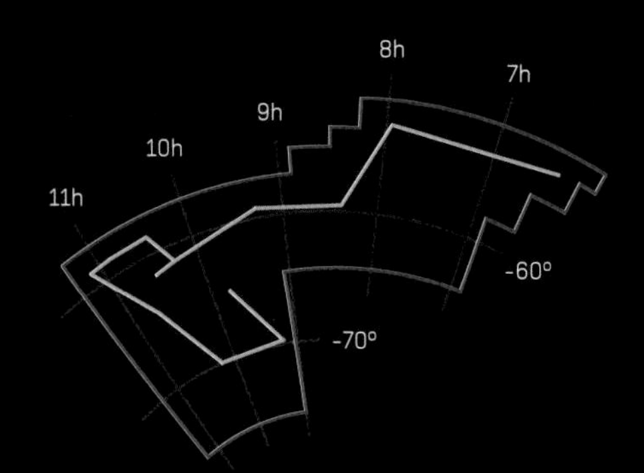

Canopus

Alpha (α) Carinae

6h24m, -52°42′	Magnitude: -0.7
Giant star	Distance: 315 light yrs

The sky's second most brilliant star is a far more formidable object than its rival Sirius, which only beats it for brightness because it is a near neighbour to the Sun. Canopus is more than 30 times further away, and is a brilliant yellow-white supergiant. The colour is unusual for a star of this type – supergiants are usually either relatively small, hot and bluish, or bloated, cool, and properly yellow, orange or red. The crucial factor is the star's gravity – Canopus is thought to weigh as much as eight Suns, which should not be enough to prevent it expanding to a larger size, so we are probably seeing it in transition. Its current diameter is roughly half the size of Mercury's orbit in our solar system, but astronomers aren't sure whether it is currently growing into a cooler orange supergiant, or shrinking as it taps into new energy sources at its core.

Appropriately for a star named after a famous helmsman of Greek myth, Canopus has proved useful for navigation. In earlier times, it acted as a useful pointer to the south for European vessels venturing towards the equator, while today its brightness and location in the sky mean it is often used to help the navigational instruments on spaceprobes find their bearings.

The Southern Pleiades

IC 2602

10h43m, -64°24′	Magnitude: 1.9
Open cluster	Distance: 480 light yrs

This bright and beautiful cluster, first catalogued by Nicolas de Lacaille from Cape Town in 1751, is rightly compared to the 'true' Pleiades of Taurus. It contains some sixty or more stars, of which the brightest is Theta Carinae at magnitude 2.7. However, several other individual stars of around magnitude 5 can be picked out individually by the sharp-eyed.

The cluster is thought to be a relatively youthful 30 million years old, and coincidentally lies just 40 light years further away than the Pleiades themselves.

The Wishing Well Cluster

NGC 3532

11h6m, -58°40′	Magnitude: 3.0
Open cluster	Distance: 1,300 light yrs

This rich open cluster is easily visible to the naked eye as a fuzzy patch of light twice the size of the full Moon, set against the Milky Way. Binoculars or telescopes on low magnification will display a field that contains dozens of stars of around magnitude 7 and lower – there are thought to be 150 in total, and their twinkling appearance when low on the horizon and filtered through turbulent air is said to make them resemble coins at the bottom of a wishing well.

On the cluster's eastern edge lies a bright yellow star known as x Carinae. However, its position is a simple line-of-sight effect – the star is not associated with the cluster and is in fact a yellow supergiant about three times further away.

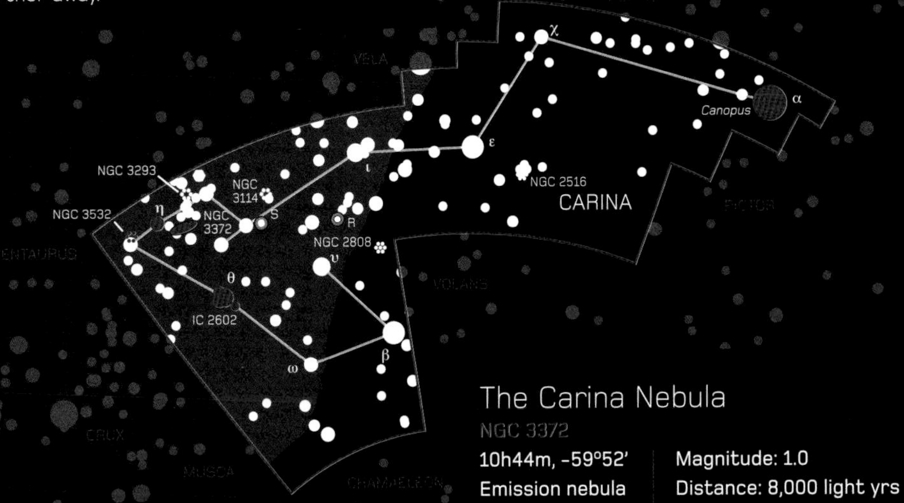

The Carina Nebula

NGC 3372

10h44m, -59°52′	Magnitude: 1.0
Emission nebula	Distance: 8,000 light yrs

Spread across two degrees of the Milky Way around Eta Carinae lies a bright, star-forming nebula best viewed through binoculars or with a very wide field and low magnification. This cloud of gas and dust gave birth to Eta Carinae itself about a million years ago, and stars are still forming here today – the nebula contains several bright knots of stars, whose hot ultraviolet radiation is exciting the gas and causing it to glow. The Homunculus Nebula around Eta Carinae forms one bright patch within the wider nebula, while another famous feature is the Keyhole Nebula – a distinctive patch of darker gas within which new stars may be forming, silhouetted against the brighter background.

Eta (η) Carinae

10h45m, -59°41′	Magnitude: 6.2 (var)
Binary star	Distance: 8,000 light yrs

Although it currently lies just beyond the limit of naked-eye visibility, this highly unpredictable variable star is well worth tracking down with binoculars, for it is one of the true monsters of our galaxy. Eta Carinae lies within the large Carina Nebula, surrounded by its own clump of bright nebulosity known as the Homunculus Nebula. This cloud of expanding gas obscures much of the star's visible light, but allows infrared radiation to shine through, making Eta the brightest infrared object in the entire sky.

While the star's visible brightness today varies unpredictably over fairly long periods, the 1840s saw a much more violent eruption, in which Eta outshone Canopus and was briefly the second brightest star in the sky, before slowly fading away. The Homunculus Nebula seems to have originated in this enormous stellar explosion. Larger amateur telescopes can identify its twin lobes pinched around the star in an hourglass shape (perhaps by the action of a companion star, since there is good evidence that Eta may be a binary).

Estimates of the distance to Eta Carinae have to be made from the stars around it in the Keyhole Nebula. They put it a colossal 8,000 light years away, suggesting that it is 1.5 million times as luminous as the Sun in visible light, and perhaps 5 million times as bright overall.

To produce such stupendous amounts of energy, Eta must be a truly monstrous star or stellar pair – current estimates suggest it weighs about 100 times the mass of the Sun, close to the theoretical upper limit for a star, where it will burst apart due to the pressure from its own intense radiation. As a double system, each star might be slightly smaller, but still with the mass of 60-80 Suns.

Whatever its true nature, Eta Carinae is clearly heading rapidly towards a violent end. In other galaxies 'supernova impostor events' such as the eruption of 1843 have presaged the final death of a star in a supernova within a few years. No one knows quite how much time Eta has left – it could be anything between a few months and several thousand years – but when it finally destroys itself, the dying Eta will go out in a blaze of glory, outshining even Sirius to become the brightest star in the sky.

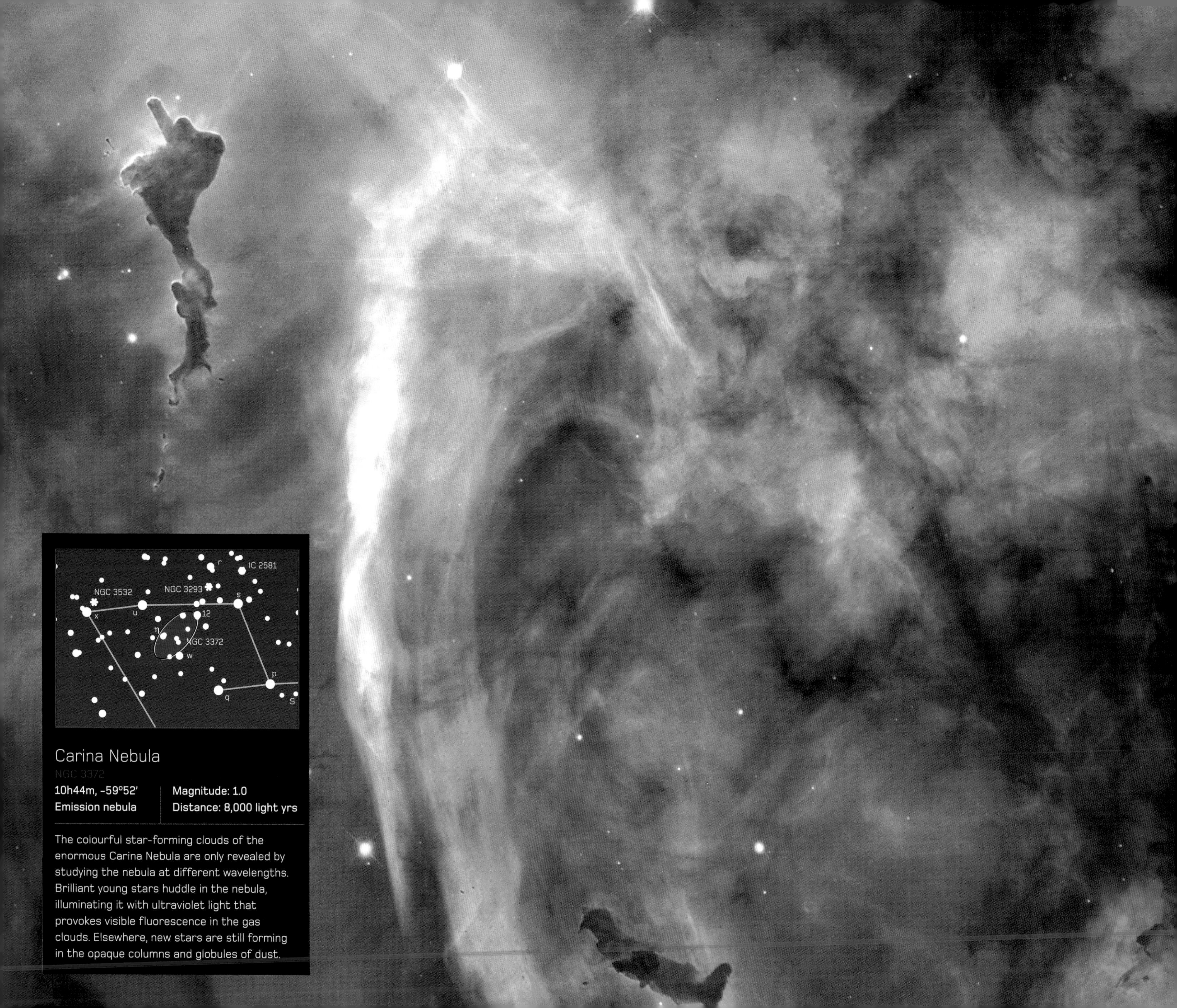

Carina Nebula

NGC 3372

10h44m, -59º52′
Emission nebula

Magnitude: 1.0
Distance: 8,000 light yrs

The colourful star-forming clouds of the enormous Carina Nebula are only revealed by studying the nebula at different wavelengths. Brilliant young stars huddle in the nebula, illuminating it with ultraviolet light that provokes visible fluorescence in the gas clouds. Elsewhere, new stars are still forming in the opaque columns and globules of dust.

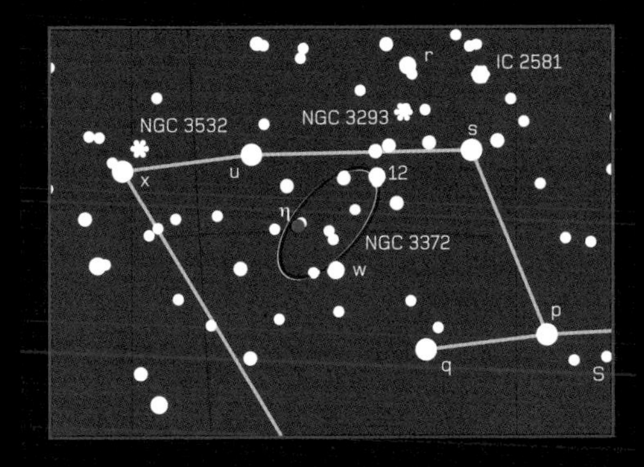

Eta (η) Carinae

10h45m, -59°41′	Magnitude: 6.2 (var)
Binary star	Distance: 8,000 light yrs

The double-lobed Homunculus Nebula that lurks within the large star-forming cloud of the Carina Nebula conceals at its heart Eta Carinae, one of the most massive stars known. During the outburst that threw off these great bubbles of gas, Eta shone briefly with the light of 11 million Suns. It seems miraculous that the star actually survived this violent 'supernova impostor' event.

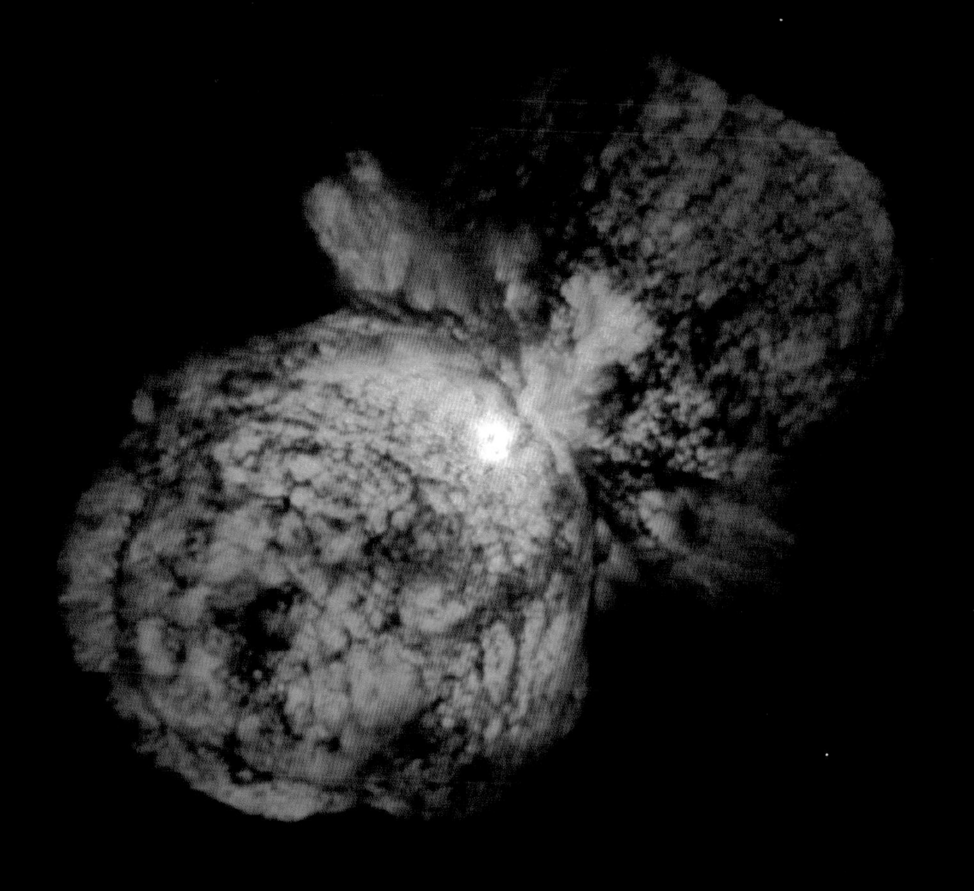

Crux & Musca

The Southern Cross and the Fly

The sky's smallest constellation is also one of the most famous, represented on several national flags. Crux is a tight grouping of four bright stars embedded deep in the southern Milky Way, invented by early European navigators journeying into the southern hemisphere. The ancient Greeks knew of its stars before precession carried them out of sight from the Mediterranean, but they considered them as part of Centaurus.

Though very distinctive, Crux is not entirely unmistakable to novice stargazers – a little to its west stars in Carina form a quite similar grouping, known as the 'false cross'.

South of Crux lies Musca, a fairly jumbled group of stars in the Milky Way. When first invented by Dutch sailors in the late 1600s, this was Apis, the bee, but its identity changed in the 17th century, possibly to avoid confusion with Apus, the Bird of Paradise.

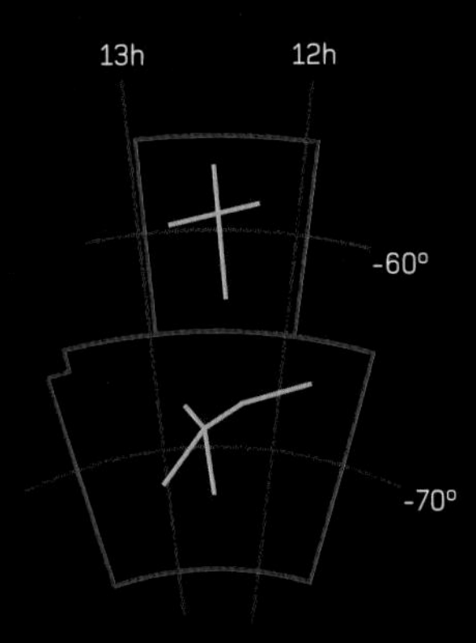

Theta (θ) Muscae

13h8m, -65°18′ | **Magnitude: 5.7**
Multiple star | **Distance: 10,000 light yrs**

Small telescopes will resolve this blue-white star into two distinct components of magnitude 5.7 and 7.3. The brighter component is a hot blue star, pumping out most of its energy in the ultraviolet, but about 50,000 times as luminous as the Sun in visible light (if the rough measures of the system's distance are correct). The fainter companion is a stellar oddity called a Wolf-Rayet star – an object that was originally so massive and hot (with perhaps 40 or 50 times the mass of the Sun) that strong stellar winds began to blow away its outer layers. Over the star's few million years of life, this effect has intensified, so that now the star has lost much of its outer envelope of hydrogen, leaving its intensely hot inner layers, rich in the products of nuclear fusion, exposed to view. Theta Muscae is among the brightest Wolf-Rayets in the sky – another is a component of Gamma Velorum, but in that case, it lies so close to its binary companion that the two stars cannot be distinguished.

The Jewel Box

NGC 4755

12h54m, -60°20′ | **Magnitude: 4.2**
Open cluster | **Distance: 7,600 light yrs**

This beautiful open star cluster is rightly considered a highlight of the southern skies. Positioned neatly between Mimosa and the Coalsack, it is visible to the naked eye as a blur of light around magnitude 5.9 Kappa Crucis, and is a truly spectacular sight through binoculars or a telescope on low magnification. The cluster contains several dozen stars, most of which are blue and white, but one of which, a brilliant red giant of magnitude 7.6, forms a strong contrast.

Because many of its most massive and short-lived stars are still on their hydrogen-burning 'main sequence' lifetimes, the cluster must be fairly young, and astronomers have calculated that it is actually among the youngest known, just 7.1 million years old.

Acrux

Alpha (α) Crucis
12h27m, -63°06′ | **Magnitude: 0.8**
Multiple star | **Distance: 320 light yrs**

The southern end of the cross is marked by a dazzling double star. What appears to the naked eye as a single blue-white star of magnitude 0.8 is transformed by the smallest telescope into a double of magnitudes 1.3 and 1.7. These near-twins are more than 300 light years away, generating 25,000 and 16,000 times as much energy as the Sun respectively. Spectroscopy has revealed that the brighter star is itself a double, containing stars of 14 and 10 solar masses in a 76-day orbit around each other. The fainter, single star has the mass of 13 Suns on its own, so at least two of the stars in this system will one day end their lives as supernovae.

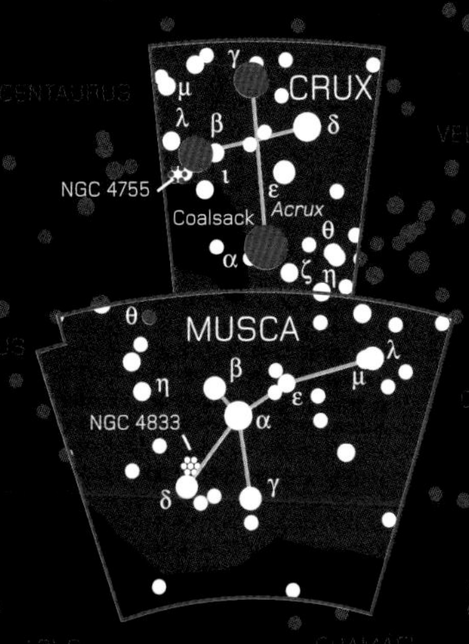

The Coalsack

12h52m, -63°18′ | **Magnitude: -**
Dark nebula | **Distance: 2,000 light yrs**

The southeastern quadrant of Crux is dominated by a striking hole in the sky – a black gulf in the Milky Way created by a dark cloud of dust that intervenes between Earth and the more distant star clouds. The Coalsack is thought to be about 70 light years across and perhaps 2,000 light years away – only a few stars lie on our line of sight towards it, and none of them are visible with the naked eye, but it is not quite as black as it appears – the dust does in fact glow weakly with reflected starlight.

Mimosa

Beta (β) Crucis
12h48m, -59°41′ | **Magnitude: 1.3**
Variable star | **Distance: 350 light yrs**

This brilliant blue-white giant is the hottest star of 'first magnitude' (i.e. one that has a magnitude of brighter than 1.5). Its surface temperature is more than 27,000°C, and this means that it radiates more than 90 percent of its energy as invisible ultraviolet light. To the naked eye, it is 3,000 times more luminous than the Sun, and matches the brightness of Deneb in Cygnus, but if all Mimosa's energy were poured out as visible light, it would rival Sirius as the brightest star in the sky.

Mimosa is also a variable similar to Beta Cephei, wracked with multiple oscillations that cause its brightness to fluctuate by about 0.1 magnitudes in a few hours.

Gacrux

Gamma (γ) Crucis
12h31m, -57°07′ | **Magnitude: 1.6**
Giant star | **Distance: 88 light yrs**

Contrasting obviously with the bright blue-white stars that form the rest of Crux, Gacrux is a brilliant red giant – a star with about three times the mass of the Sun, swollen to half the diameter of Earth's orbit, or about 150 million km. With comparatively little material spread over a huge volume of space, the star's outer layers are extremely tenuous and its surface gravity extremely weak, so the pressure of radiation emitted from the surface creates strong stellar winds that cause the star to vary slightly in brightness. The cool surface temperature means that the star emits much of its energy in the infrared, so its visible luminosity, 140 times that of the Sun, represents about 10 percent of its total energy output

Binoculars or a small telescope will reveal what appears to be a wide companion star of magnitude 6.5, but this is a mere chance alignment. However, there is some evidence that Gacrux may have a white dwarf companion – rather like Alphard in Hydra, its upper atmosphere seems to have been 'polluted' with barium and other elements from a companion star that long ago shed its outer layers in a planetary nebula.

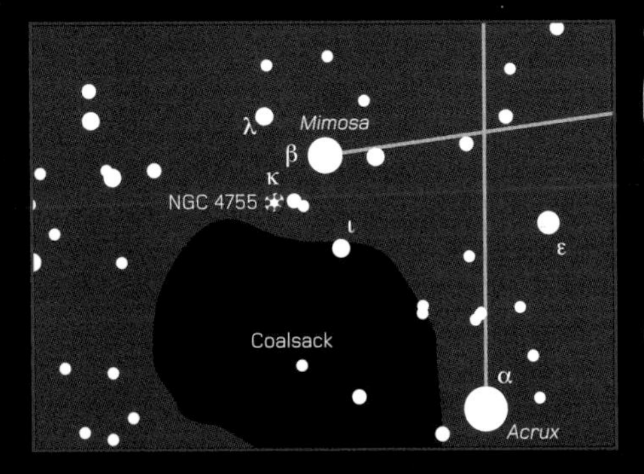

Jewel Box Cluster

NGC 4755

12h54m, -60°20′
Open cluster

Magnitude: 4.2
Distance: 7,600 light yrs

Studded with heavyweight blue stars, Crux's famous Jewel Box cluster is nevertheless dominated by the bright red Kappa Crucis. A ruby set among diamonds, this star is so massive that it has already exhausted its main fuel supplies and entered its death throes.

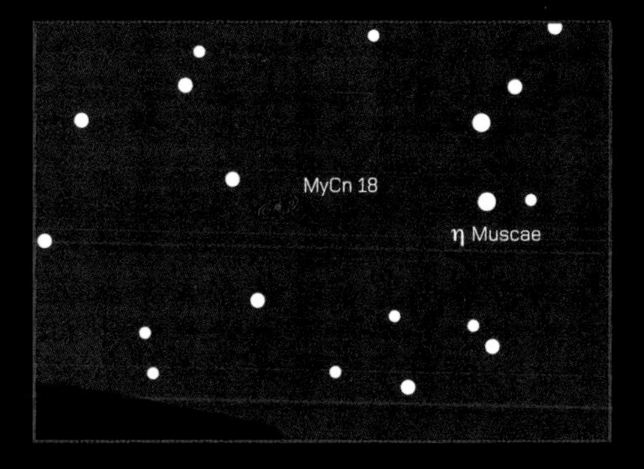

Hourglass Nebula

MyCn18

13h39m, -67°23′
Planetary nebula

Magnitude: 13.0
Distance: 8,000 light yrs

This stunning 'eye' in space is a planetary nebula caught in the act of formation – an exhausted red giant shrugging off its outer layers, shortly to become a dead white dwarf star. Colours trace the presence of different elements, including nitrogen (red), hydrogen (green), and oxygen (blue). The pinched waistline of this celestial hourglass is due to the interaction of a fast-growing shell of hot gas, blown out on strong stellar winds, with a cooler ring of slower-moving gas already surrounding the star's equator.

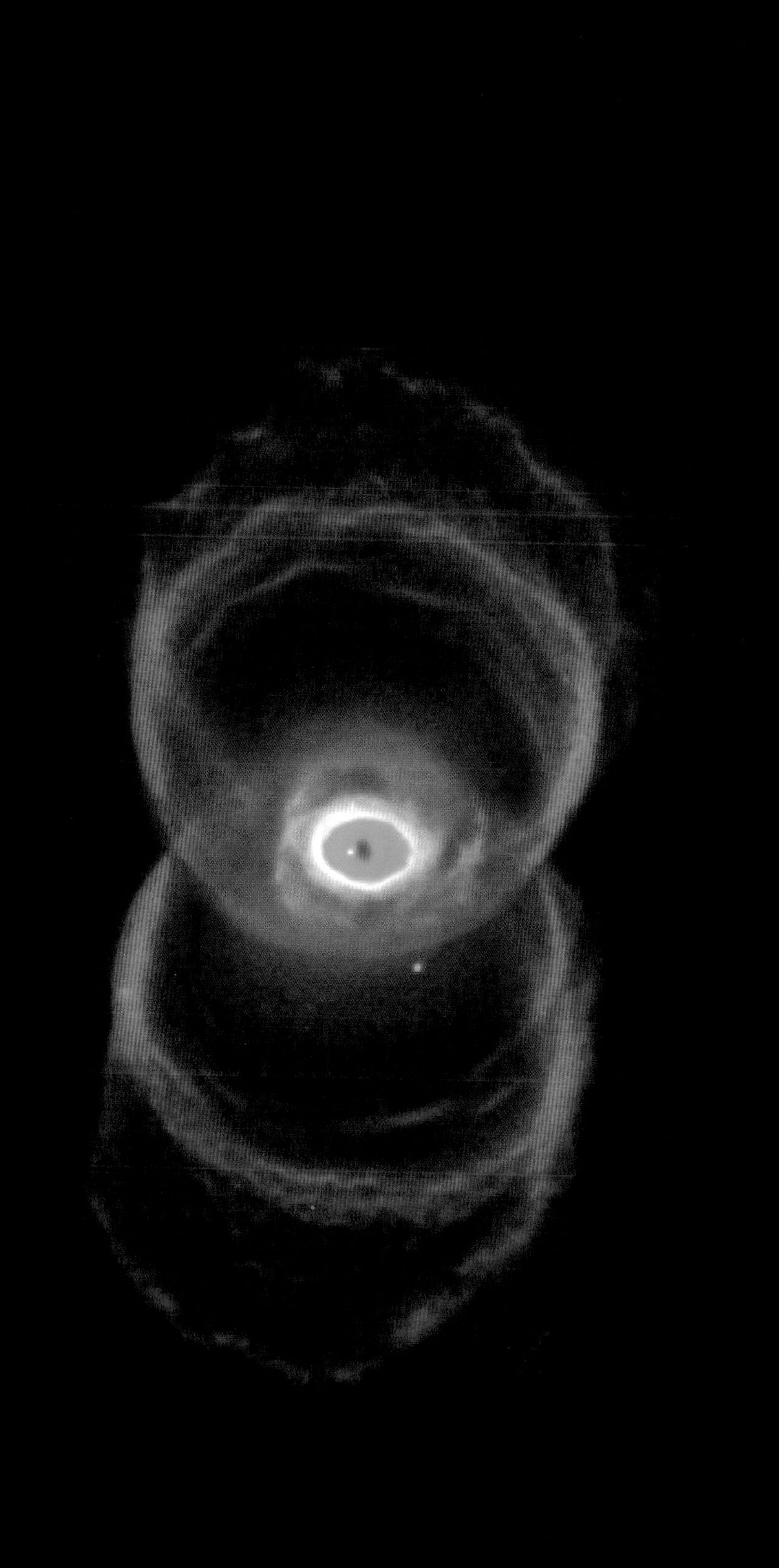

Circinus & Triangulum Australe

The Compasses and the Southern Triangle

Directly to the east of Centaurus's bright southern stars Hadar and Rigil Kentaurus lie two less well-known constellations. Circinus is a random scattering of unimpressive stars, though its three brightest members form a long, thin triangle. This was the shape that doubtless inspired Nicolas de Lacaille when he named the area after the compasses used by the surveyors of his day, and placed it near another surveying tool, Norma (the level), in the sky.

East of Circinus is another triangle, though this one is far broader – almost equilateral. The Southern Triangle's first appearance on a European star map dates to 1603, when German astronomer Johann Bayer included it in his *Uranometria*. However, it was apparently invented by Pieter Dirkszoon Keyser a few years before, and the pattern was probably already recognized as a constellation in the Middle East, where the Arabs called it by a similar name, and the three bright stars were known as the 'Three Patriarchs'.

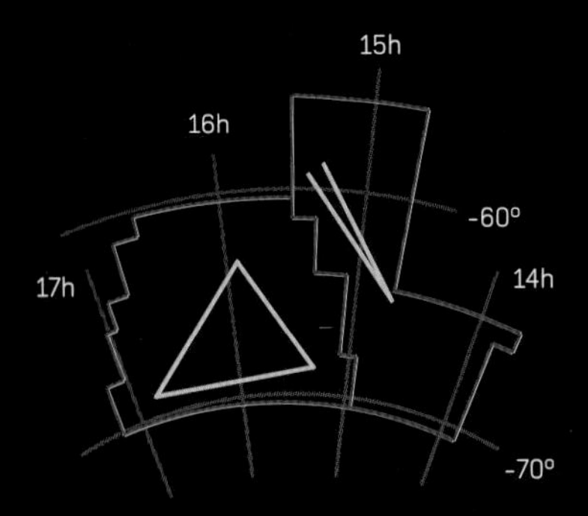

Theta (θ) Circini

14h57m, -62°47' | **Magnitude: 5.1 (var)**
Variable star | **Distance: 830 light yrs**

This hot and brilliant blue-white star varies unpredictably between magnitudes 5.0 and 5.5. It has a mass around ten times that of the Sun – large enough to ensure that when it eventually dies, it will end its life in a huge supernova explosion. Theta is thought to be a rapidly rotating, unstable variable of the same type as Gamma Cassiopeiae – a so-called 'Be' or emission-line star. It is pulsating unpredictably and occasionally throwing off rings of material around its equator.

Atria

Alpha (α) Trianguli Australis
16h49m,-69°02' | **Magnitude: 1.9**
Giant star | **Distance: 415 light yrs**

With a name that is simply an abbreviation of its formal designation, Atria is an orange giant that is currently shining by the nuclear fusion of helium in its core into heavier elements. It pumps out about 5,000 times the energy of the Sun, but since much of this is released as infrared radiation, it is roughly 2,500 times as luminous as the Sun in visible light.

Alpha (α) Circini

14h43m, -64°59' | **Magnitude: 3.2**
Binary star | **Distance: 53 light yrs**

A small telescope will reveal that this naked-eye white star is accompanied by an orange dwarf that orbits it in about 3,000 years. The unremarkable-looking primary is actually interesting in its own right – it combines two different peculiarities. On the one hand, it is a rapidly pulsating variable similar to Delta Scuti. On the other, it has a remarkably strong magnetic field, which creates patches of different chemical composition in the atmosphere, similar to sunspots. These in turn cause the chemical 'fingerprints' in the star's spectrum to shift as it rotates, revealing that the star rotates in roughly 4.5 days.

Lambda (λ) Trianguli Australis

NGC 6025
16h4m, -60°30' | **Magnitude: 5.1**
Open cluster | **Distance: 2,700 light yrs**

Like several other star clusters on the edge of naked-eye visibility, NGC 6025 was classified as a star before its true nature was established by French astronomer Nicolas de Lacaille in the 1750s. Binoculars reveal that this starlike blob is in fact composed of thirty or more stars of around 7th magnitude.

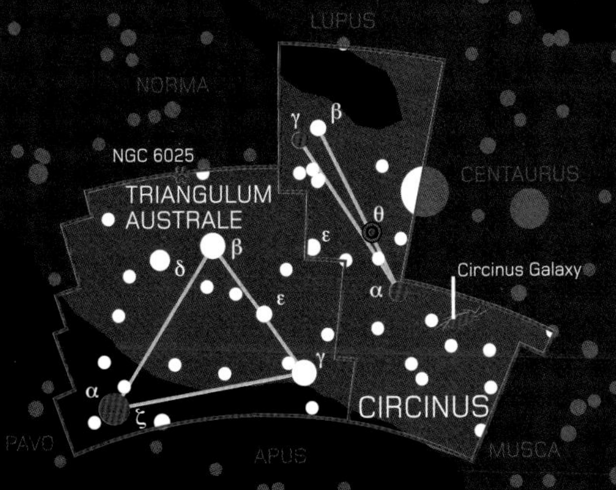

Gamma (γ) Circini

15h23m, -59°19' | **Magnitude: 5.1, 5.5**
Optical double? | **Distance: 510 light yrs**

This attractive contrasting double star is so tightly bound that it can be divided only through medium-sized telescopes – to the naked eye or binoculars it appears as a single star of magnitude 4.5. High magnifications will reveal that the stars are a magnitude 5.1 blue-white star and a magnitude 5.5 white star that sometimes appears yellow in contrast. There is some evidence that these stars are a mere chance alignment, and the white star is considerably closer than the blue one.

The Circinus Galaxy

ESO 97-G13
14h13m, -65°20' | **Magnitude: 12.1**
Active galaxy | **Distance: 13M light yrs**

As the star clouds of the southern Milky Way run through Circinus, they block our view of objects behind them, such as those outside our own galaxy. As a result, the so-called Circinus Galaxy was only discovered in the 1970s. It proved to be a fascinating object, however, since it is the nearest example of an active galaxy. Its unusually bright central region marks it out as a Seyfert Galaxy – the least violent type of active galaxy – but astronomers have also discovered lobes of hot gas above and below the disc probably ejected in a similar way to the radio-emitting lobes spat from the centres of much more violent radio galaxies. It seems the central 'engine' of all active galaxies is essentially the same – a supermassive black hole at their heart, consuming material from a surrounding region that has been disrupted, perhaps by a collision or close encounter with another galaxy. The major difference between the Circinus Galaxy and the most violent of distant quasars is essentially a matter of scale.

Corona Australis, Telescopium & Pavo

The Southern Crown, the Telescope and the Peacock

Running south of Sagittarius lie three constellations of varying prominence. Corona Australis is an easily identifiable arc of stars, while Telescopium may rank as the most obscure constellation of all, and Pavo lies somewhere between the two, with moderately bright stars but no really obvious pattern.

Corona Australis rises high enough in the sky to have been known to the ancient Greeks. It is an obvious counterpart to the northern Corona Borealis, although its shape is not quite so perfect. Mythologically, the constellation was sometimes seen as a crown, and sometimes as a wreath. It was associated with the Bacchus, the god of wine, but also with the centaur to its north. As a result it was sometimes known as Corona Sagittarii. Eastern astronomers, however, saw these same stars as a celestial tortoise.

Next in antiquity comes Pavo, one of the southern birds invented by Dutch navigators in the late 17th century. Telescopium, meanwhile, is another invention of Nicolas de Lacaille. It is a paltry tribute to the astronomer's most useful tool, largely because Lacaille 'stole' stars from other constellations to make his figure. When the stars were returned to their rightful owners, all that remained was this barren area of sky.

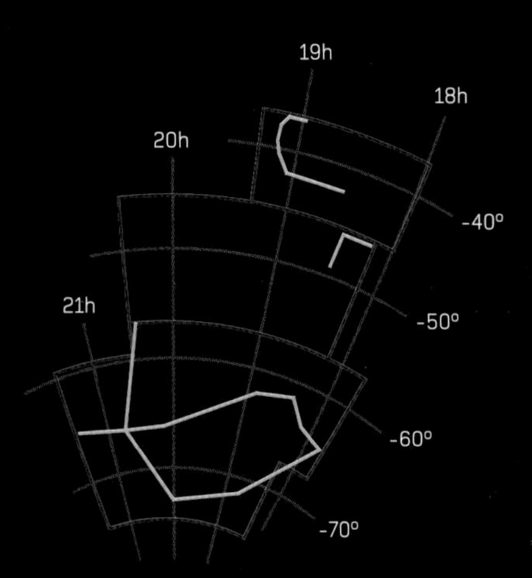

Delta (δ) Telescopium

18h32m, -45°55′	Magnitude: 4.9, 5.1
Optical double star	Distance: 800 light yrs 1,080 light yrs

The only notable star in Telescopium is a mere chance alignment between two blue-white stars at different distances. This 'optical double', easily separated with binoculars, consists of Delta 1 (southernmost of the pair), magnitude 4.9 and about 800 light years from Earth, and Delta 2, magnitude 5.1 and about 1080 light years away. In reality, the more distant Delta 2 is the more luminous of the two stars – while Delta 1 has just started to evolve off the 'main sequence' of stellar evolution, Delta 2 has already reached the 'giant' phase of its life.

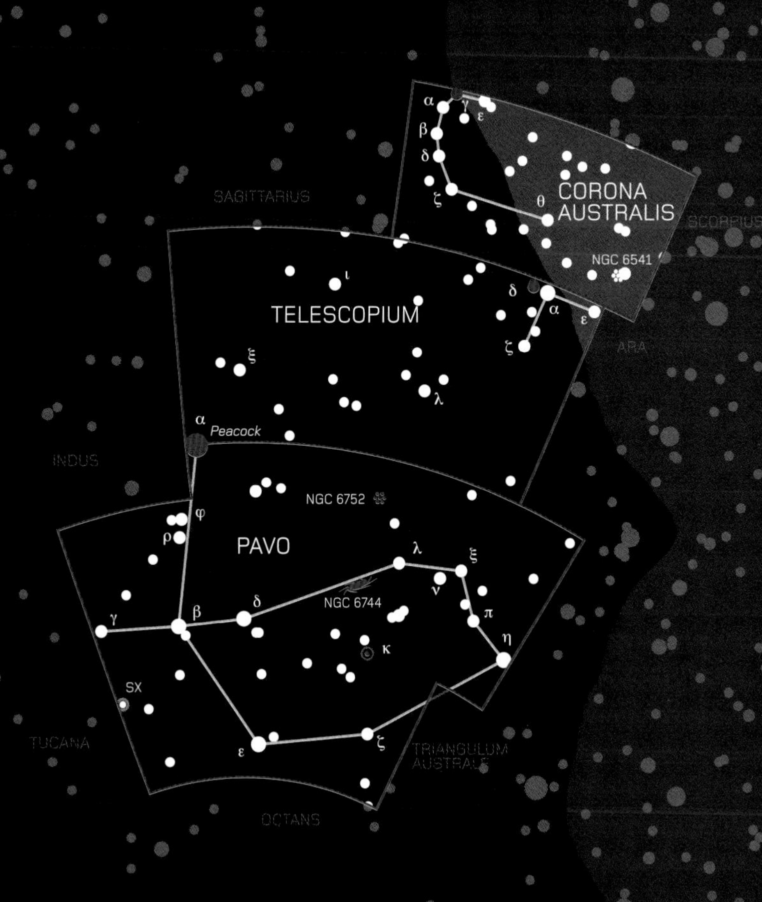

Peacock

Alpha (α) Pavonis

20h26m, -56°44′	Magnitude: 1.9
Multiple star	Distance: 183 light yrs

The name of Pavo's brightest star is one of the most recent to be widely adopted – it was invented when the star was selected by Britain's Royal Air Force for inclusion as a navigational aid in an almanac of the 1930s. The star itself is one of the brightest in far southern skies – a hot blue star that is probably still burning hydrogen in its core. Peacock has the mass of five or more Suns, and produces about 2,000 times the energy of our star, though about three quarters of this is released as ultraviolet radiation, and only a quarter as visible light. Although it appears as one star through even the most powerful telescopes, Peacock is a spectroscopic binary – two stars locked in orbit around each other with a period of less than 12 days.

NGC 6744

19h10m, -63°51′	Magnitude: 9.1
Spiral galaxy	Distance: 25M light yrs

The core of this beautiful barred spiral galaxy can be picked out with a small telescope, but larger instruments or long-exposure photographs are needed to capture its full structure, since it lies almost face on to us, and so its light is well spread out. Nevertheless, NGC 6744 is one of the finest galaxies in southern skies, and is also thought to bear a striking resemblance to our own Milky Way galaxy. Its arms are comparatively loose and fluffy, and it even has a companion galaxy similar to our own Magellanic Clouds.

NGC 6752

19h11m, -59°59′	Magnitude: 5.4
Globular cluster	Distance: 13,000 light yrs

This bright and nearby globular star cluster is visible to the naked eye as a fuzzy 'star', and a beautiful sight through binoculars or a small telescope. It appears half the size of the full Moon in Earth's skies, and is comparatively loose, making it easy to resolve the stars around its perimeter and a good way toward its centre. Studies of the motion of stars in NGC 6752 have provided some of the best evidence that clusters like this have 'intermediate mass' black holes, with masses of perhaps several thousand Suns at their centres.

Gamma (γ) Coronae Australis

19h6m, -37°04′	Magnitude: 4.7
Binary star	Distance: 58 light yrs

Small telescopes will show this star, at the northern end of the constellation's arc of bright stars, as a double – a pair of yellow stars of magnitudes 4.8 and 5.1. The stars are both in the 'main sequence' phase of stellar evolution, brighter and hotter than the Sun because they have a little more mass – perhaps 1.5 times the mass of the Sun. The two elements orbit each other in 120 years, and are currently drawing apart from a close alignment in the late 1990s.

Kappa (κ) Pavonis

18h57m, -67°14′	Magnitude: 4.4 (var)
Variable star	Distance: 540 light yrs

Kappa Pavonis is among the brightest Cepheid variables – pulsating stars similar to Delta Cephei. It varies in brightness between magnitudes 3.9 and 4.8 in a cycle that takes 9.1 days to repeat. Because the period of a Cepheid is related to its true luminosity, stars like Kappa Pavonis are valuable 'standard candles' for measuring the distance of galaxies across the nearby Universe.

Tucana

The Toucan

Best found by tracking from the brighter stars of Grus towards the South Celestial Pole, or looking to the southwest of brilliant Achernar in Eridanus, Tucana is a group of stars with moderate brightness. Fortunately, on dark nights it draws attention to itself through the hazy patch of the Small Magellanic Cloud on its southern edge.

The constellation was first recorded in Johann Bayer's *Uranometria* of 1603, and is another of the 'southern birds' invented by the Dutch navigators Keyser and de Houtman in the late 16th century.

The Magellanic Clouds today bear the name of Ferdinand Magellan, but they were known to other cultures long before the first European set eyes on them. Polynesian islanders called them the Upper and Lower Mists, while some Aboriginal tribes of northern Australia believed that they came down to Earth to choke people as they slept. Even Arab astronomers were aware of the larger cloud (see Dorado p208).

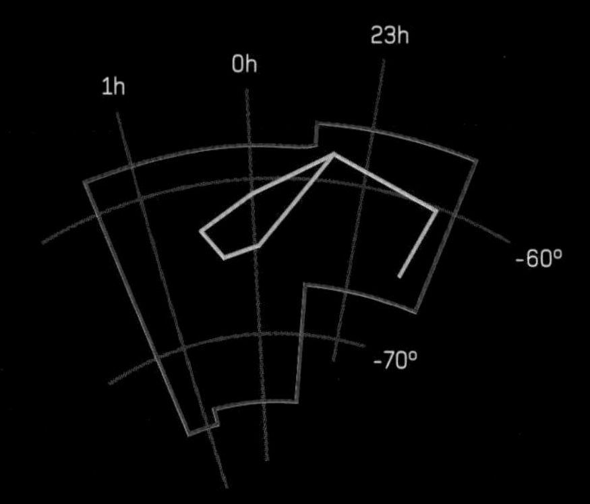

Small Magellanic Cloud

SMC

0h53m, -72°50′	**Magnitude: 2.3**
Irregular galaxy	**Distance: 210,000 light yrs**

The Milky Way's second major satellite galaxy, the Small Magellanic Cloud moves on almost the same orbit as the larger LMC in Dorado, but appears smaller in the sky partly because it is further away, but also because it is genuinely the lesser galaxy. As the clouds orbit the Milky Way, they are gradually stripped of stars, gas and dust with each close encounter, leaving a trail of debris, the 'Magellanic Stream' across the southern sky. In a few billion years time, they will be completely absorbed into the Milky Way.

For the moment, though, the SMC is still some 10,000 light years across, and contains some huge regions of starbirth, including the bright star cluster and nebula NGC 346, which can be picked up with a small telescope.

Although they were obvious features of the sky to the peoples of the southern hemisphere, the clouds bear the name of the first European to report them – Portuguese explorer Ferdinand Magellan, whose vessels were the first to circle the globe in 1519-21.

The SMC has also played an important role in 20th-century history – by assuming that all its stars were at the same distance, astronomer Henrietta Leavitt established in 1921 that the period and true luminosity of Cepheid variable stars (yellow supergiants similar to Delta Cephei) were closely linked. This link allowed other astronomers to map the Milky Way and even establish the distance to other galaxies.

Alpha (α) Tucanae

22h19m, -60°16′	**Magnitude: 2.9**
Binary star	**Distance: 199 light yrs**

Tucana's brightest star is an orange giant of a type very common among naked-eye stars. At a distance of just under 200 light years, it is more than 200 times as luminous as the Sun at visible wavelengths, but thanks to its cool surface, it produces just as much energy at infrared wavelengths. The visible star has a mass of a little under three Suns, but it is also a spectroscopic binary, with a low-mass companion that orbits it every 11.5 years.

Kappa (κ) Tucanae

1h15m, -68°53′	**Magnitude: 4.3**
Multiple star	**Distance: 67 light yrs**

Slightly simpler than Beta Tucanae, Kappa is nevertheless another fascinating multiple star – a 'double double' system. Binoculars will show the star as a well-separated pair with magnitudes 5.0 and 7.2. A small telescope will reveal that the brighter element is in fact a double of magnitudes 5.1 and 7.3, taking 1200 years to orbit each other, while a larger instrument will show that the faint component is also two stars – these with magnitudes 7.8 and 8.2, orbiting each other in 86 years.

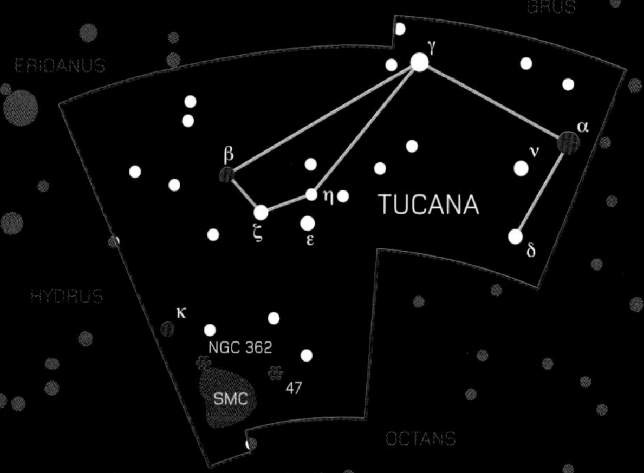

ERIDANUS
GRUS
HYDRUS
OCTANS
TUCANA
β
γ
α
ν
η
ζ
ε
δ
κ
NGC 362
SMC
47

47 Tucanae

NGC 104

0h24m, -72°05′	**Magnitude: 4.9**
Globular cluster	**Distance: 13,400 light yrs**

Second only to Omega Centauri among the sky's globular clusters, 47 Tucanae is a beautiful stellar glitterball easily visible to the naked eye as a slightly out-of-focus 'star'. It was first identified as a 'nebulous' object by the French astronomer Nicolas de Lacaille in 1751. Binoculars will show the true extent of the cluster – a ball of light the size of the full Moon – but its stars are so densely packed that a moderate-sized telescope is needed to resolve even its outer edges. In total, it is thought to have about 100,000 stars, packed into a volume of space 120 light years across.

Beta (β) Tucanae

0h32m, -62°57′	**Magnitude: 4.3, 5.1**
Optical double star?	**Distance: 144 light yrs**
	151 light yrs

This complex multiple star consists of stars so far apart that they may not actually be a single group. Naked eye observers can usually identify two components of magnitudes 4.3 and 5.1, and binoculars or a small telescope will split the brighter star into two elements of near-identical brightness – a blue star of magnitude 4.4 and a white one of 4.5, known as Beta 1 and Beta 2 respectively. The fainter and more distant white star is known as Beta 3 – it lies seven light years beyond the close pair – too far to be gravitationally bound if current measurements are correct. However, the stars do share a common motion through space, and certainly originated very recently as part of the same open cluster, so if they are not siblings they are certainly cousins.

All this is just one level of the system's complexity – more powerful telescopes will show that Beta 2 is itself a double of white stars with magnitudes 4.8 and 6.0, in an eccentric 43-year orbit around each other. And Beta 1 also has a binary companion of its own – a faint red dwarf star of magnitude 13.5 that orbits in about 700 years. The two pairs in this quadruple system take at least 150,000 years to orbit one another.

NGC 362

1h3m, -70°51′	**Magnitude: 6.4**
Globular cluster	**Distance: 30,000 light yrs**

Far more distant than 47 Tucanae, this globular cluster happens to lie on the edge of the Small Magellanic Cloud as seen from the Earth. Easily spotted in binoculars, the cluster's location has allowed some interesting comparisons between its stars and the more remote stars of the SMC beyond it. The most obvious difference is that, typically for a globular cluster, NGC 362 is almost entirely composed of old, slow-burning yellow and red stars. This fits with the idea that the stars in the Milky Way's globular clusters all formed at roughly the same time, long ago, so that any hotter and more massive stars have long since died.

Horologium & Reticulum

The Clock and the Reticle

Nestling beneath the curve of southern Eridanus, Horologium and Reticulum are typical of the faint and obscure constellations that fill much of the southern sky, but at least they stand as testimony to the imagination of their inventor, French astronomer Nicolas de Lacaille.

Horologium, the clock, is particularly imaginative – a scattering of faint stars in which Lacaille somehow perceived the form of a pendulum clock. Most interpretations of the constellation pattern strip it to its bare minimum – a swinging pendulum suspended at Alpha Horologii.

Reticulum is a little easier to spot – a compact diamond or kite-shaped group of five stars tucked beneath the jagged diagonal of Horologium. Although its name is the Latin for 'net', Lacaille intended it to represent a reticle – the crosshair sometimes used in telescope eyepieces to aid precise measurements.

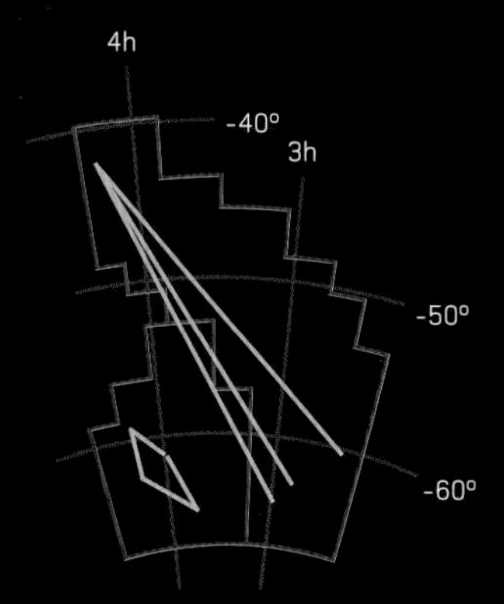

Zeta (ζ) Reticuli

3h18m, -62°31′	Magnitude: 5.0
Binary star	Distance: 39 light yrs

The twin stars of this binary system are well separated, and close enough to Earth that they can usually be distinguished with the naked eye (and easily split with binoculars). The stars are Sunlike yellow dwarfs of individual magnitudes 5.5 and 5.2, with masses and other properties within a few percent of our own star's. They are separated by more than a hundred times the diameter of Pluto's orbit, circling each other in a million or more years. This should allow ample room for each to support an independent solar system of its own, and makes the Zeta Reticuli system an obvious target in the search for extrasolar planets. Since the stars are somewhat older than our Sun at around 8 billion years old, there might even have been time for life and intelligence to evolve on a suitable planet. But as yet, no evidence for a planetary system has been found around either star.

Intriguingly for such an apparently hospitable system, Zeta Reticuli has another unusual significance – since the 1960s, UFO enthusiasts have claimed that this star system is the home of the big-eyed 'grey' aliens beloved of abduction stories and popular culture.

R Horologii

2h54m, -49°53′	Magnitude: 9.5 (var)
Variable star	Distance: 1,000 light yrs

Located in the same binocular field as Iota Horologii, this star is an extreme example of the Mira-type long period variables. It is a pulsating red giant with a 407-day period, and usually lurks beyond the range of even small telescopes at around magnitude 14. However, around maximum brightness it can be nine magnitudes (about 4,000 times) brighter, easily outshining Iota and reaching magnitude 4.0.

NGC 1261

3h12m, -55°13′	Magnitude: 8.4
Globular cluster	Distance: 54,000 light yrs

This globular cluster, first spotted by James Dunlop in 1826, is a particularly rewarding sight for medium-sized telescopes, since its stars are loosely packed and can be resolved almost to its centre. Through a smaller telescope or binoculars, it appears as a diffuse starlike blob. Infrared studies of this cluster reveal that it has unusually large amounts of gas in between its stars – astronomers think that, in larger globulars, such gas is literally blown away by the force of countless supernova explosions.

NGC 1512

4h4m, -43°21′	Magnitude: 11.1
Spiral galaxy	Distance: 30M light yrs

This 11th-magnitude barred spiral is among the finest galaxies of its type in the sky, although it needs a moderate-sized telescope to be properly appreciated. This will reveal the elongated bar, 2,400 light years wide, across its centre, and perhaps the traces of a surrounding ring of new starbirth – where brilliant, short-lived stars are making the galaxy's structure far more visible.

Ultraviolet images have shown that the galaxy's spiral arms extend much further into space – almost to the nearby small elliptical galaxy NGC 1510. Overall, the spiral has a diameter of 70,000 light years, but the spiral arms are far harder to see in visible light because they are lit mainly by the fainter light of more sedate, longer-lived stars. It is possible that the gravitational tides raised by a close encounter with NGC 1510 have triggered the wave of star formation in its larger neighbour.

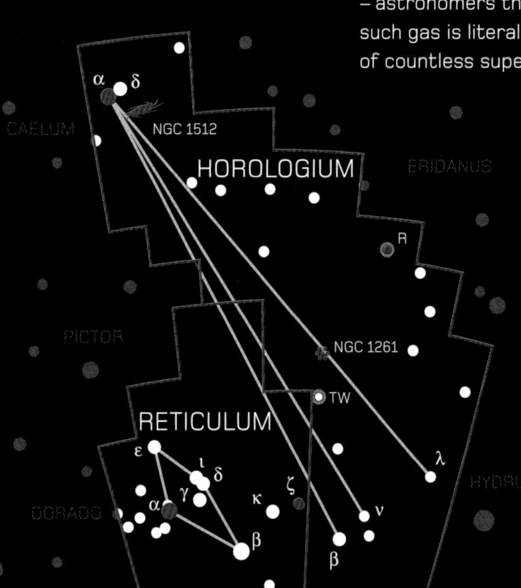

Alpha (α) Horologii

4h14m, -42°17′	Magnitude: 3.9
Giant star	Distance: 117 light yrs

The brightest star in Horologium is hard to spot at magnitude 3.9. It marks the pivot of the pendulum, well to the northeast of the constellation's other naked-eye stars. It is a fairly typical orange giant, twice the mass of the Sun and 30 times more luminous in visible light (though it produces about 50 times the Sun's energy overall, since much of it is released as low-energy invisible infrared). Alpha is thought to be passing through the second fairly settled period of its life, shining largely through fusion of the helium produced in its core during its main-sequence youth.

Alpha (α) Reticuli

4h14m, -62°28′	Magnitude: 3.4
Binary star	Distance: 165 light yrs

The brightest star in Reticulum, like so many other stars in the sky, is a binary system – though in this case only one star of the pair contributes much light to the overall appearance. This primary is a yellow giant with more than three times the mass of the Sun. It is thought to be passing through the same helium-burning phase as Alpha Horologii, though because it is more massive, this star produces about five times the energy (its overall luminosity, including infrared radiation, is around 240 times that of the Sun).
The second star in the system is a dim, slow-burning red dwarf that takes tens of thousands of years to orbit the primary.
At magnitude 12.0, it can be spotted through medium-sized telescopes.

Dorado, Pictor & Mensa

The Dorado, the Painter's Easel and Table Mountain

Southwest of brilliant Canopus lie three fairly faint and amorphous constellations with a bright, starry mist at their heart – the Large Magellanic Cloud. Long before Magellan journeyed to the southern hemisphere, Arab astronomers knew this distant satellite of our own galaxy, just visible from their latitudes, as the White Ox.

The cloud's host constellation, Dorado, was invented by the Dutch sailors Keyser and de Houtman in the 1590s, and has something of an identity crisis. The literal translation of its name is 'goldfish', but it was never intended to show the ornamental fish of garden ponds. Often it is seen as a swordfish, but it seems the original intention was to show a mahi-mahi – a type of fish native to waters around Hawaii.

Mensa and Pictor are late additions to the sky, added by Lacaille during his studies from the Cape of Good Hope in the 1750s. Stationed as he was at the foot of Table Mountain, it seems likely that he was inspired by the sight of its flat top, shrouded in cloud, to invent something similar in the sky.

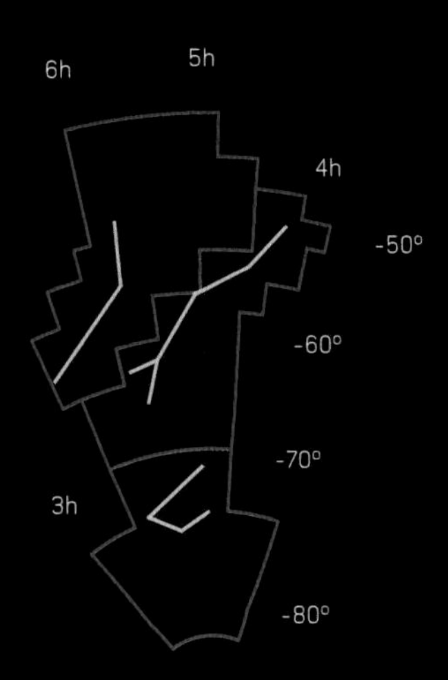

Beta (β) Pictoris

5h47m, -51°04'	Magnitude: 3.9
Extrasolar planetary system	Distance: 63 light yrs

Although this average white star has little to catch the eye of an observer, it was an early success in the quest for planets beyond our solar system. The launch of the first infrared space telescope, IRAS, in the 1980s revealed that, for a fairly hot white star, it is an unusual emitter of infrared radiation. The source of these emissions was eventually traced to a broad disc of gas and dust surrounding the star and extending up to 40 times Pluto's distance from the Sun. More recent and detailed images have revealed the presence of gaps and distortions in the disc, perhaps caused by larger planets moving through it.

Unfortunately, Beta Pictoris is not a strong candidate for life beyond the solar system, since it is a young and relatively short-lived hot star – a close relative of the bright stars Vega (in Lyra) and Fomalhaut (in Piscis Austrinus), both of which have similar discs. It has a mass roughly 1.7 times that of the Sun, and shines about 10 times as brightly.

The Tarantula Nebula

NGC 2070

5h39m, -69°06'	Magnitude: 8.0
Emission nebula	Distance: 180,000 light yrs

Appearing through binoculars as a bright knot in the Magellanic Cloud (with its own star designation, 30 Doradus), binoculars or a small telescope will reveal the long tendrils of gas that give NGC 2070 its popular name. The Tarantula Nebula is a huge region of star formation, about 1,000 light years across and larger than any such area in the Milky Way. At the cluster's heart, and visible through a small telescope, is a star cluster called NGC 2070, which contains some of the brightest and most massive stars known. Radiation from these stars blasts out through the nebula, exciting gas atoms and causing them to glow. According to some estimates, if the Tarantula were at the same distance as our own galaxy's Orion Nebula, it would cover some 30 degrees of the sky, and be bright enough to cast shadows at night.

Alpha (α) Mensae

6h10m, -74°45'	Magnitude: 5.1
Main sequence star	Distance: 33 light yrs

Of all the naked-eye stars, Alpha Mensae is one of the most similar to our Sun. It has a mass around 90 percent of the Sun's and shines with 80 percent of its luminosity (the brilliance of stars is highly dependent on their mass, so drops away sharply as stars get lighter). It also has a high metal content, which (based on examples elsewhere in the sky) should predispose it to forming a planetary system – but no such worlds have so far been detected.

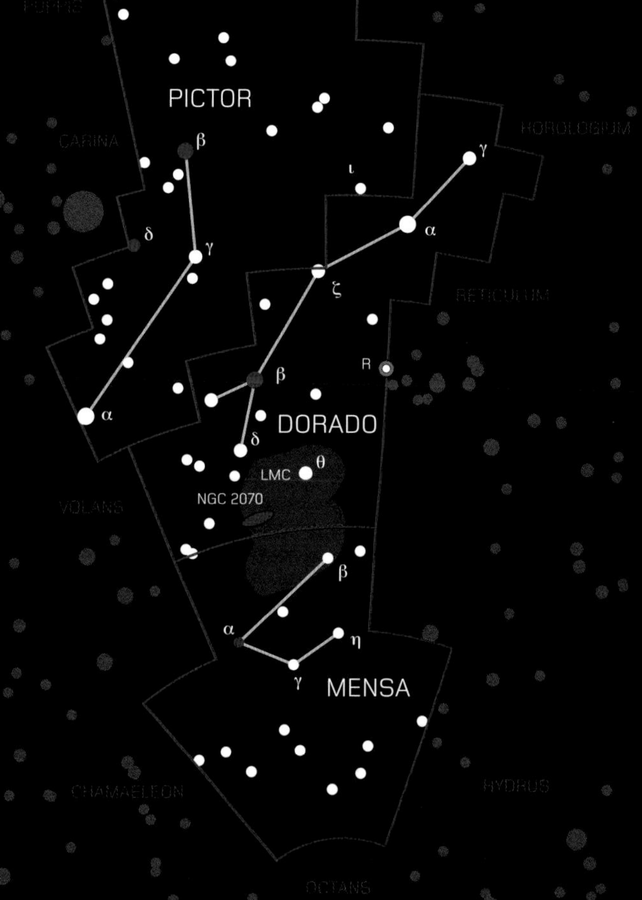

Delta (δ) Pictoris

6h10m, -54°58'	Magnitude: 4.8 (var)
Binary star	Distance: 1,650 light yrs

What seems with even the largest telescope to be a single blue-white star is in fact an eclipsing binary that normally shines at magnitude 4.7, but dips to magnitude 4.9 (a difference just detectable to the human eye) every 40 hours. The small size of the dip indicates that the system's eclipses are 'grazing' ones in which one star just clips the edge of the other one as seen from Earth, blocking out a little of its light. Studies of the star's light suggest the components are both hot and blue-white, and that one star is probably twice as massive as the other, though each must have several times the Sun's mass.

Large Magellanic Cloud

LMC

5h24m, -69°45'	Magnitude: 0.1
Irregular galaxy	Distance: 179,000 light yrs

Dominating the far south of Dorado and extending across the border into Mensa is a cloud of stars that resembles a detached part of the Milky Way. This is the Large Magellanic Cloud, a satellite galaxy in a 1.5-billion-year orbit around our own. Despite its relatively small size (roughly 20,000 light years in diameter), it contains roughly 10 billion stars, and is rich in other features such as clusters and nebulosity. Its shape is usually considered irregular, but there are intriguing hints of spiral-like features at its heart, and traces of a single 'arm' of stars.

Scanning the LMC with binoculars reveals countless individual stars, but these are only the galaxy's brightest giants. To give some idea of the distance involved, a star cluster just to the north of the galaxy's central 'bar' is home to S Doradus, one of the most luminous stars known. This star is a blue supergiant similar to P Cygni, over a million times more luminous than the Sun and erratically variable, yet across the gulf of 170,000 light years, it only reaches magnitude 8.6 at best – barely bringing it within the reach of binoculars.

Other features of the LMC include the wreckage of the most recent supernova visible to the naked eye. Supernova 1987A was formed by a blue supergiant that blew itself to pieces in February 1987, briefly shining almost as brightly as its entire galaxy.

Beta (β) Doradus

5h34m, -62°29'	Magnitude: 3.8 (var)
Variable star	Distance: 1,040 light yrs

One of the sky's brightest variable stars, Beta Doradus is a pulsating yellow supergiant – a southern equivalent of the famous Delta Cephei. It varies in brightness between magnitude 3.5 and 4.1 in a 9.9 day period, and its state can easily be tracked by comparison with the other stars in the constellation – notably Alpha (magnitude 3.3) and Delta (magnitude 4.3). Beta has an estimated mass of more than six Suns, and pumps out almost 3,000 times our Sun's energy. Its similar colour means that, like our star, it emits most of its energy as visible light.

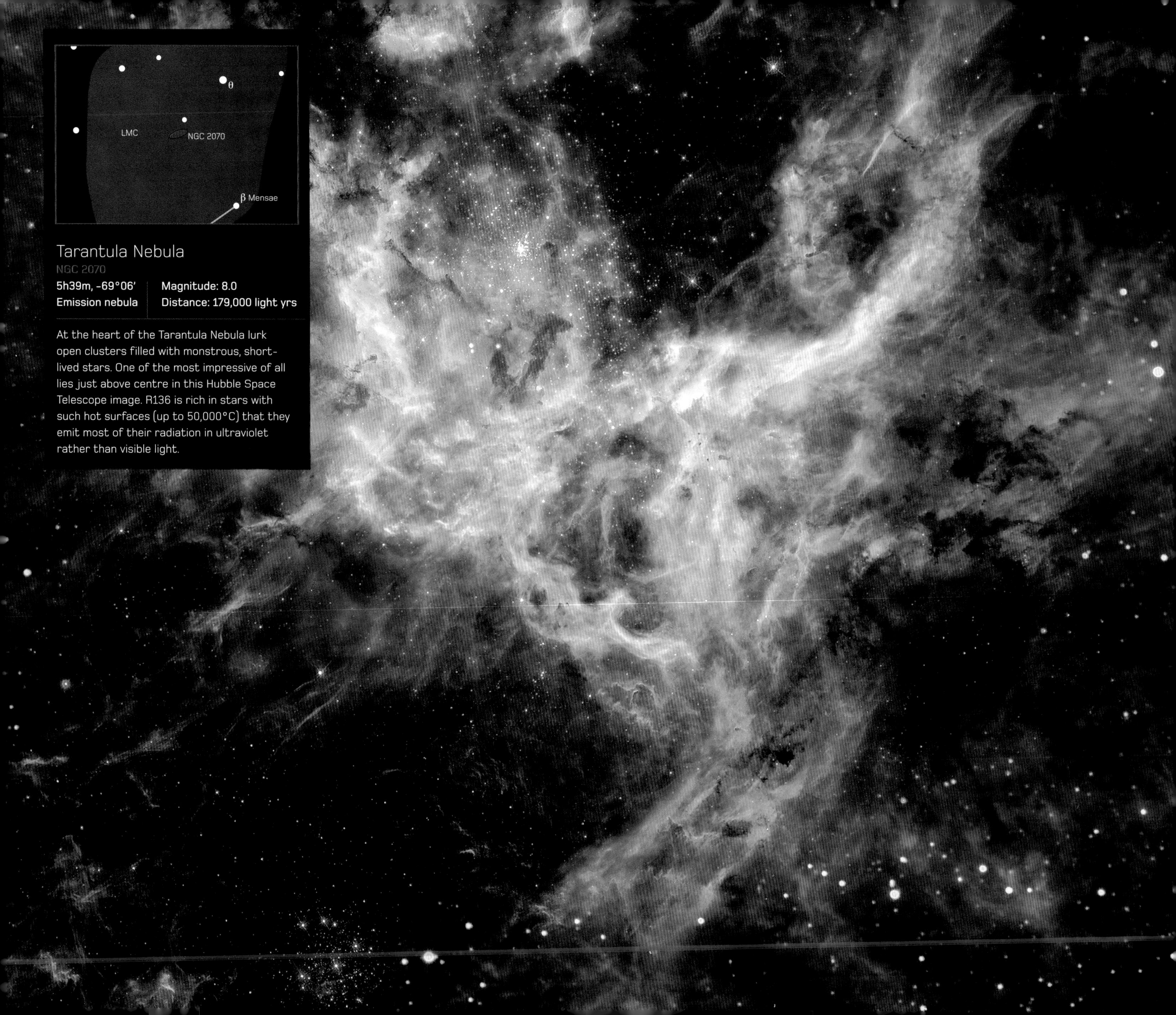

Tarantula Nebula

NGC 2070

5h39m, -69°06′	Magnitude: 8.0
Emission nebula	Distance: 179,000 light yrs

At the heart of the Tarantula Nebula lurk open clusters filled with monstrous, short-lived stars. One of the most impressive of all lies just above centre in this Hubble Space Telescope image. R136 is rich in stars with such hot surfaces (up to 50,000°C) that they emit most of their radiation in ultraviolet rather than visible light.

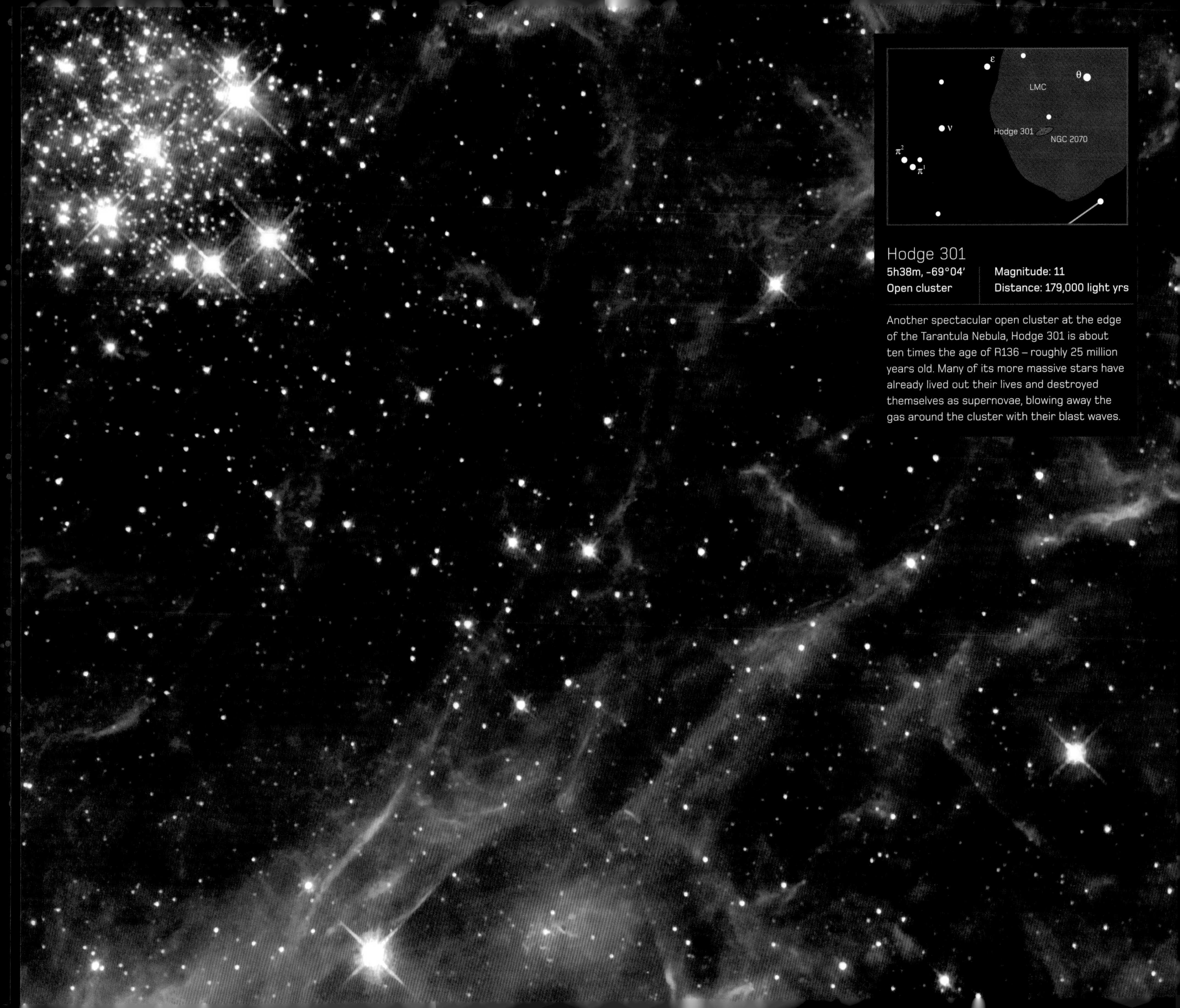

Hodge 301

5h38m, -69°04′
Open cluster

Magnitude: 11
Distance: 179,000 light yrs

Another spectacular open cluster at the edge of the Tarantula Nebula, Hodge 301 is about ten times the age of R136 – roughly 25 million years old. Many of its more massive stars have already lived out their lives and destroyed themselves as supernovae, blowing away the gas around the cluster with their blast waves.

Wega
Drachen
Leier
Cepheus
Deneb
Schwan
Eidechse
Stier des Poniatowsky
Adler
Antinous
Altair
Fuchs
Delphin
Friedrichsehre
Sirrah
Krebses
Pegasus
Marcab
Kleines Pferd
AEQUATOR
Declination
285
XX
300
XX
315
XXI

4 Monthly sky guide

As our planet travels around the Sun in its ceaseless orbit, the night sky is constantly shifting in position and orientation. The following charts offer a guide to the key constellations and stars visible on any night of the year, from most inhabited latitudes.

20°

WE

February

Northern hemisphere

20° 40° 60° Looking North

The charts on these pages show the northern hemisphere skies during February. They are correct for 11pm local time (midnight Daylight Saving Time) at the start of the month, 10pm (11pm DST) in the middle of the month, and 9pm (10pm DST) at the end of the month.

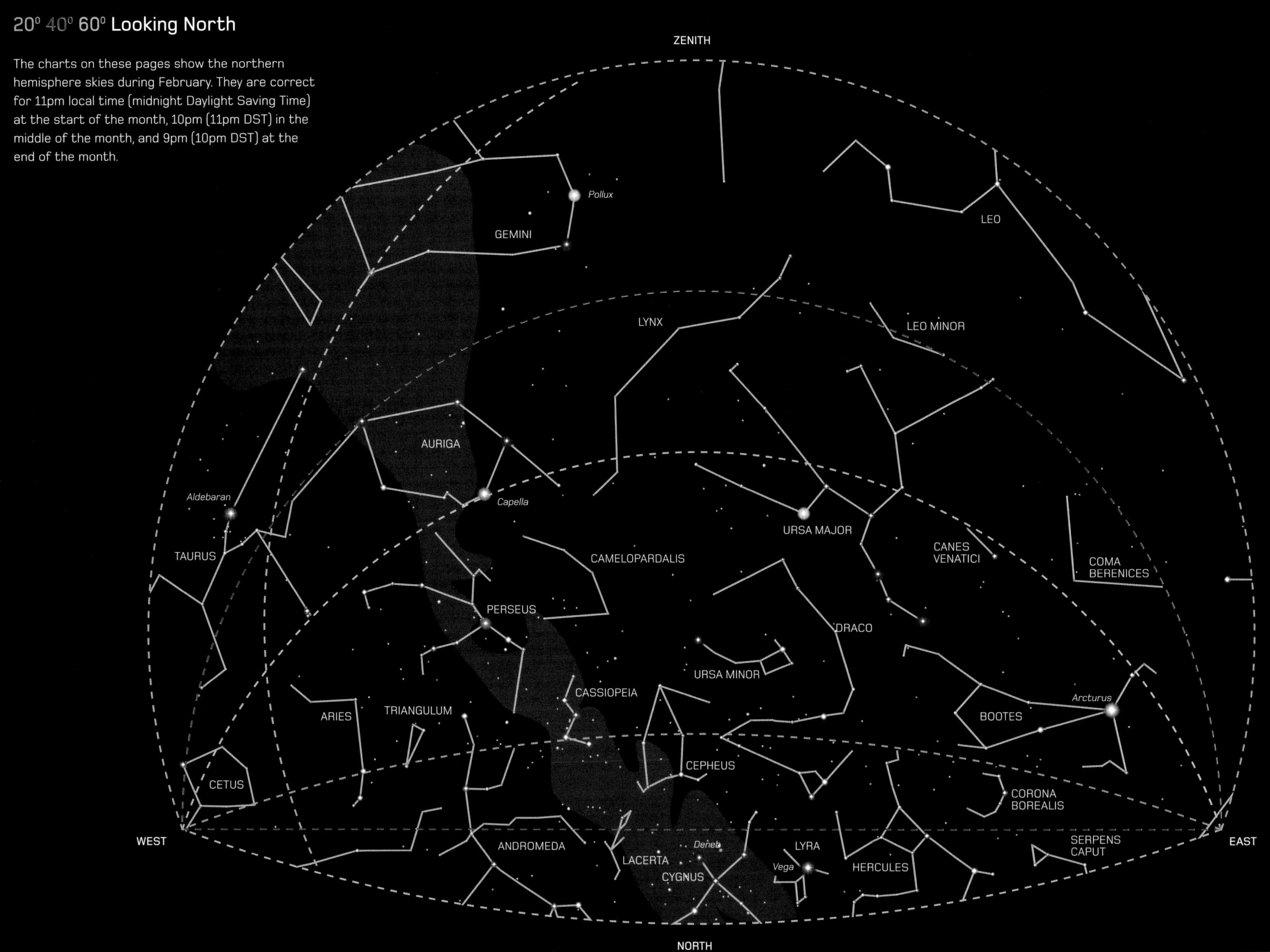

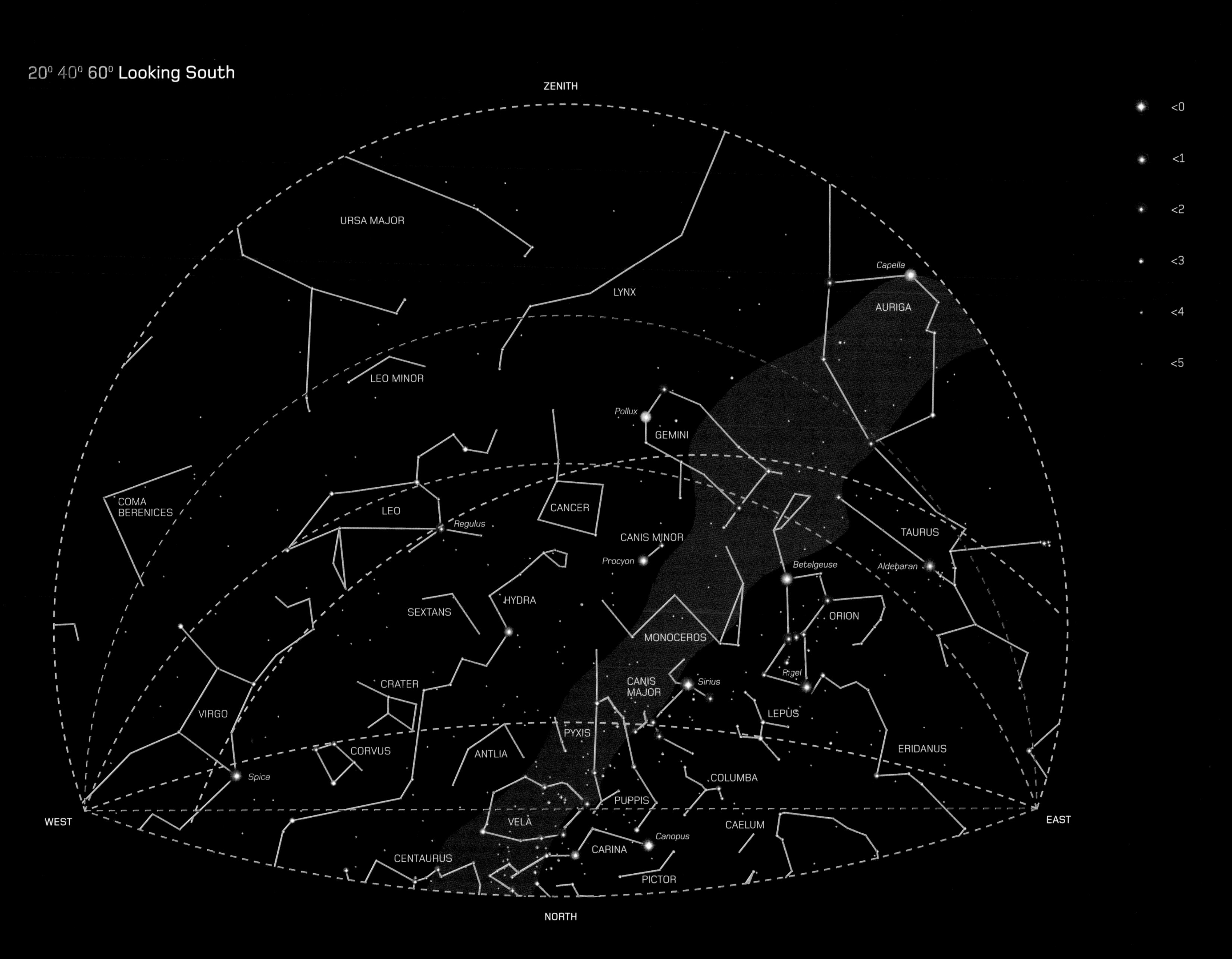

20° 40° 60° Looking South
ZENITH
<0
<1
<2
<3
<4
<5
URSA MAJOR
LYNX
Capella
AURIGA
LEO MINOR
Pollux
GEMINI
COMA
BERENICES
LEO
Regulus
CANCER
CANIS MINOR
TAURUS
Procyon
Betelgeuse
Aldebaran
SEXTANS
HYDRA
ORION
MONOCEROS
CRATER
CANIS
MAJOR
Sirius
Rigel
VIRGO
LEPUS
PYXIS
CORVUS
ANTLIA
ERIDANUS
Spica
COLUMBA
PUPPIS
WEST
VELA
EAST
CAELUM
Canopus
CARINA
CENTAURUS
PICTOR
NORTH

February

Southern hemisphere

-20° -40° -60° Looking North

The charts on these pages show the southern hemisphere skies during February. They are correct for 11pm local time (midnight Daylight Saving Time) at the start of the month, 10pm (11pm DST) in the middle of the month, and 9pm (10pm DST) at the end of the month.

20° 40° 60° Looking South

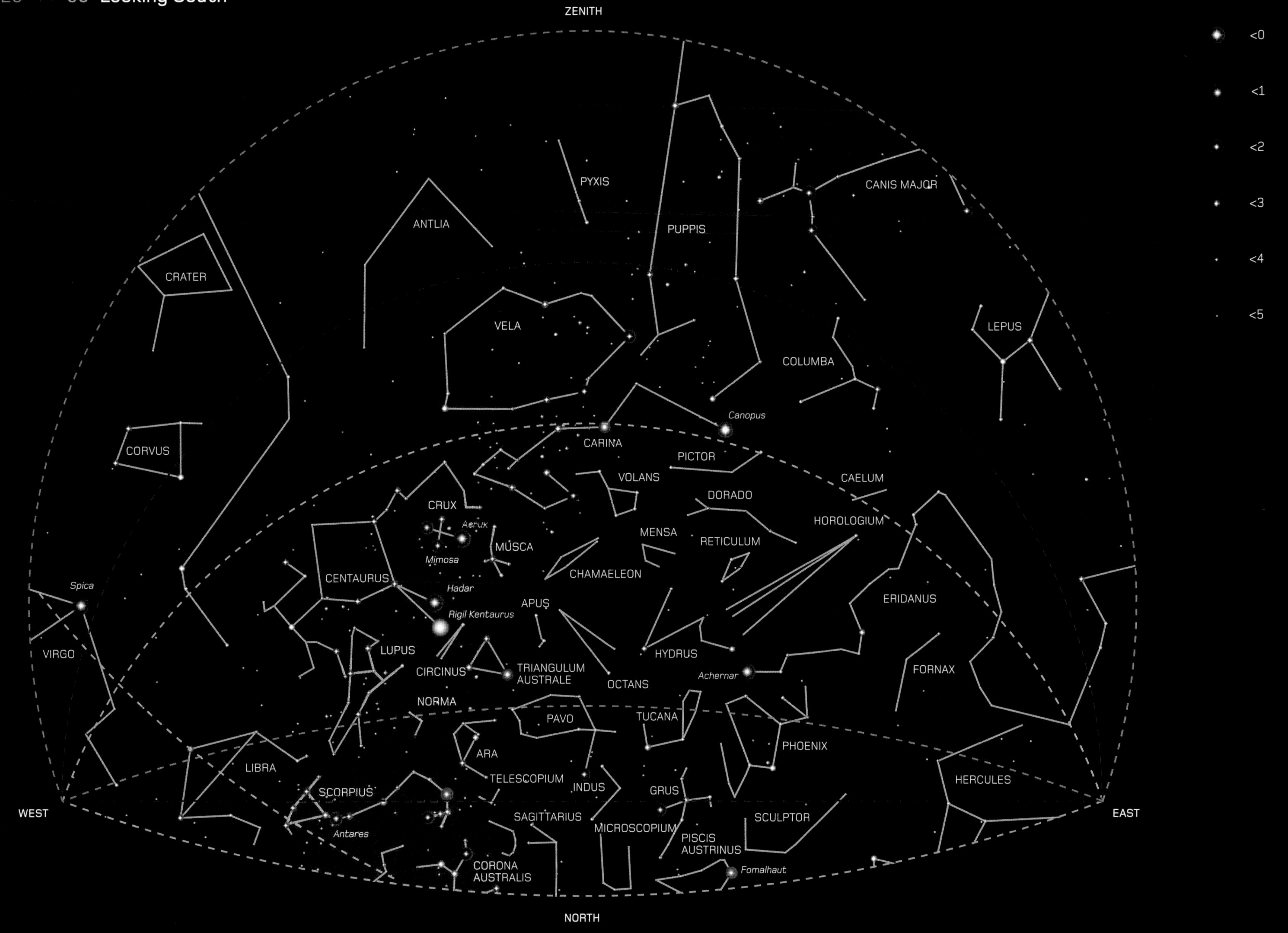

ZENITH

<0
<1
<2
<3
<4
<5

PYXIS
ANTLIA
PUPPIS
CANIS MAJOR
CRATER
VELA
LEPUS
COLUMBA
Canopus
CORVUS
CARINA
PICTOR
CAELUM
VOLANS
DORADO
CRUX
Acrux
HOROLOGIUM
MUSCA
MENSA
RETICULUM
Mimosa
CENTAURUS
CHAMAELEON
Spica
Hadar
ERIDANUS
APUS
Rigil Kentaurus
VIRGO
LUPUS
HYDRUS
CIRCINUS
TRIANGULUM AUSTRALE
FORNAX
Achernar
OCTANS
NORMA
PAVO
TUCANA
PHOENIX
ARA
LIBRA
TELESCOPIUM
HERCULES
INDUS
SCORPIUS
GRUS
SAGITTARIUS
SCULPTOR
WEST
EAST
MICROSCOPIUM
Antares
PISCIS AUSTRINUS
Fomalhaut
CORONA AUSTRALIS
NORTH

March

Northern hemisphere

20° 40° 60° Looking North

The charts on these pages show the northern hemisphere skies during March. They are correct for 11pm local time (midnight Daylight Saving Time) at the start of the month, 10pm (11pm DST) in the middle of the month, and 9pm (10pm DST) at the end of the month.

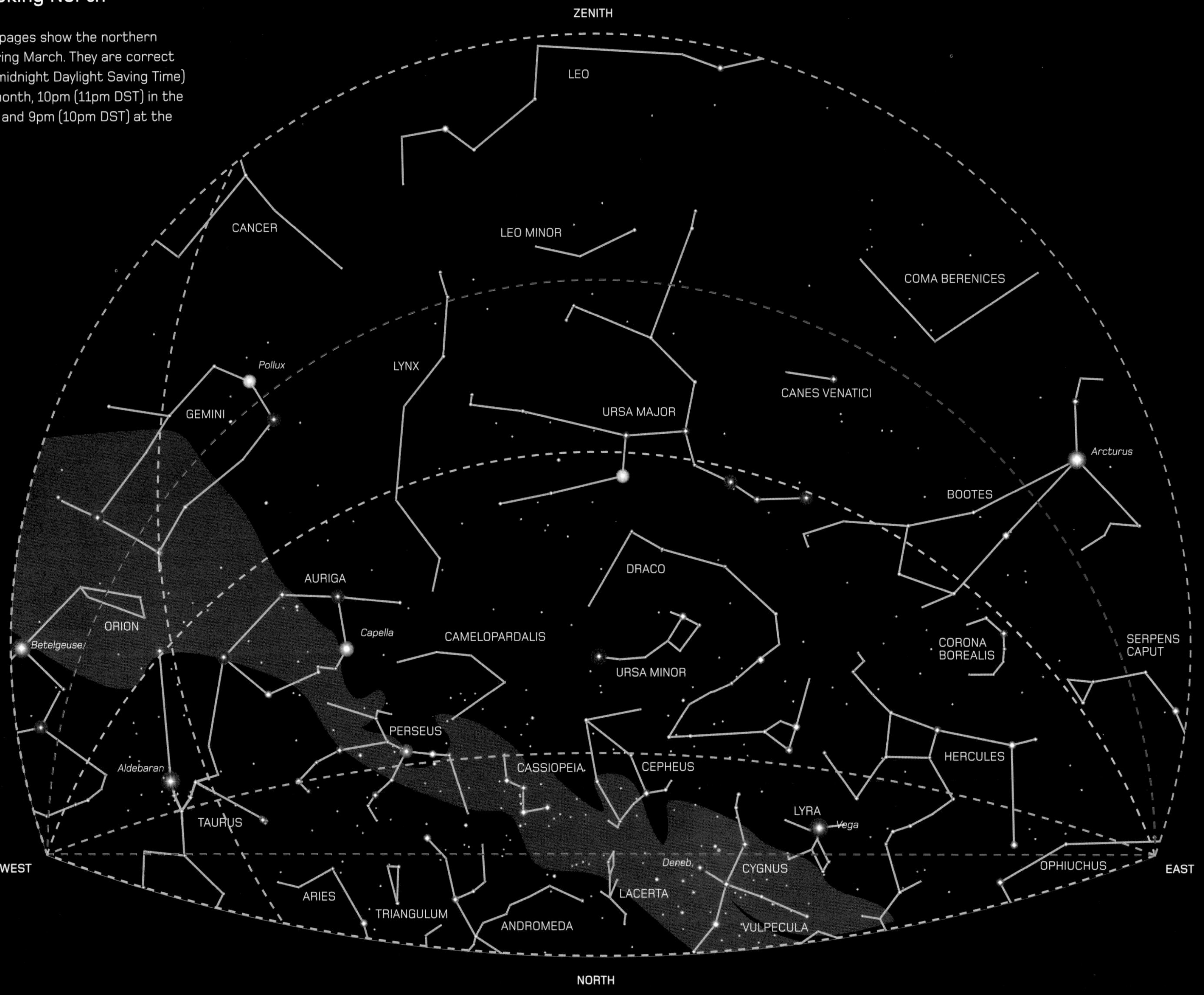

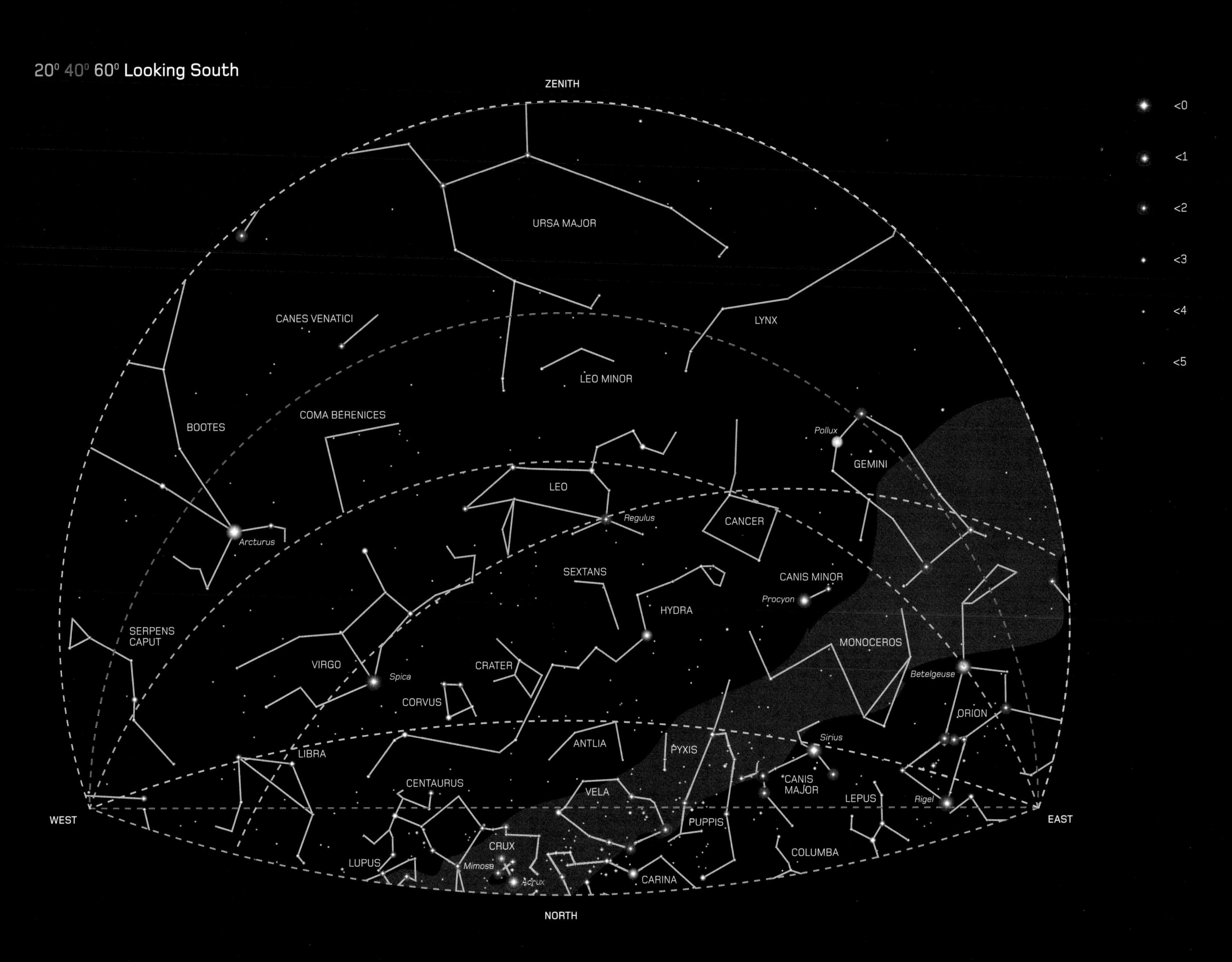

20° 40° 60° Looking South
ZENITH
<0
<1
<2
<3
<4
<5
URSA MAJOR
CANES VENATICI
LYNX
LEO MINOR
BOOTES
COMA BERENICES
Pollux
GEMINI
LEO
Regulus
CANCER
Arcturus
SEXTANS
CANIS MINOR
Procyon
HYDRA
SERPENS CAPUT
MONOCEROS
VIRGO
Spica
CRATER
Betelgeuse
CORVUS
ORION
ANTLIA
Sirius
LIBRA
PYXIS
CANIS MAJOR
CENTAURUS
VELA
LEPUS
Rigel
WEST
PUPPIS
EAST
CRUX
COLUMBA
LUPUS
Mimosa
Acrux
CARINA
NORTH

March

Southern hemisphere

-20° -40° -60° **Looking North**

The charts on these pages show the southern hemisphere skies during March. They are correct for 11pm local time (midnight Daylight Savings Time) at the start of the month, 10pm (11pm DST) in the middle of the month, and 9pm (10pm DST) at the end of the month.

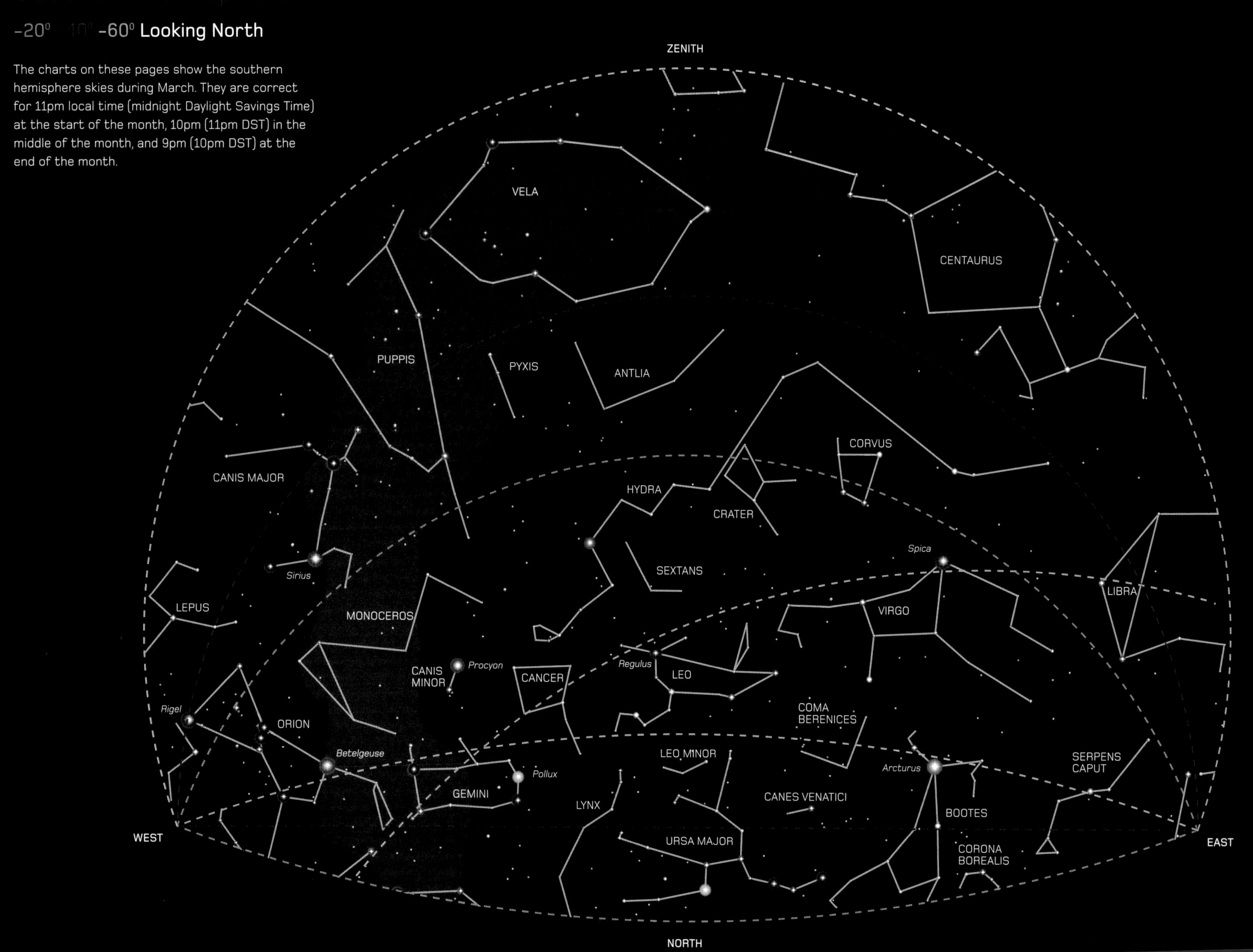

20° 40° 60° Looking South

ZENITH

<0
<1
<2
<3
<4
<5

CORVUS
ANTLIA
HYDRA
PYXIS
VELA
PUPPIS
CENTAURUS
CANIS MAJOR
CRUX
Acrux
Mimosa
Sirius
Hadar
MUSCA
VOLANS
Rigil Kentaurus
CARINA
Canopus
LUPUS
CIRCINUS
CHAMAELEON
MENSA
PICTOR
COLUMBA
LIBRA
APUS
NORMA
TRIANGULUM AUSTRALE
OCTANS
DORADO
LEPUS
RETICULUM
SCORPIUS
CAELUM
ARA
HYDRUS
Antares
Rigel
PAVO
HOROLOGIUM
TELESCOPIUM
Achernar
CORONA AUSTRALIS
ERIDANUS
OPHIUCHUS
INDUS
TUCANA
WEST
FORNAX
EAST
SERPENS CAUDA
SAGITTARIUS
MICROSCOPIUM
PHOENIX
GRUS
SCULPTOR

NORTH

April

Northern hemisphere

20° 40° 60° Looking North

The charts on these pages show the northern hemisphere skies during April. They are correct for 11pm local time (midnight Daylight Saving Time) at the start of the month, 10pm (11pm DST) in the middle of the month, and 9pm (10pm DST) at the end of the month.

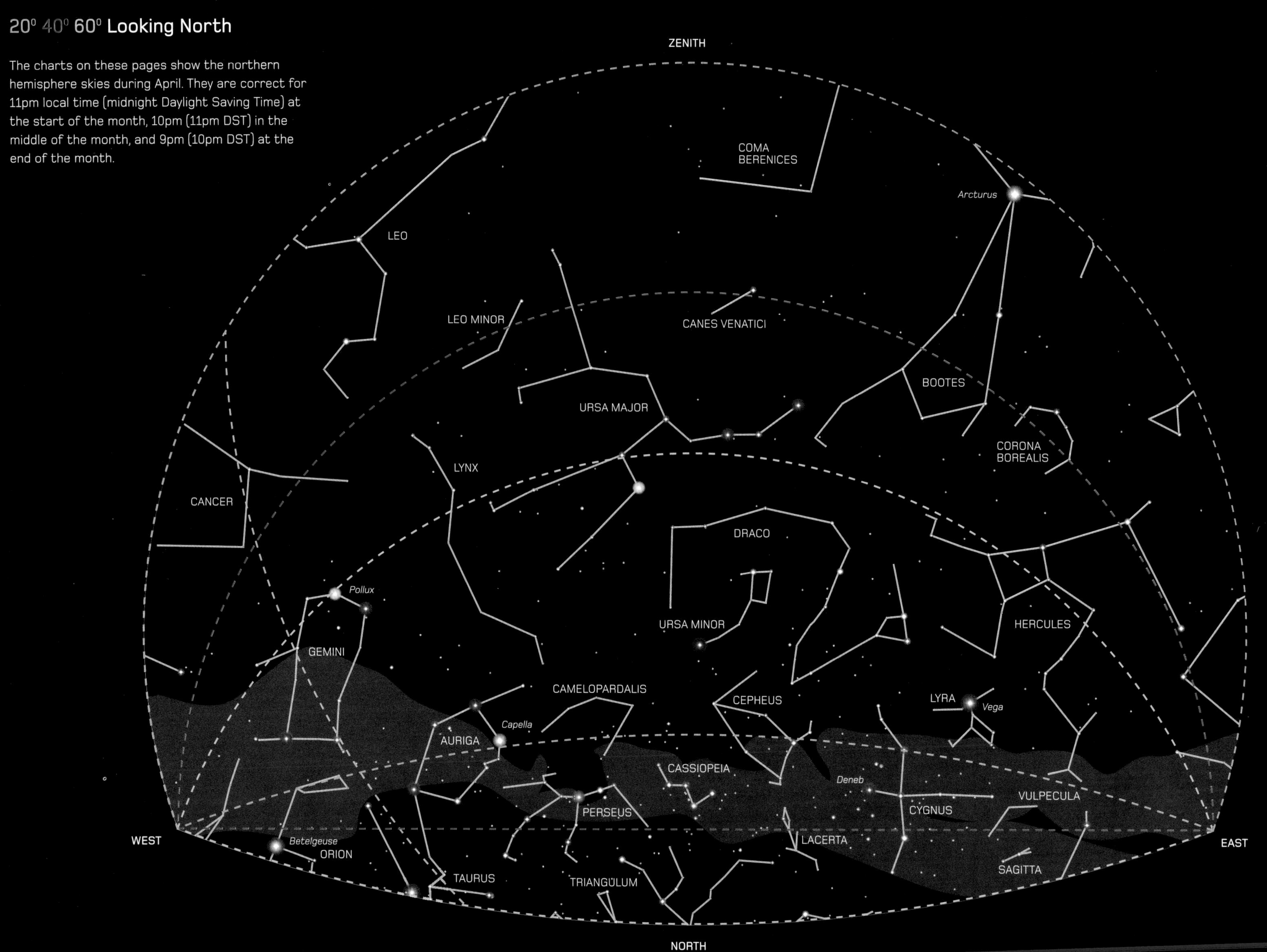

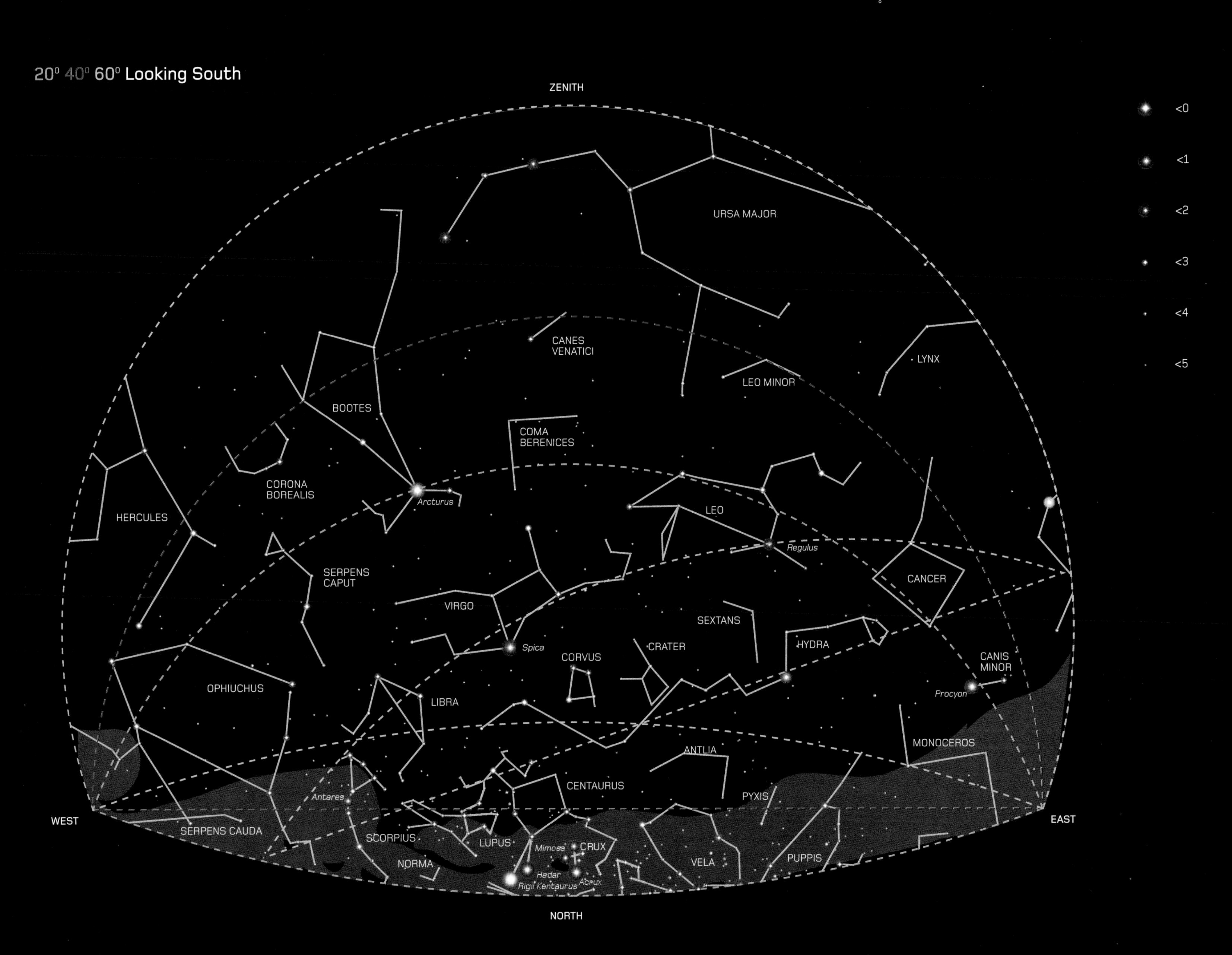
20° 40° 60° Looking South
ZENITH
<0
<1
<2
<3
<4
<5
URSA MAJOR
CANES
VENATICI
LYNX
LEO MINOR
BOOTES
COMA
BERENICES
CORONA
BOREALIS
HERCULES
Arcturus
LEO
Regulus
SERPENS
CAPUT
CANCER
VIRGO
SEXTANS
HYDRA
Spica
CORVUS
CRATER
CANIS
MINOR
OPHIUCHUS
LIBRA
Procyon
MONOCEROS
ANTLIA
CENTAURUS
PYXIS
Antares
WEST
SERPENS CAUDA
SCORPIUS
LUPUS
Mimosa
CRUX
VELA
PUPPIS
EAST
NORMA
Hadar
Rigil Kentaurus
Acrux
NORTH

April

Southern hemisphere

-20° -40° -60° Looking North

The charts on these pages show the southern hemisphere skies during April. They are correct for 11pm local time (midnight Daylight Saving Time) at the start of the month, 10pm (11pm DST) in the middle of the month, and 9pm (10pm DST) at the end of the month.

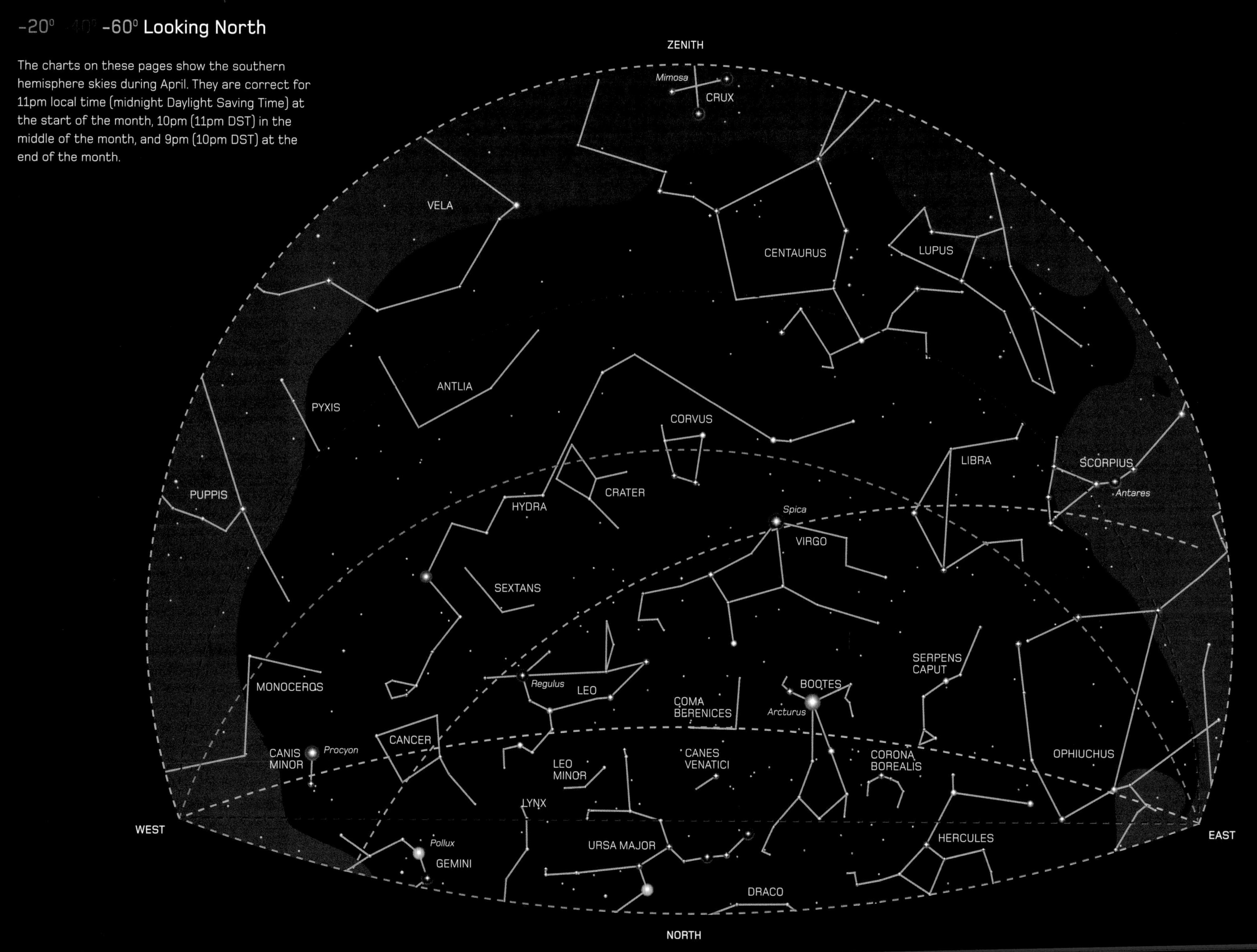

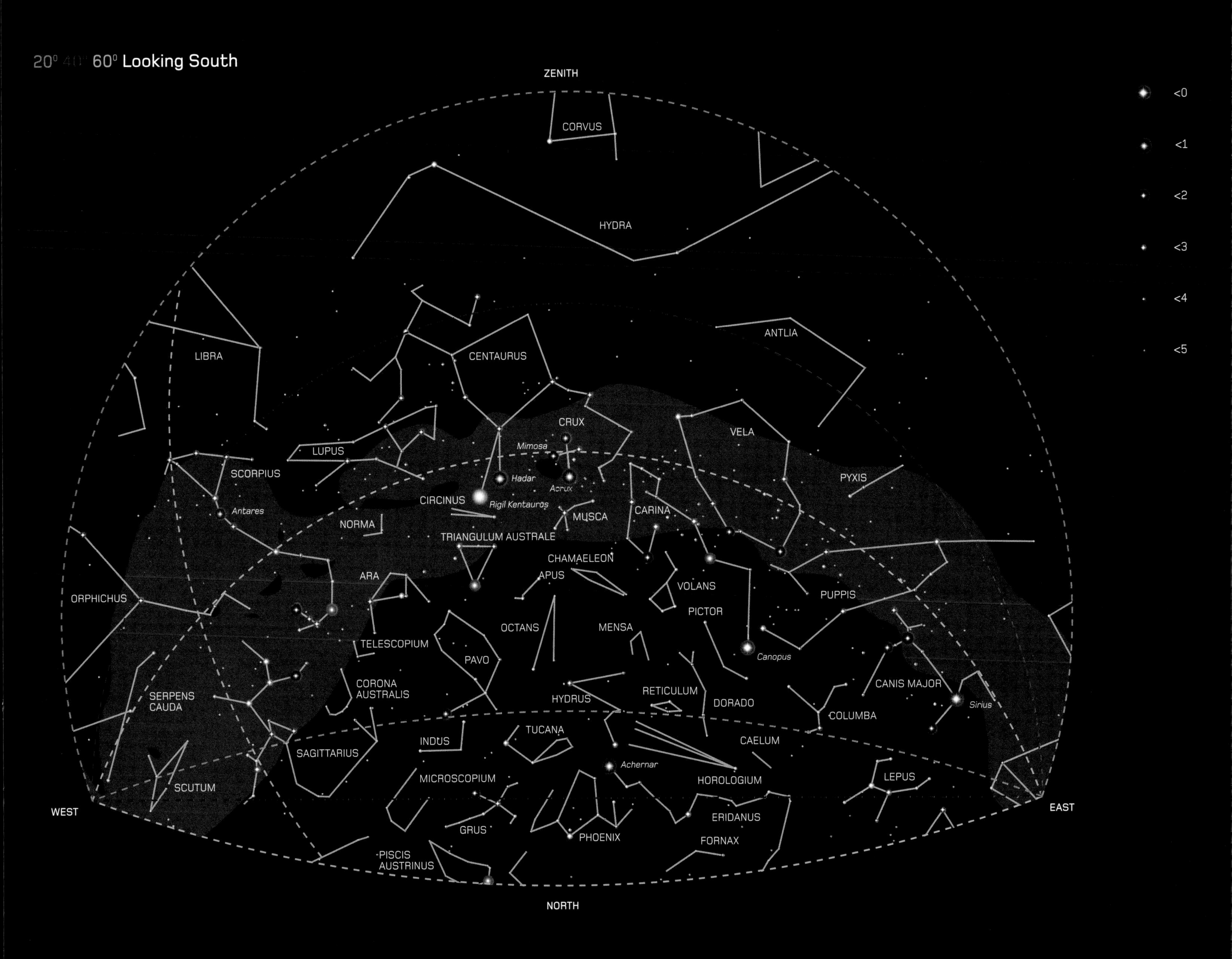

20° 40° 60° Looking South
ZENITH
<0
<1
<2
<3
<4
<5
CORVUS
HYDRA
ANTLIA
LIBRA
CENTAURUS
CRUX
Mimosa
VELA
LUPUS
SCORPIUS
Hadar
Acrux
PYXIS
CIRCINUS
Rigil Kentauros
Antares
CARINA
MUSCA
NORMA
TRIANGULUM AUSTRALE
CHAMAELEON
APUS
ARA
VOLANS
PUPPIS
ORPHICHUS
PICTOR
OCTANS
MENSA
TELESCOPIUM
Canopus
PAVO
CANIS MAJOR
CORONA AUSTRALIS
SERPENS CAUDA
RETICULUM
HYDRUS
DORADO
Sirius
COLUMBA
TUCANA
INDUS
CAELUM
SAGITTARIUS
Achernar
MICROSCOPIUM
HOROLOGIUM
LEPUS
SCUTUM
WEST
EAST
ERIDANUS
GRUS
PHOENIX
FORNAX
PISCIS AUSTRINUS
NORTH

May

Northern hemisphere

20° 40° 60° Looking North

The charts on these pages show the northern hemisphere skies during May. They are correct for 11pm local time (midnight Daylight Saving Time) at the start of the month, 10pm (11pm DST) in the middle of the month, and 9pm (10pm DST) at the end of the month.

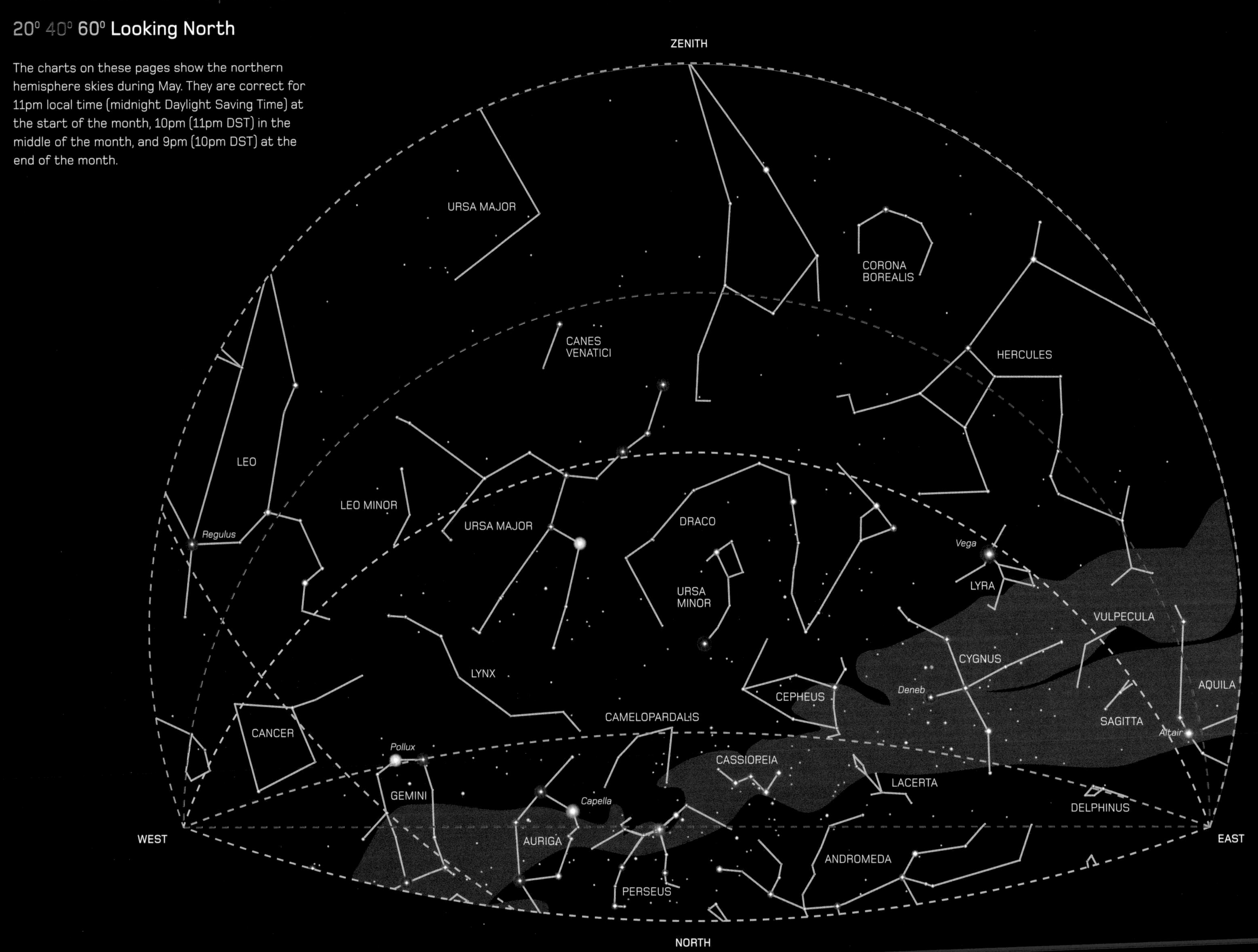

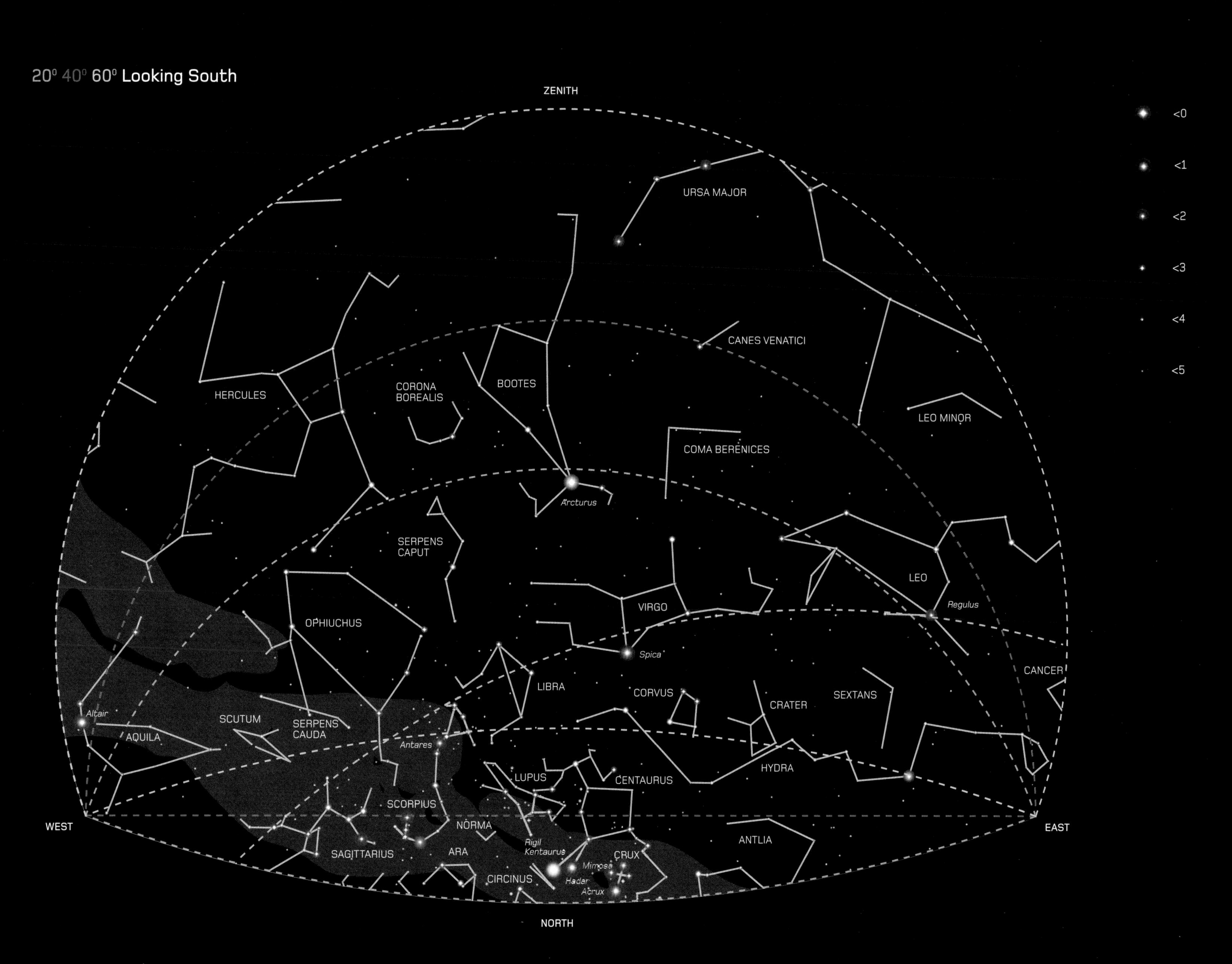
20° 40° 60° Looking South
ZENITH
<0
<1
<2
<3
<4
<5
URSA MAJOR
CANES VENATICI
HERCULES
CORONA BOREALIS
BOOTES
LEO MINOR
COMA BERENICES
Arcturus
SERPENS CAPUT
LEO
OPHIUCHUS
VIRGO
Regulus
Spica
LIBRA
CORVUS
CANCER
CRATER
SEXTANS
Altair
SCUTUM
SERPENS CAUDA
AQUILA
Antares
LUPUS
CENTAURUS
HYDRA
SCORPIUS
WEST
NORMA
EAST
Rigil Kentaurus
ANTLIA
SAGITTARIUS
ARA
CRUX
Mimosa
CIRCINUS
Hadar
Acrux
NORTH

May

Southern hemisphere

-20° -40° -60° Looking North

The charts on these pages show the southern hemisphere skies during May. They are correct for 11pm local time (midnight Daylight Saving Time) at the start of the month, 10pm (11pm DST) in the middle of the month, and 9pm (10pm DST) at the end of the month.

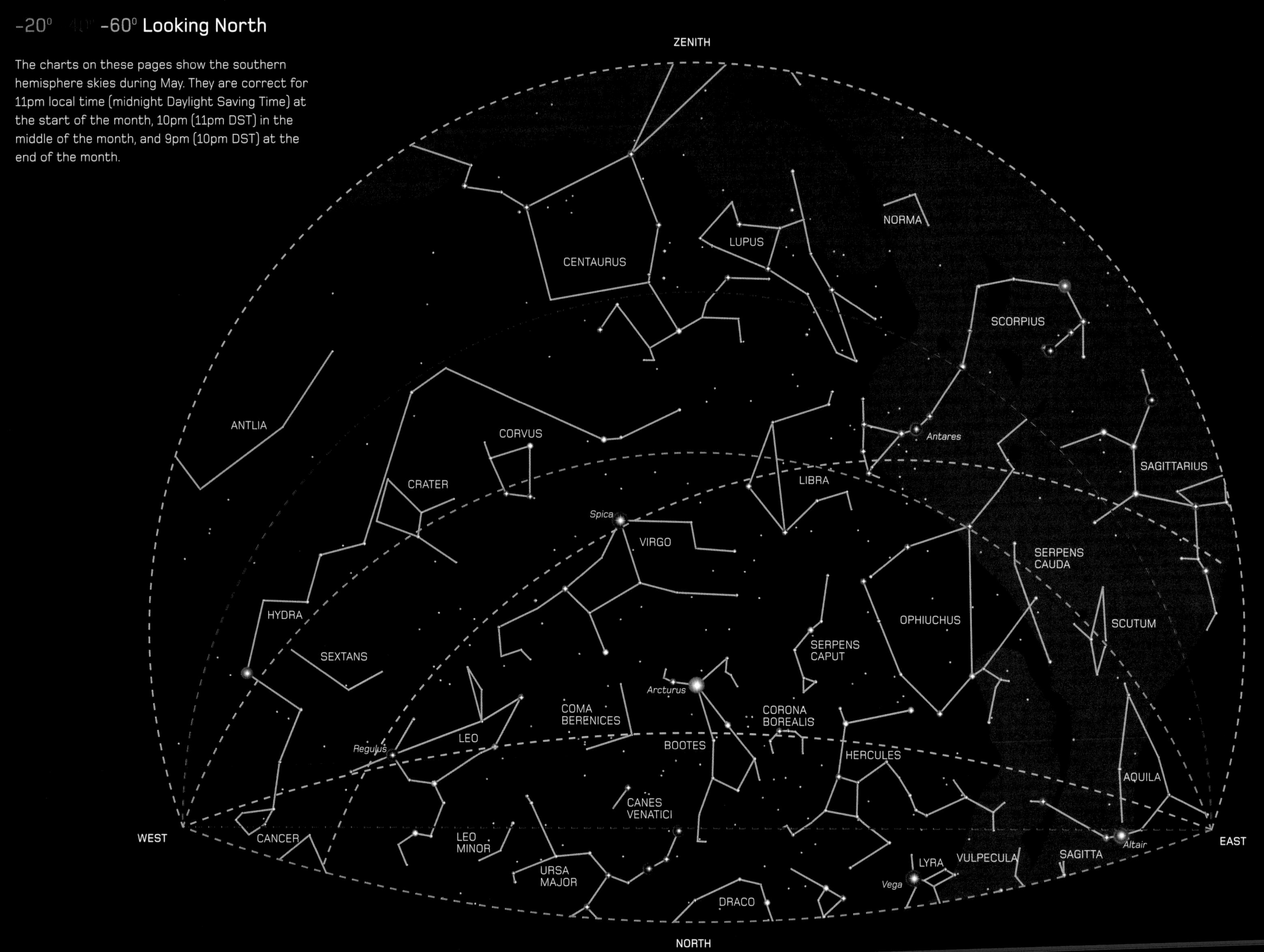

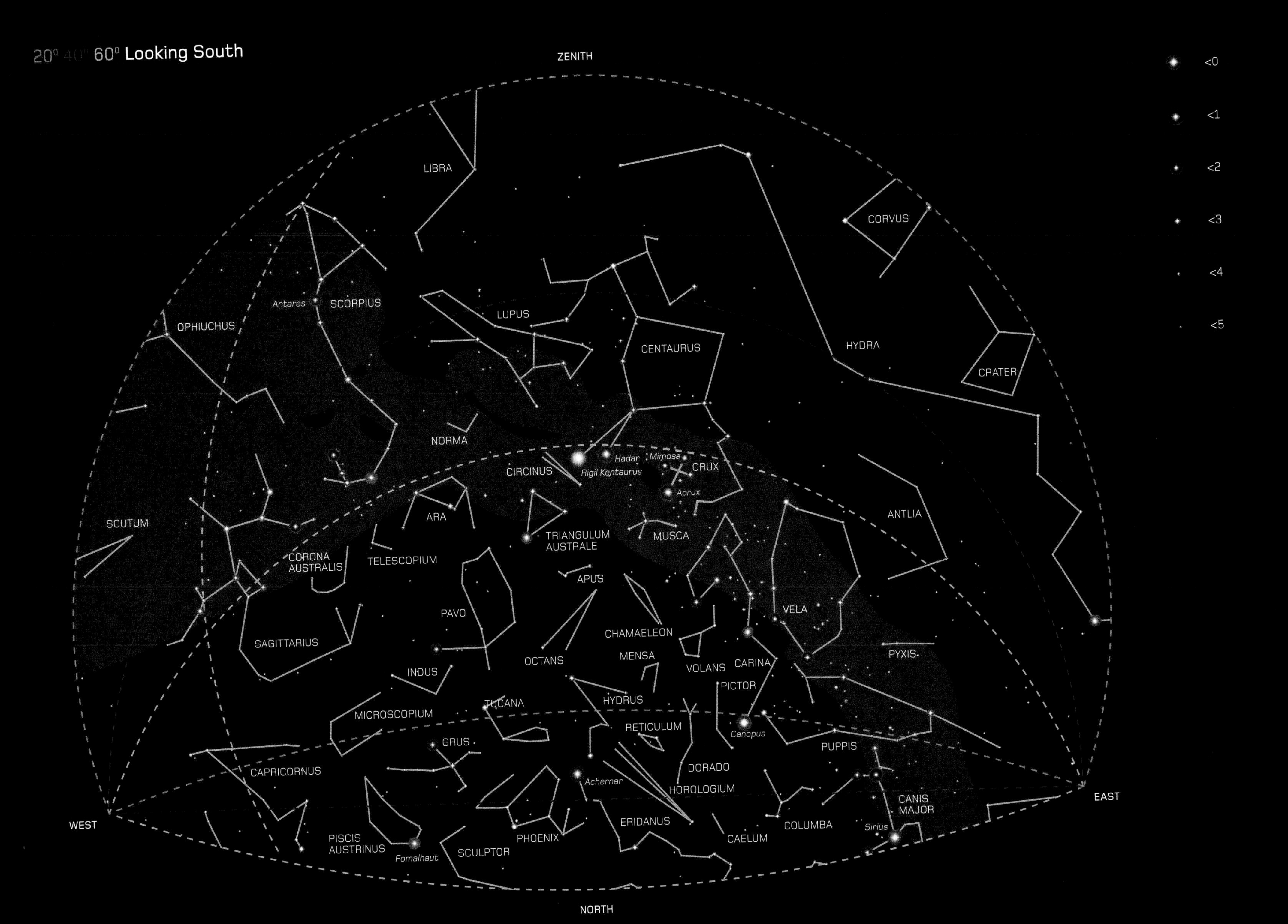

20° 40° 60° Looking South
ZENITH
<0
<1
<2
<3
<4
<5
LIBRA
CORVUS
SCORPIUS
Antares
LUPUS
OPHIUCHUS
CENTAURUS
HYDRA
CRATER
NORMA
Hadar
Mimosa
CRUX
CIRCINUS
Rigil Kentaurus
Acrux
ANTLIA
ARA
SCUTUM
TRIANGULUM AUSTRALE
MUSCA
CORONA AUSTRALIS
TELESCOPIUM
APUS
VELA
PAVO
SAGITTARIUS
CHAMAELEON
PYXIS
MENSA
OCTANS
INDUS
VOLANS
CARINA
PICTOR
HYDRUS
TUCANA
MICROSCOPIUM
RETICULUM
Canopus
GRUS
PUPPIS
DORADO
CAPRICORNUS
Achernar
HOROLOGIUM
EAST
CANIS MAJOR
WEST
ERIDANUS
COLUMBA
PHOENIX
Sirius
PISCIS AUSTRINUS
CAELUM
Fomalhaut
SCULPTOR
NORTH

June

Northern hemisphere

20° 40° 60° Looking North

The charts on these pages show the northern hemisphere skies during June. They are correct for 11pm local time (midnight Daylight Saving Time) at the start of the month, 10pm (11pm DST) in the middle of the month, and 9pm (10pm DST) at the end of the month.

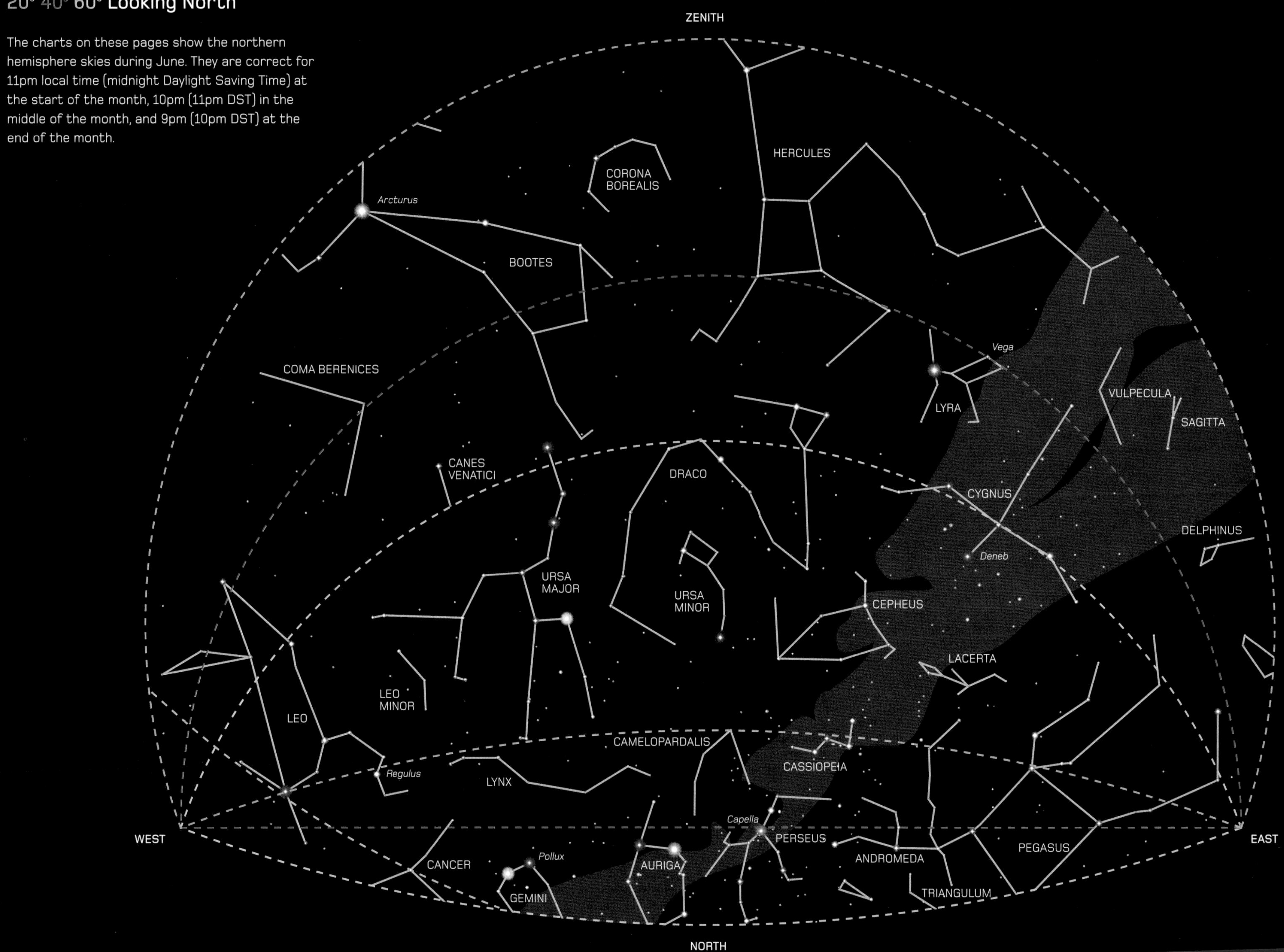

20° 40° 60° Looking South

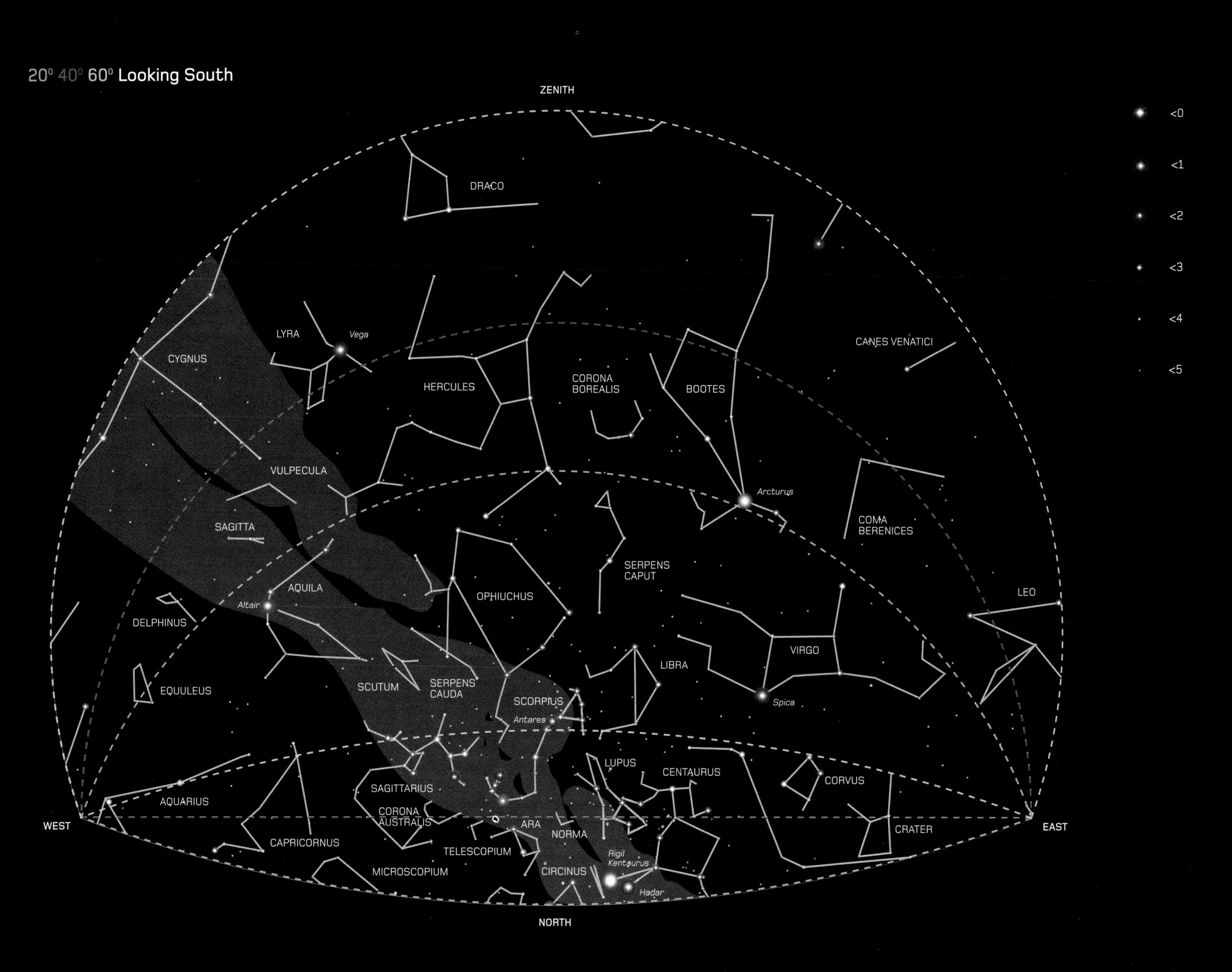

June

Southern hemisphere

-20° -40° -60° **Looking North**

The charts on these pages show the southern hemisphere skies during June. They are correct for 11pm local time (midnight Daylight Saving Time) at the start of the month, 10pm (11pm DST) in the middle of the month, and 9pm (10pm DST) at the end of the month.

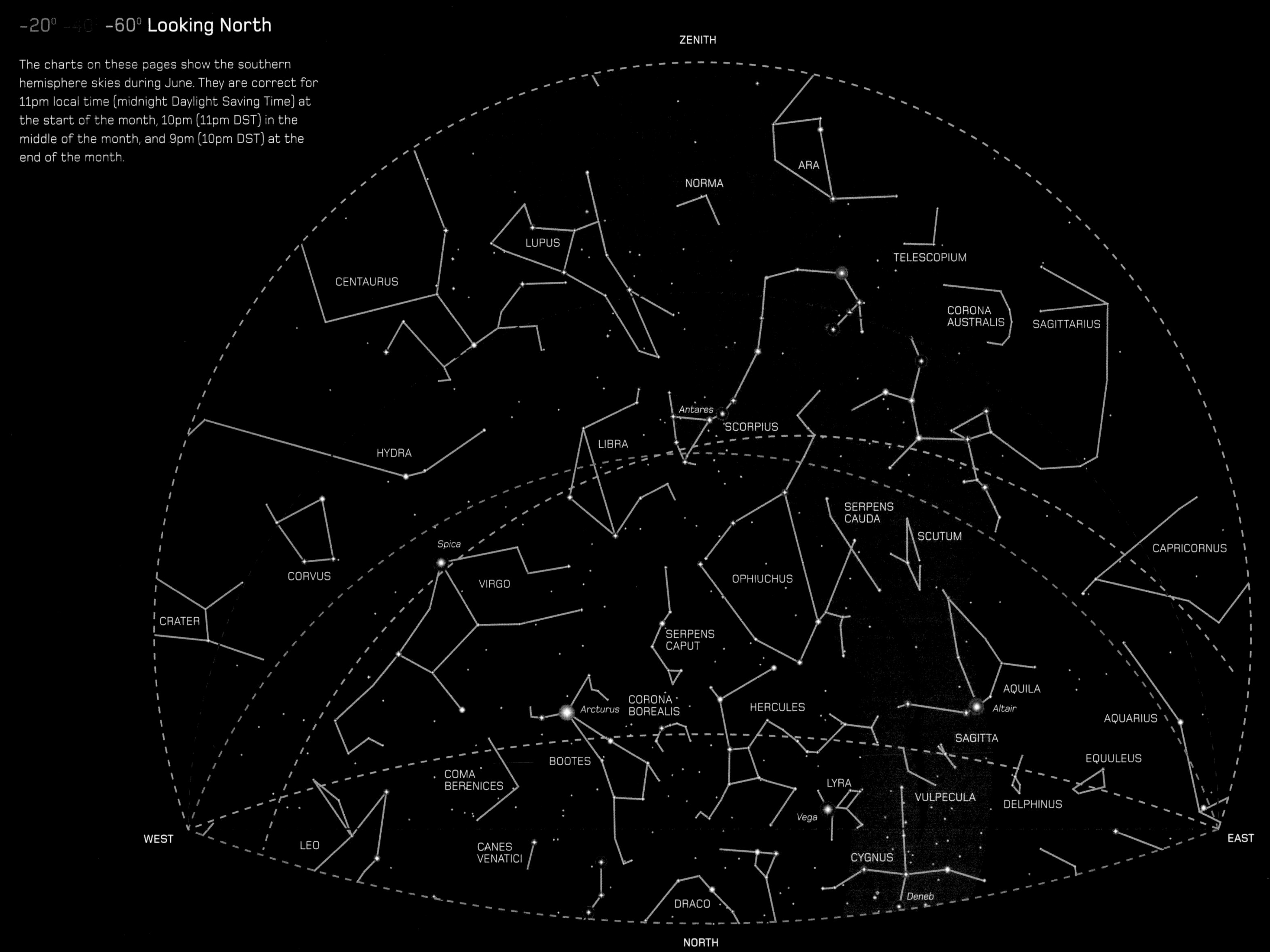

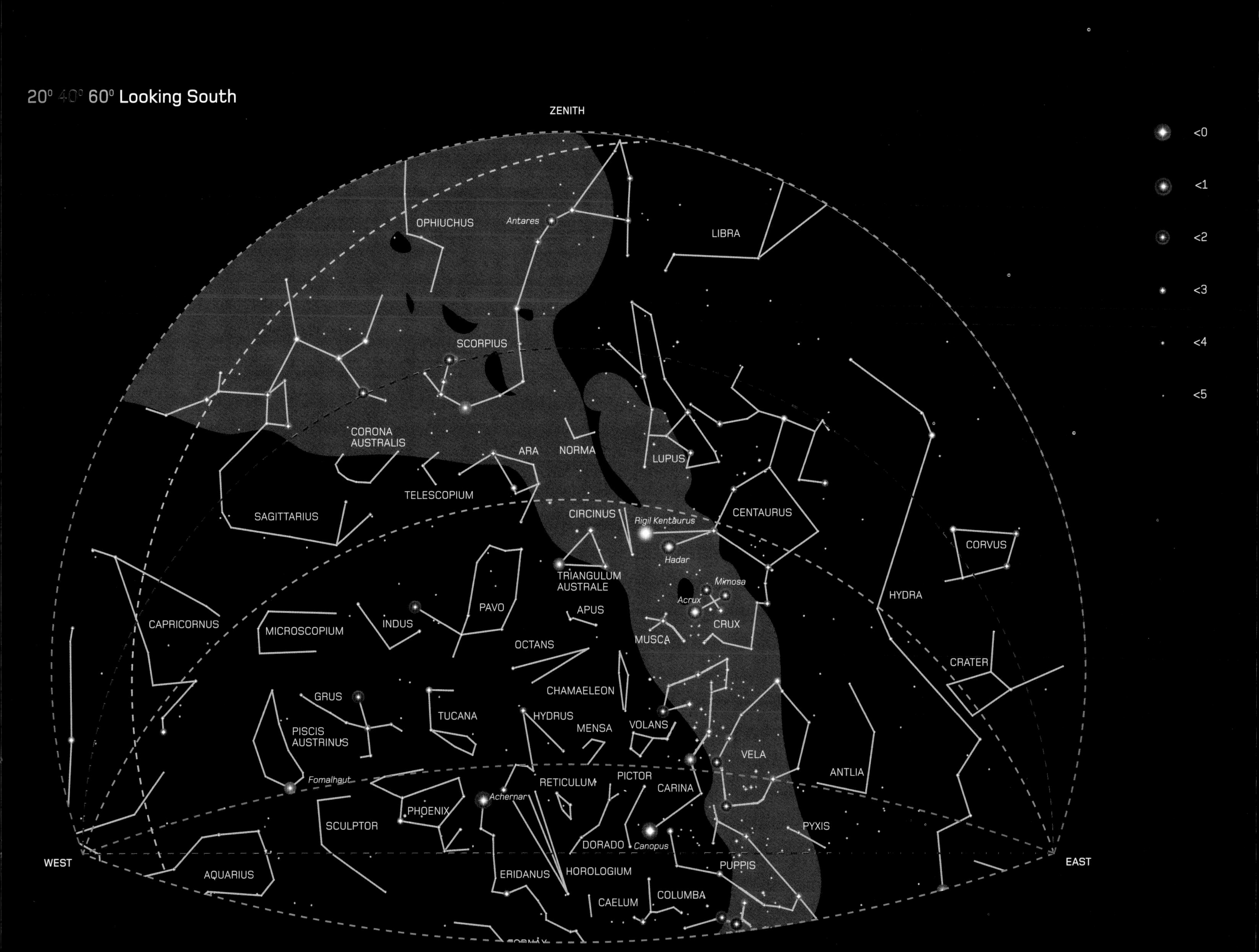

20° 40° 60° Looking South
ZENITH
<0
<1
<2
<3
<4
<5
OPHIUCHUS
Antares
LIBRA
SCORPIUS
CORONA AUSTRALIS
ARA
NORMA
LUPUS
TELESCOPIUM
SAGITTARIUS
CIRCINUS
Rigil Kentaurus
CENTAURUS
Hadar
CORVUS
TRIANGULUM AUSTRALE
Mimosa
Acrux
HYDRA
PAVO
APUS
CRUX
CAPRICORNUS
MICROSCOPIUM
INDUS
MUSCA
OCTANS
CRATER
CHAMAELEON
GRUS
TUCANA
HYDRUS
VOLANS
MENSA
PISCIS AUSTRINUS
VELA
ANTLIA
Fomalhaut
PICTOR
RETICULUM
CARINA
Achernar
PHOENIX
SCULPTOR
PYXIS
DORADO
Canopus
WEST
EAST
PUPPIS
AQUARIUS
ERIDANUS
HOROLOGIUM
CAELUM
COLUMBA

July

Northern hemisphere

20° 40° 60° Looking North

The charts on these pages show the northern hemisphere skies during July. They are correct for 11pm local time (midnight Daylight Saving Time) at the start of the month, 10pm (11pm DST) in the middle of the month, and 9pm (10pm DST) at the end of the month.

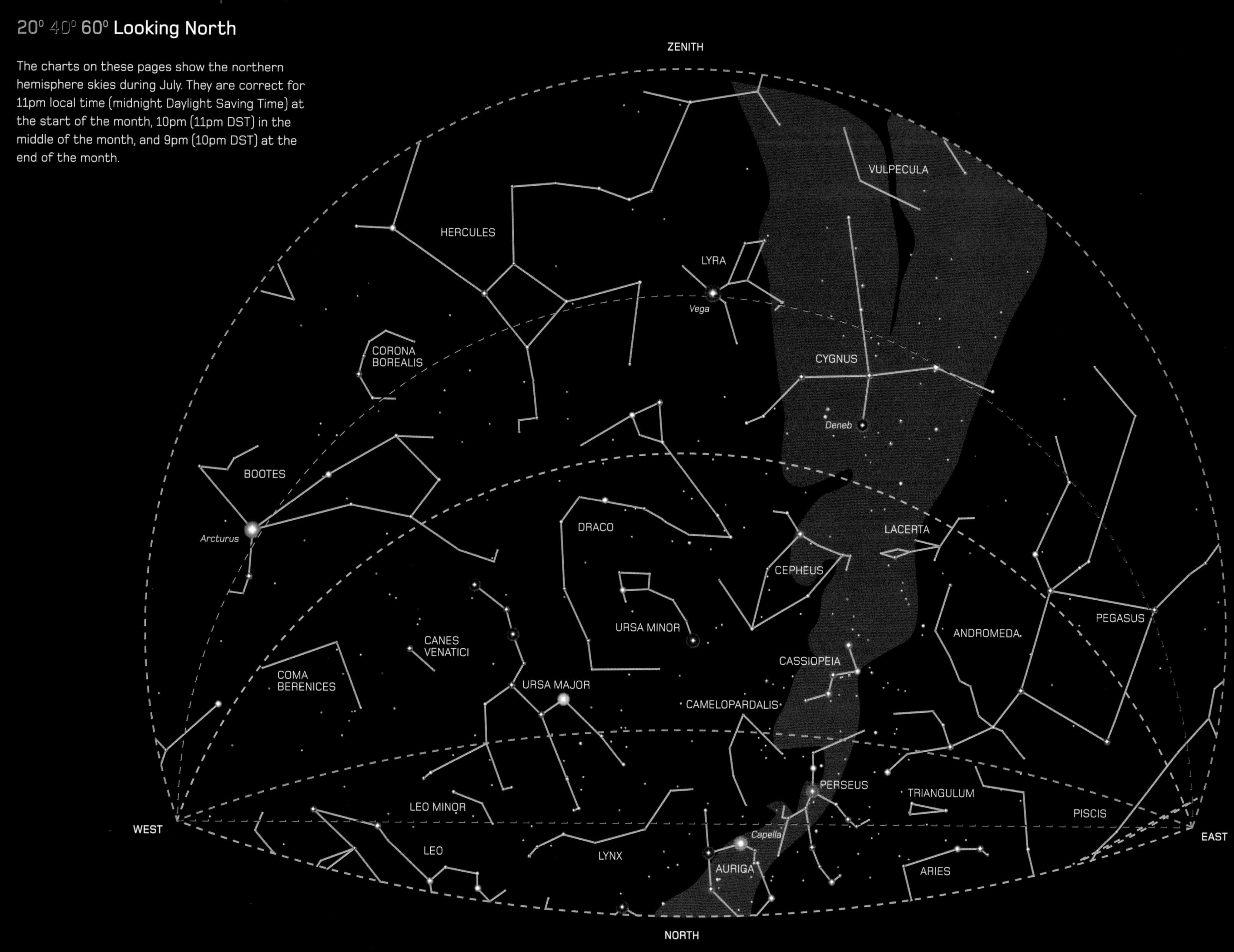

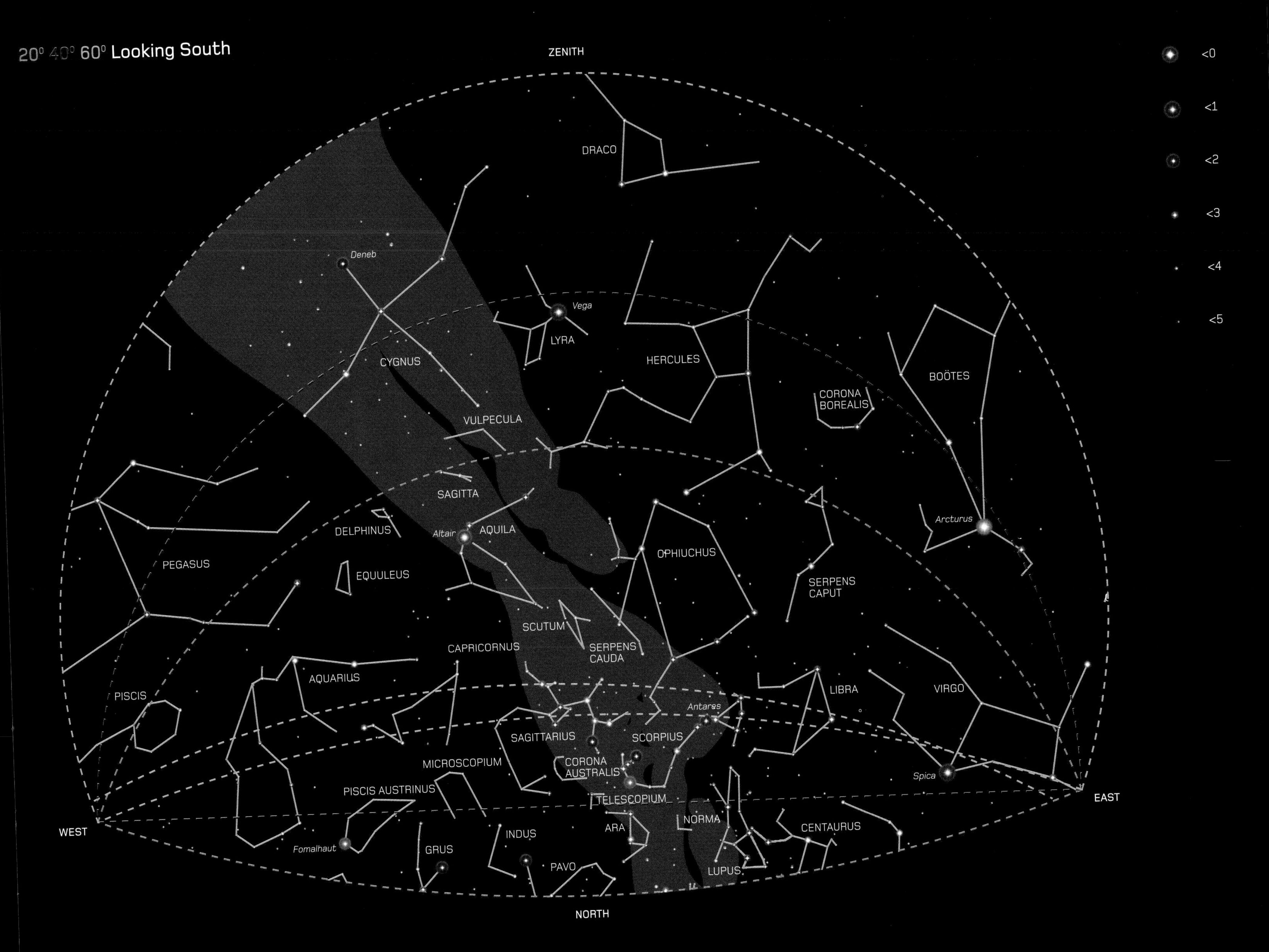
20° 40° 60° Looking South
ZENITH
<0
<1
<2
<3
<4
<5
DRACO
Deneb
Vega
LYRA
CYGNUS
HERCULES
BOÖTES
CORONA BOREALIS
VULPECULA
SAGITTA
Arcturus
DELPHINUS
Altair
AQUILA
OPHIUCHUS
PEGASUS
EQUULEUS
SERPENS CAPUT
SCUTUM
SERPENS CAUDA
CAPRICORNUS
AQUARIUS
LIBRA
VIRGO
PISCIS
Antares
SAGITTARIUS
SCORPIUS
MICROSCOPIUM
CORONA AUSTRALIS
Spica
PISCIS AUSTRINUS
TELESCOPIUM
EAST
WEST
ARA
NORMA
CENTAURUS
INDUS
Fomalhaut
GRUS
PAVO
LUPUS
NORTH

July

Southern hemisphere

-20° -40° -60° Looking North

The charts on these pages show the southern hemisphere skies during July. They are correct for 11pm local time (midnight Daylight Saving Time) at the start of the month, 10pm (11pm DST) in the middle of the month, and 9pm (10pm DST) at the end of the month.

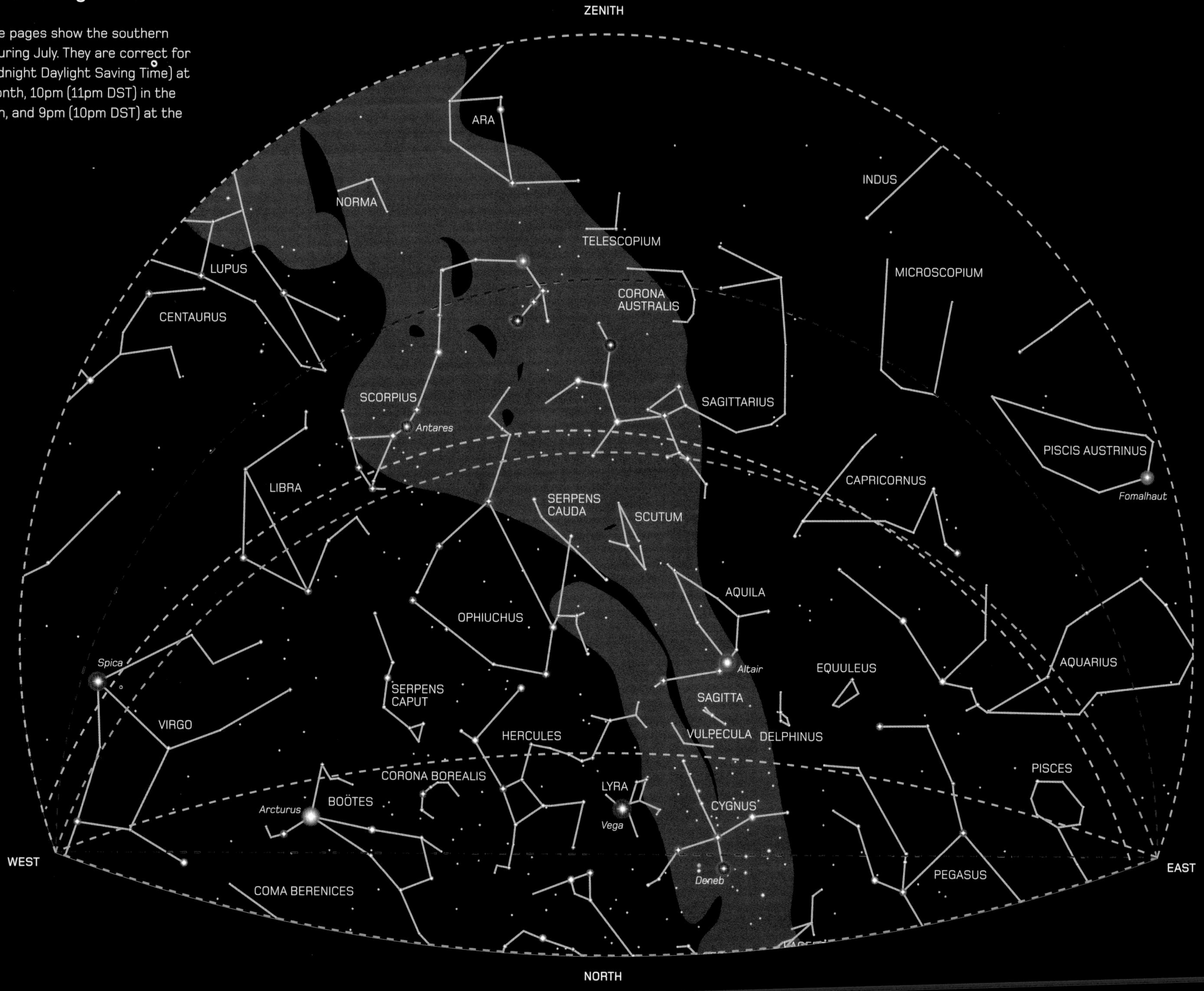

ZENITH
ARA
INDUS
NORMA
TELESCOPIUM
LUPUS
MICROSCOPIUM
CENTAURUS
CORONA AUSTRALIS
SCORPIUS
Antares
SAGITTARIUS
PISCIS AUSTRINUS
Fomalhaut
CAPRICORNUS
LIBRA
SERPENS CAUDA
SCUTUM
AQUILA
OPHIUCHUS
Spica
Altair
EQUULEUS
AQUARIUS
SERPENS CAPUT
SAGITTA
VIRGO
HERCULES
VULPECULA
DELPHINUS
CORONA BOREALIS
LYRA
PISCES
Arcturus
BOÖTES
CYGNUS
Vega
WEST
Deneb
PEGASUS
EAST
COMA BERENICES
NORTH

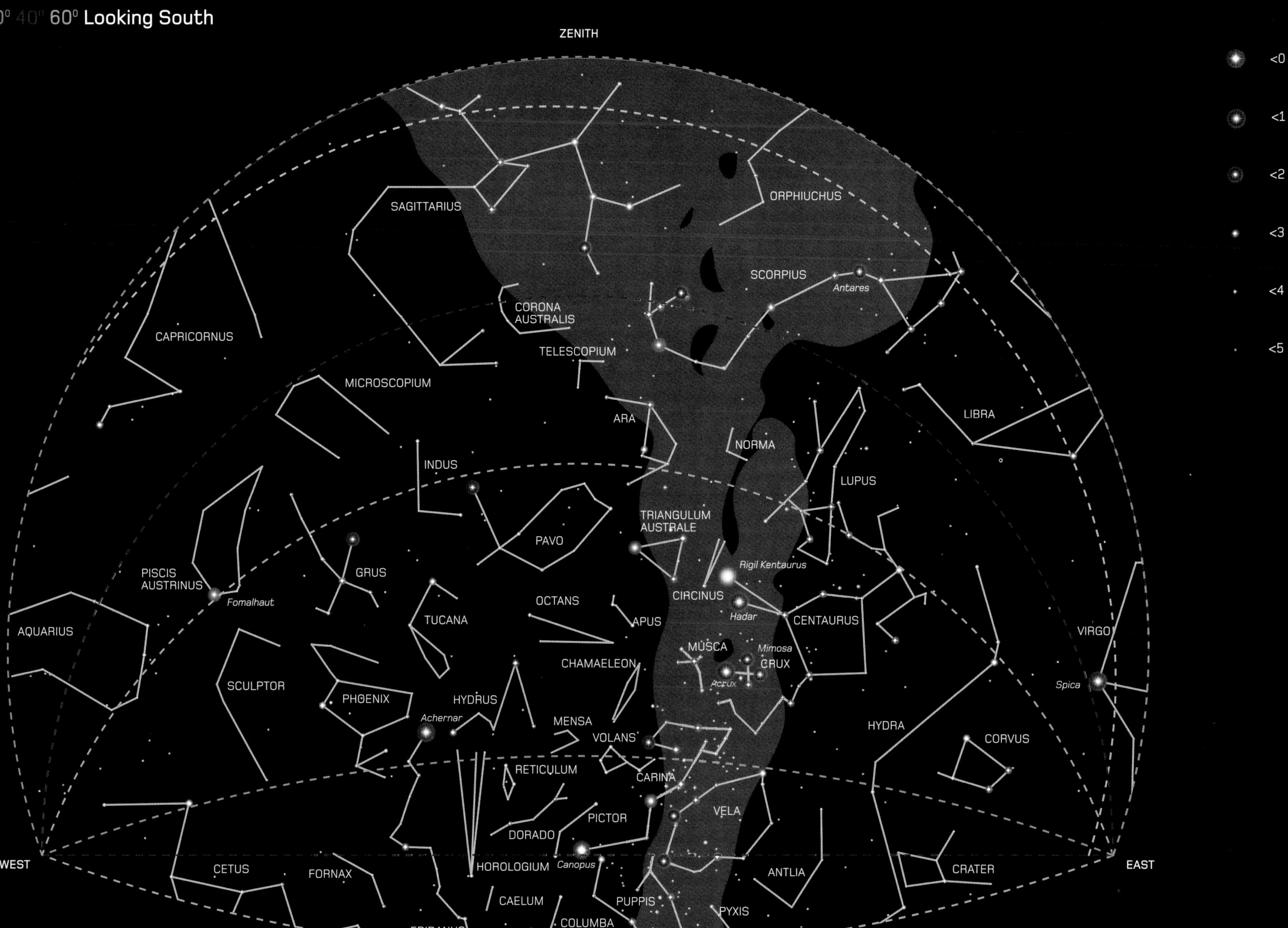

ZENITH
<0
<1
<2
<3
<4
<5
SAGITTARIUS
ORPHIUCHUS
SCORPIUS
Antares
CORONA AUSTRALIS
TELESCOPIUM
CAPRICORNUS
MICROSCOPIUM
LIBRA
ARA
NORMA
INDUS
LUPUS
TRIANGULUM AUSTRALE
PAVO
PISCIS AUSTRINUS
Fomalhaut
GRUS
Rigil Kentaurus
CIRCINUS
OCTANS
Hadar
CENTAURUS
AQUARIUS
TUCANA
APUS
VIRGO
MUSCA
Mimosa
CRUX
Acrux
CHAMAELEON
SCULPTOR
Spica
PHOENIX
HYDRUS
HYDRA
Achernar
MENSA
CORVUS
VOLANS
RETICULUM
CARINA
VELA
PICTOR
DORADO
WEST
EAST
CETUS
FORNAX
HOROLOGIUM
Canopus
CRATER
ANTLIA
CAELUM
PUPPIS
PYXIS
COLUMBA
ERIDANUS
NORTH

August

Northern hemisphere

20° 40° 60° Looking North

The charts on these pages show the northern hemisphere skies during August. They are correct for 11pm local time (midnight Daylight Saving Time) at the start of the month, 10pm (11pm DST) in the middle of the month, and 9pm (10pm DST) at the end of the month.

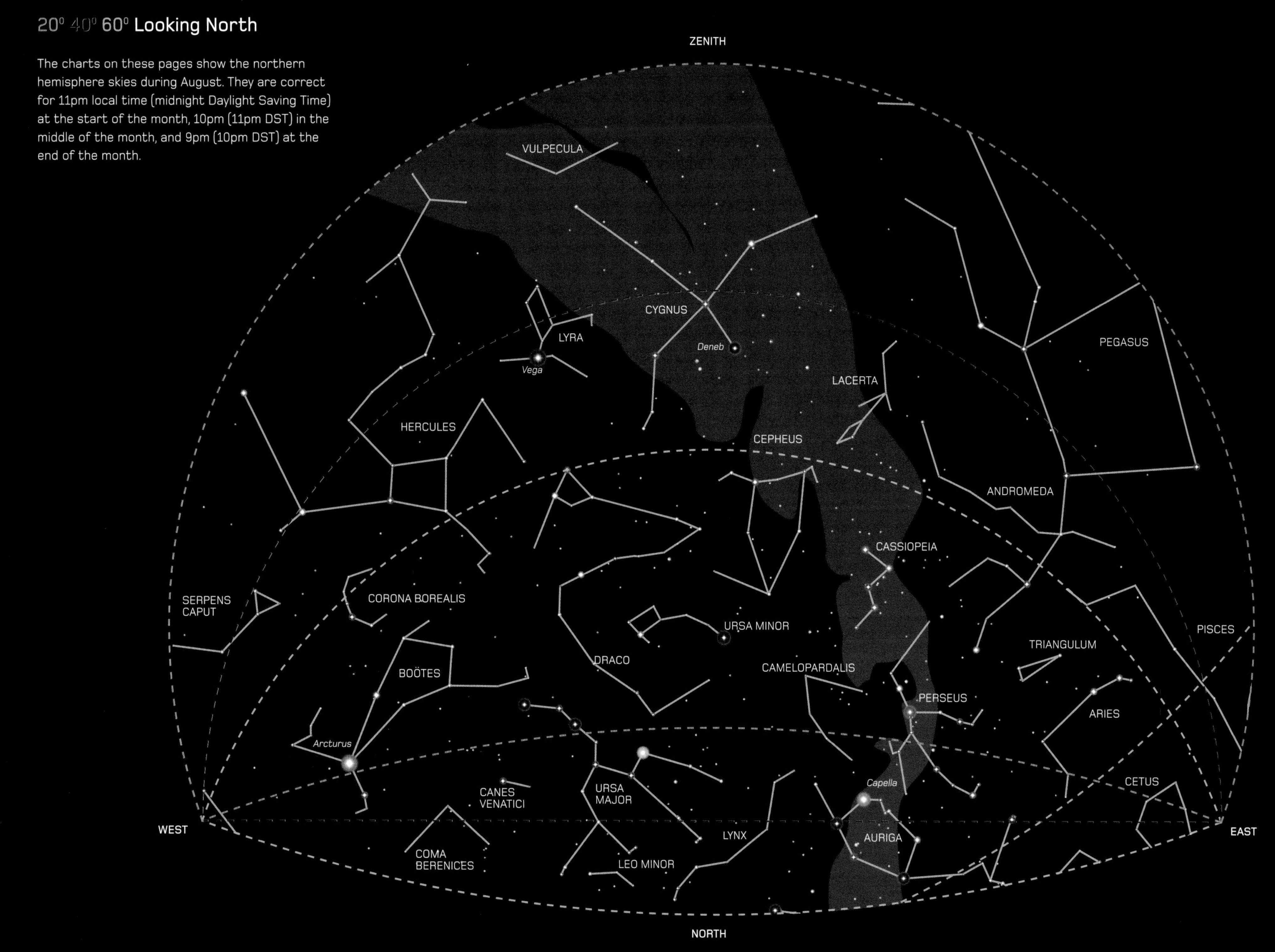

20° 40° 60° Looking South
ZENITH
<0
<1
<2
<3
<4
<5
Deneb
CYGNUS
LYRA
Vega
LACERTA
ANDROMEDA
VULPECULA
HERCULES
SAGITTA
DELPHINUS
PEGASUS
Altair
AQUILA
EQUULEUS
OPHIUCHUS
SCUTUM
AQUARIUS
PISCES
SERPENS CAUDA
SAGITTARIUS
SERPENS CAPUT
CAPRICORNUS
MICROSCOPIUM
Fomalhaut
CORONA AUSTRALIS
SCULPTOR
PISCIS AUSTRINUS
Antares
CETUS
EAST
LIBRA
WEST
GRUS
INDUS
TELESCOPIUM
SCORPIUS
PHOENIX
ARA
TUCANA
PAVO
NORTH

August

Southern hemisphere

-20° -40° -60° Looking North

The charts on these pages show the southern hemisphere skies during August. They are correct for 11pm local time (midnight Daylight Saving Time) at the start of the month, 10pm (11pm DST) in the middle of the month, and 9pm (10pm DST) at the end of the month.

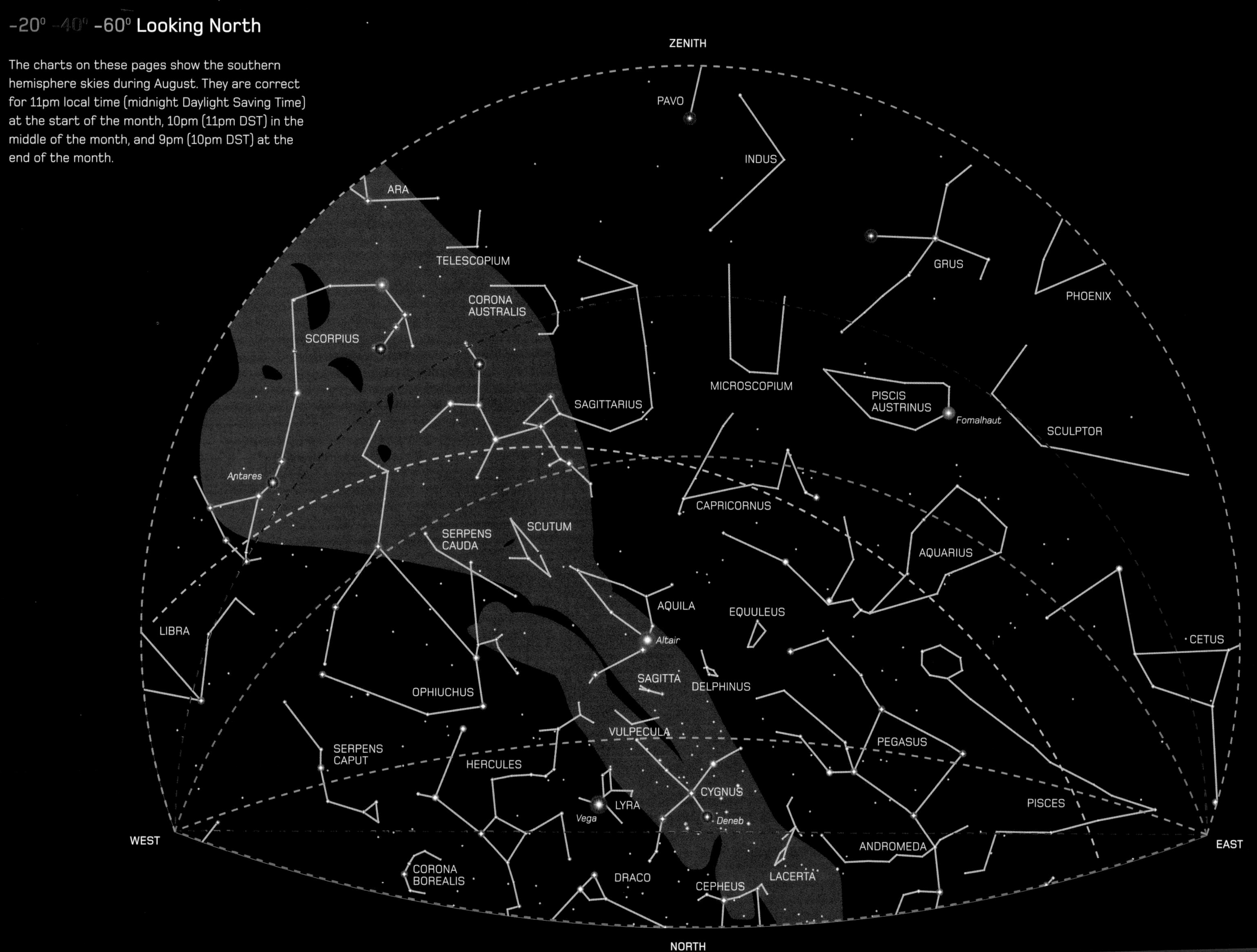

20° 40° 60° Looking South

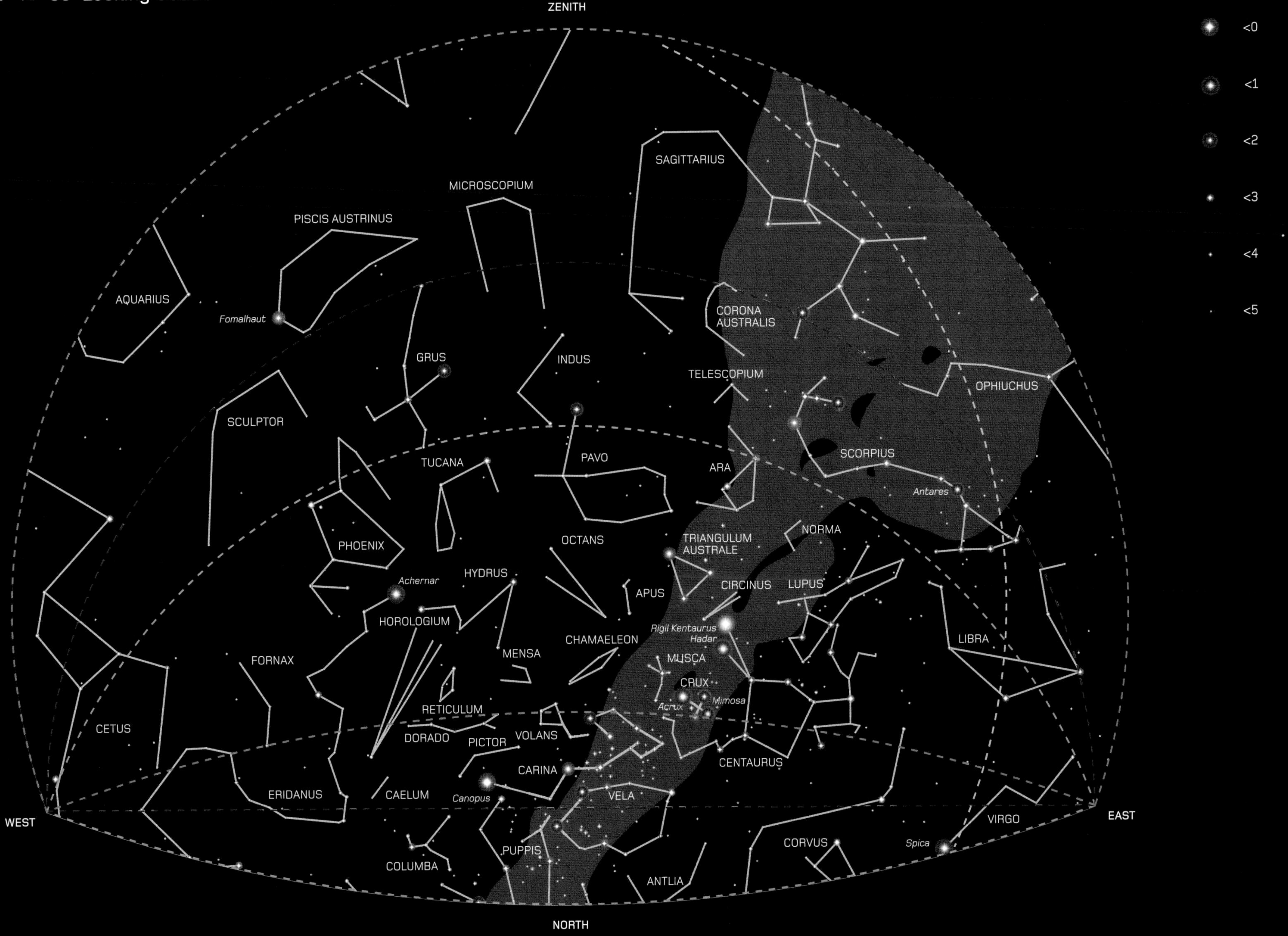

ZENITH

<0
<1
<2
<3
<4
<5

SAGITTARIUS
MICROSCOPIUM
PISCIS AUSTRINUS
AQUARIUS
Fomalhaut
CORONA AUSTRALIS
GRUS
INDUS
TELESCOPIUM
OPHIUCHUS
SCULPTOR
PAVO
ARA
SCORPIUS
TUCANA
Antares
PHOENIX
OCTANS
TRIANGULUM AUSTRALE
NORMA
Achernar
HYDRUS
APUS
CIRCINUS
LUPUS
HOROLOGIUM
Rigil Kentaurus
Hadar
CHAMAELEON
LIBRA
FORNAX
MENSA
MUSCA
CRUX
Acrux
Mimosa
RETICULUM
CETUS
DORADO
PICTOR
VOLANS
CENTAURUS
CARINA
ERIDANUS
CAELUM
Canopus
VELA
WEST
VIRGO
EAST
CORVUS
Spica
PUPPIS
COLUMBA
ANTLIA
NORTH

September

Northern hemisphere

20° 40° 60° Looking North

The charts on these pages show the northern hemisphere skies during September. They are correct for 11pm local time (midnight Daylight Saving Time) at the start of the month, 10pm (11pm DST) in the middle of the month, and 9pm (10pm DST) at the end of the month.

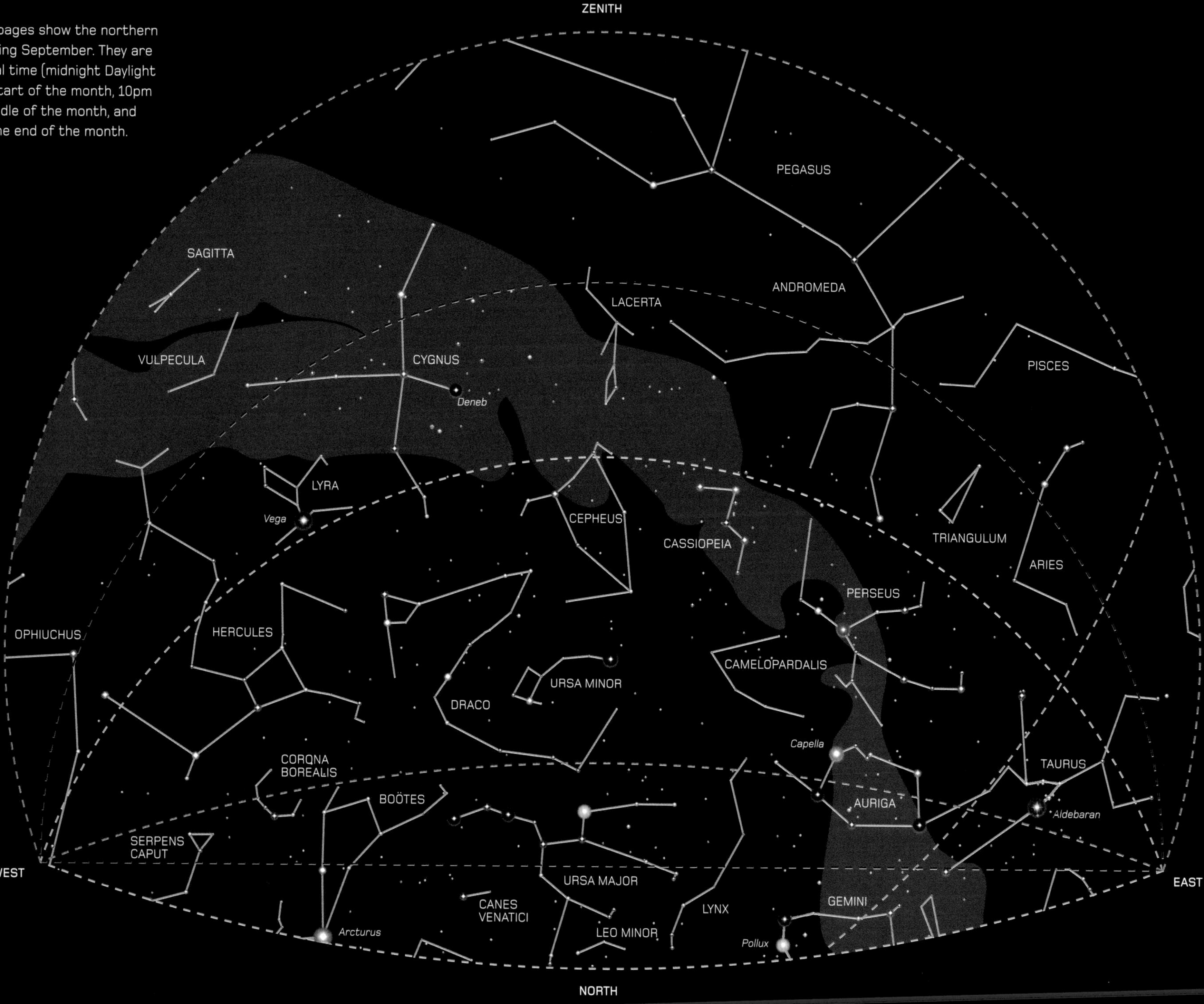

20° 40° 60° Looking South
ZENITH
<0
<1
<2
<3
<4
<5
CEPHEUS
LACERTA
Deneb
CYGNUS
LYRA
Vega
TRIANGULUM
ANDROMEDA
VULPECULA
SAGITTA
PEGASUS
ARIES
DELPHINUS
HERCULES
Altair
PISCES
EQUULEUS
AQUILA
AQUARIUS
CAPRICORNUS
CETUS
PISCIS
AUSTRINUS
Fomalhaut
SCUTUM
MICROSCOPIUM
SCULPTOR
SERPENS
CAUDA
OPHIUCHUS
GRUS
ERIDANUS
PHOENIX
SAGITTARIUS
WEST
EAST
FORNAX
INDUS
TUCANA
CORONA
AUSTRALIS
ERIDANUS
Achernar
PAVO
TELESCOPIUM
NORTH

September

Southern hemisphere

-20° -40° -60° Looking North

The charts on these pages show the southern hemisphere skies during September. They are correct for 11pm local time (midnight Daylight Saving Time) at the start of the month, 10pm (11pm DST) in the middle of the month, and 9pm (10pm DST) at the end of the month.

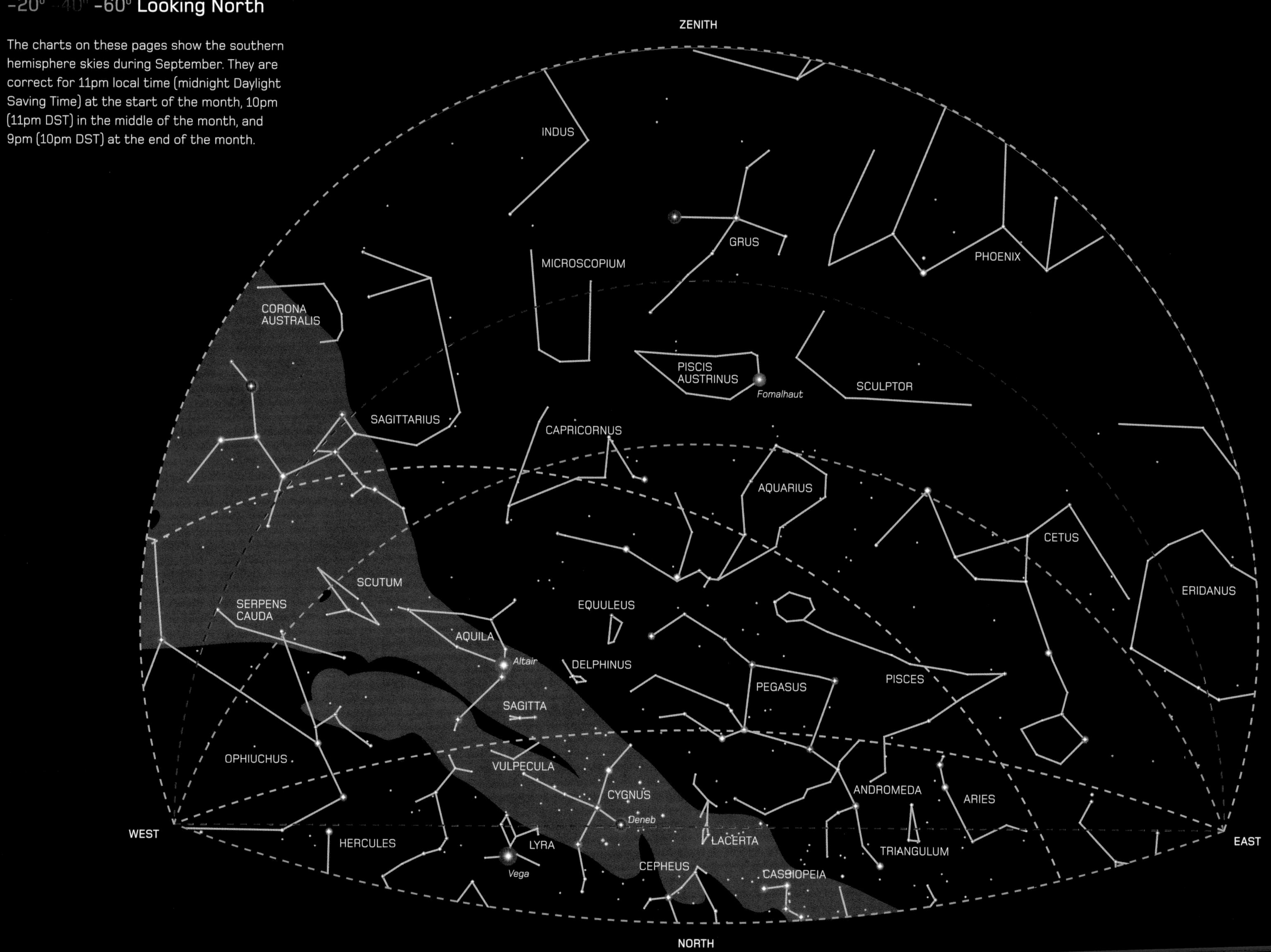

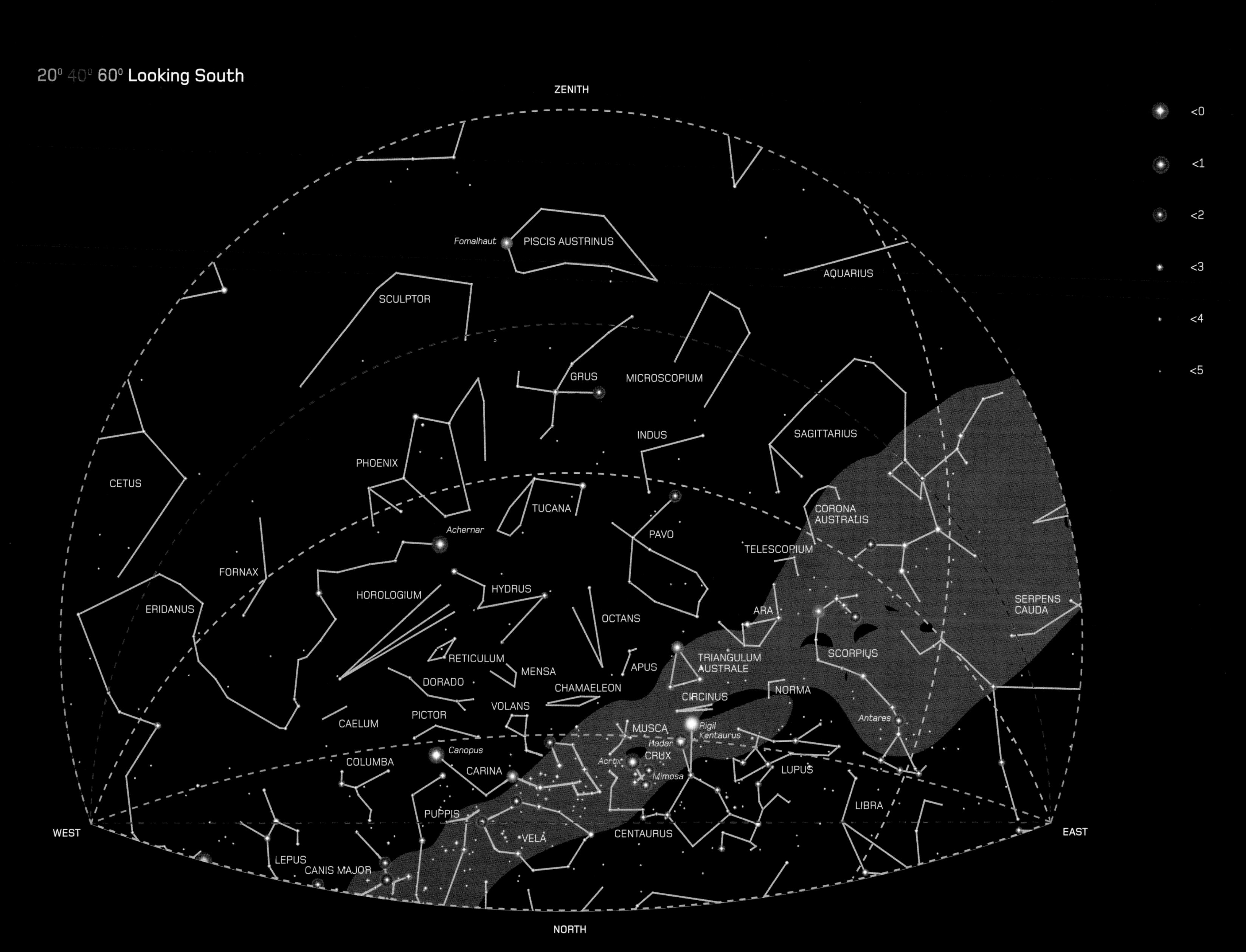

20° 40° 60° Looking South
ZENITH
<0
<1
<2
<3
<4
<5
Fomalhaut
PISCIS AUSTRINUS
AQUARIUS
SCULPTOR
GRUS
MICROSCOPIUM
CETUS
PHOENIX
INDUS
SAGITTARIUS
TUCANA
CORONA AUSTRALIS
Achernar
PAVO
FORNAX
TELESCOPIUM
HOROLOGIUM
HYDRUS
ERIDANUS
ARA
SERPENS CAUDA
OCTANS
RETICULUM
APUS
TRIANGULUM AUSTRALE
SCORPIUS
MENSA
DORADO
CHAMAELEON
CIRCINUS
NORMA
CAELUM
PICTOR
VOLANS
MUSCA
Rigil Kentaurus
Antares
Hadar
Canopus
COLUMBA
CARINA
Acrux
CRUX
Mimosa
LUPUS
LIBRA
PUPPIS
WEST
VELA
CENTAURUS
EAST
LEPUS
CANIS MAJOR
NORTH

October

Northern hemisphere

20° 40° 60° Looking North

The charts on these pages show the northern hemisphere skies during October. They are correct for 11pm local time (midnight Daylight Saving Time) at the start of the month, 10pm (11pm DST) in the middle of the month, and 9pm (10pm DST) at the end of the month.

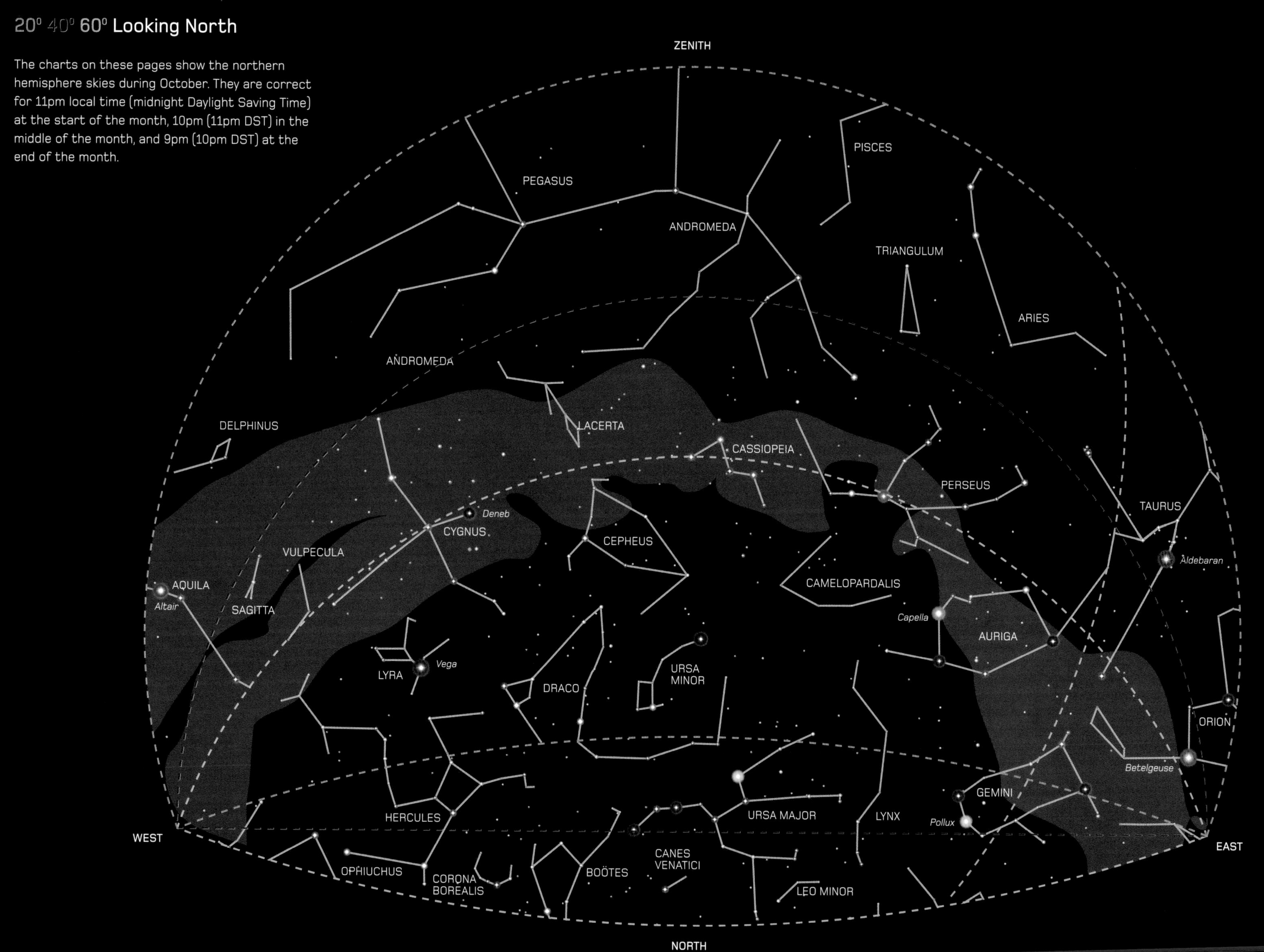

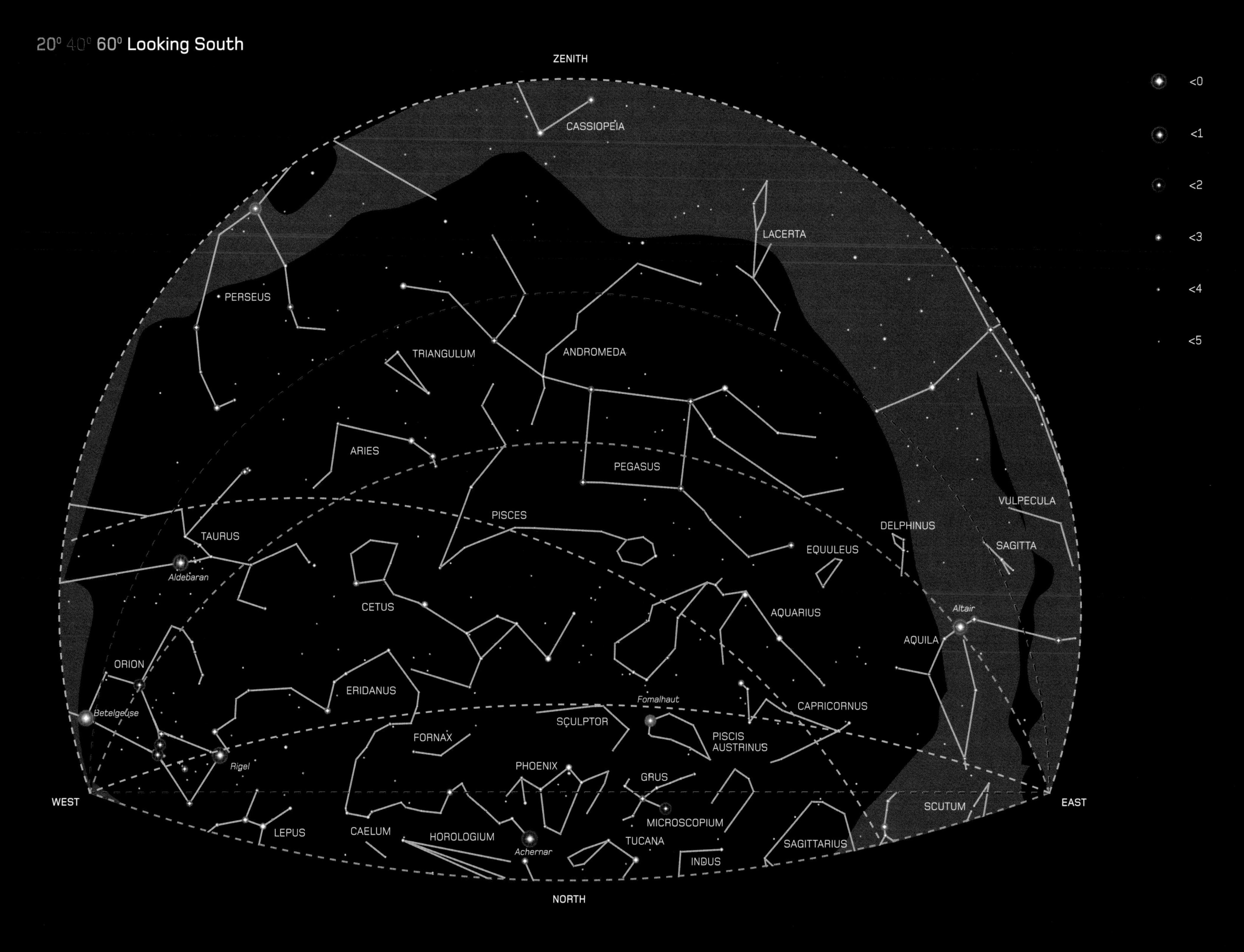
20° 40° 60° Looking South
ZENITH
<0
<1
<2
<3
<4
<5
CASSIOPEIA
LACERTA
PERSEUS
TRIANGULUM
ANDROMEDA
ARIES
PEGASUS
PISCES
TAURUS
Aldebaran
CETUS
DELPHINUS
VULPECULA
SAGITTA
EQUULEUS
AQUARIUS
Altair
AQUILA
ORION
Betelgeuse
ERIDANUS
Fomalhaut
SCULPTOR
CAPRICORNUS
FORNAX
PISCIS AUSTRINUS
Rigel
PHOENIX
GRUS
WEST
SCUTUM
EAST
MICROSCOPIUM
LEPUS
CAELUM
HOROLOGIUM
Achernar
TUCANA
SAGITTARIUS
INDUS
NORTH

October

Southern hemisphere

20° 40° 60° Looking North

The charts on these pages show the southern hemisphere skies during October. They are correct for 11pm local time (midnight Daylight Saving Time) at the start of the month, 10pm (11pm DST) in the middle of the month, and 9pm (10pm DST) at the end of the month.

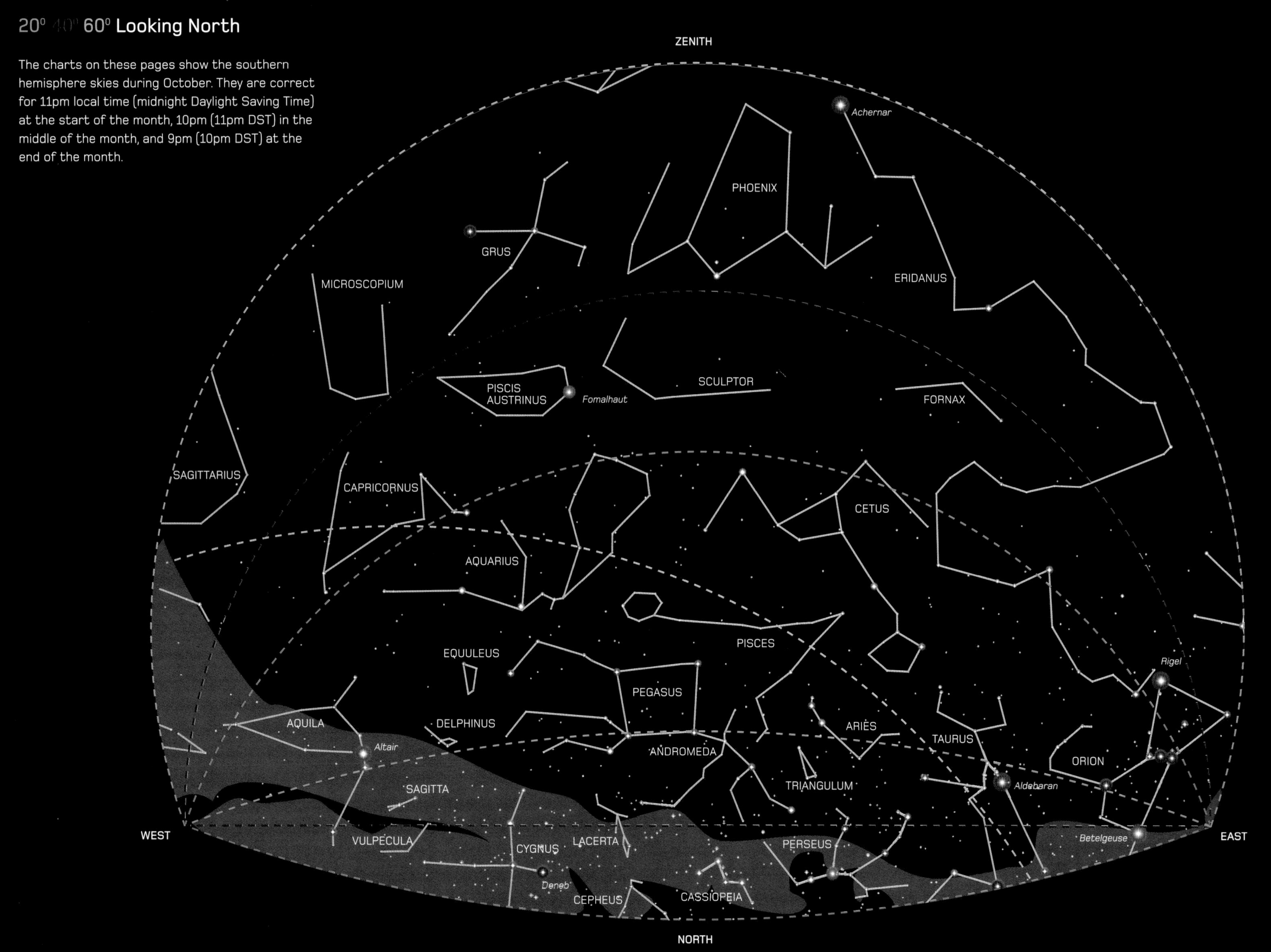

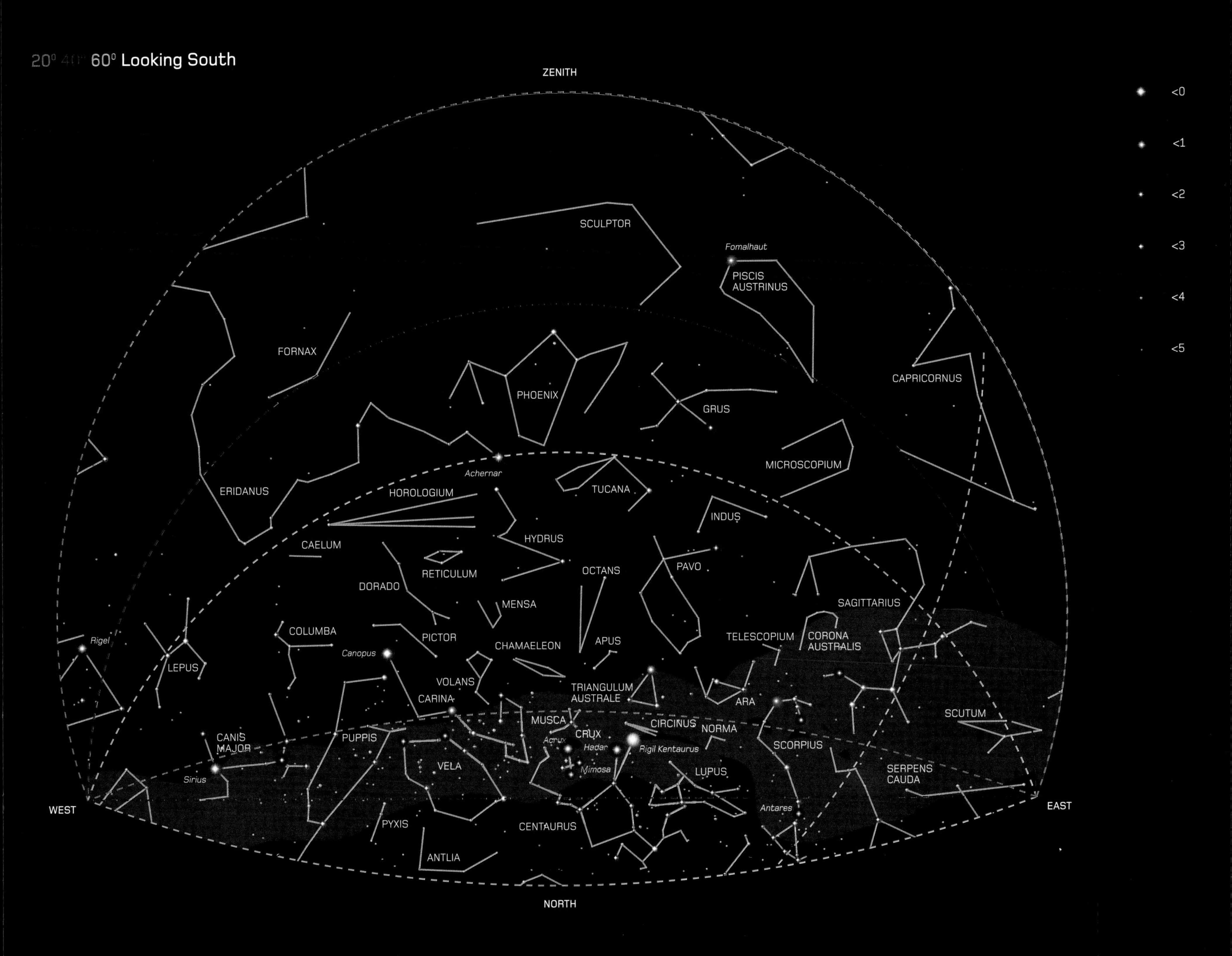
20° 40° 60° Looking South
ZENITH
<0
<1
<2
<3
<4
<5
SCULPTOR
Fomalhaut
PISCIS AUSTRINUS
FORNAX
PHOENIX
GRUS
CAPRICORNUS
MICROSCOPIUM
ERIDANUS
Achernar
HOROLOGIUM
TUCANA
INDUS
CAELUM
HYDRUS
RETICULUM
OCTANS
PAVO
DORADO
MENSA
SAGITTARIUS
COLUMBA
PICTOR
CHAMAELEON
APUS
TELESCOPIUM
CORONA AUSTRALIS
Rigel
LEPUS
Canopus
VOLANS
TRIANGULUM AUSTRALE
CARINA
ARA
SCUTUM
MUSCA
CIRCINUS
NORMA
CANIS MAJOR
PUPPIS
CRUX
Acrux
Hadar
Rigil Kentaurus
SCORPIUS
VELA
Mimosa
LUPUS
SERPENS CAUDA
Sirius
WEST
Antares
EAST
PYXIS
CENTAURUS
ANTLIA
NORTH

November

Northern hemisphere

20° 40° 60° Looking North

The charts on these pages show the northern hemisphere skies during November. They are correct for 11pm local time (midnight Daylight Saving Time) at the start of the month, 10pm (11pm DST) in the middle of the month, and 9pm (10pm DST) at the end of the month.

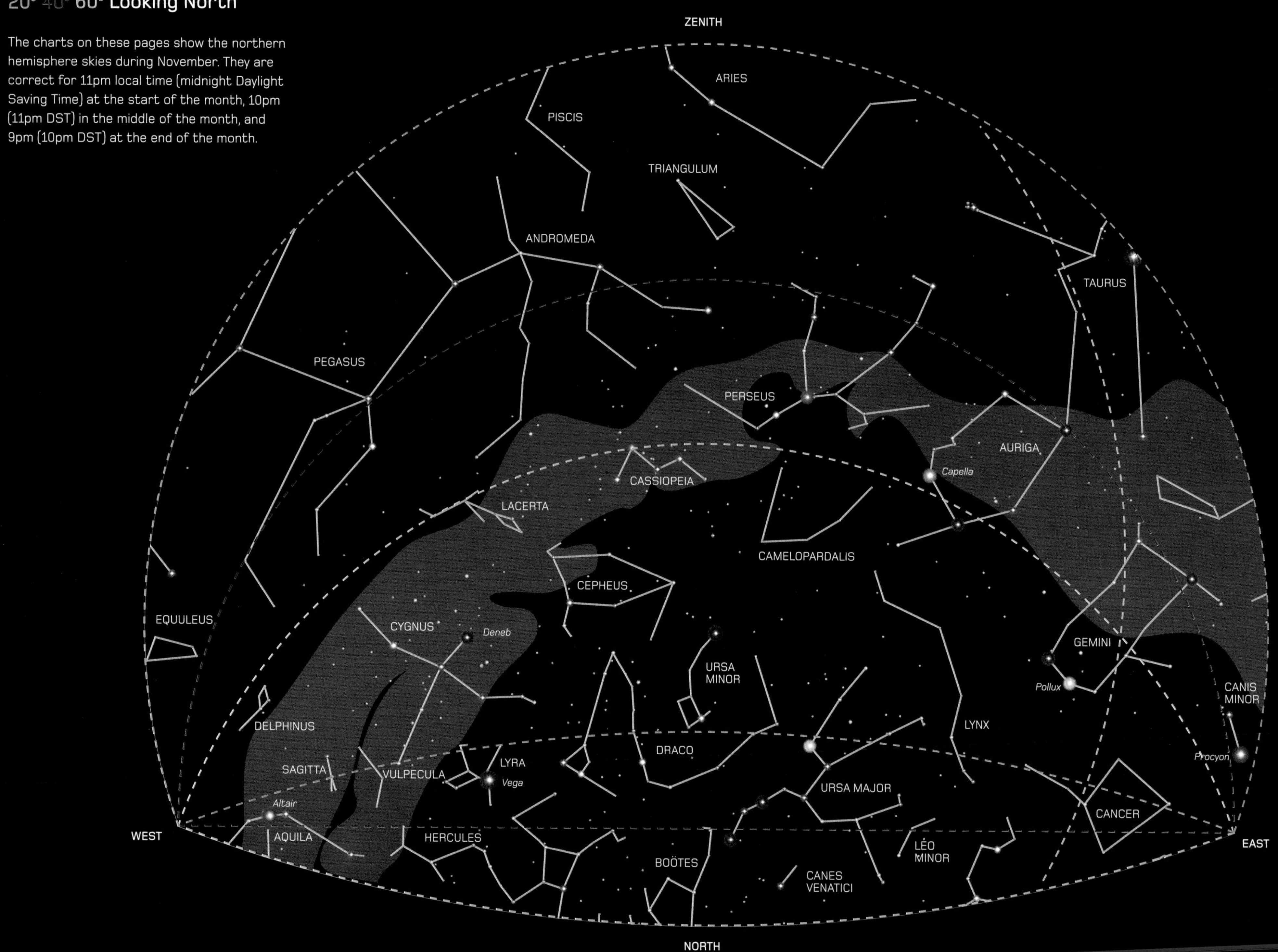

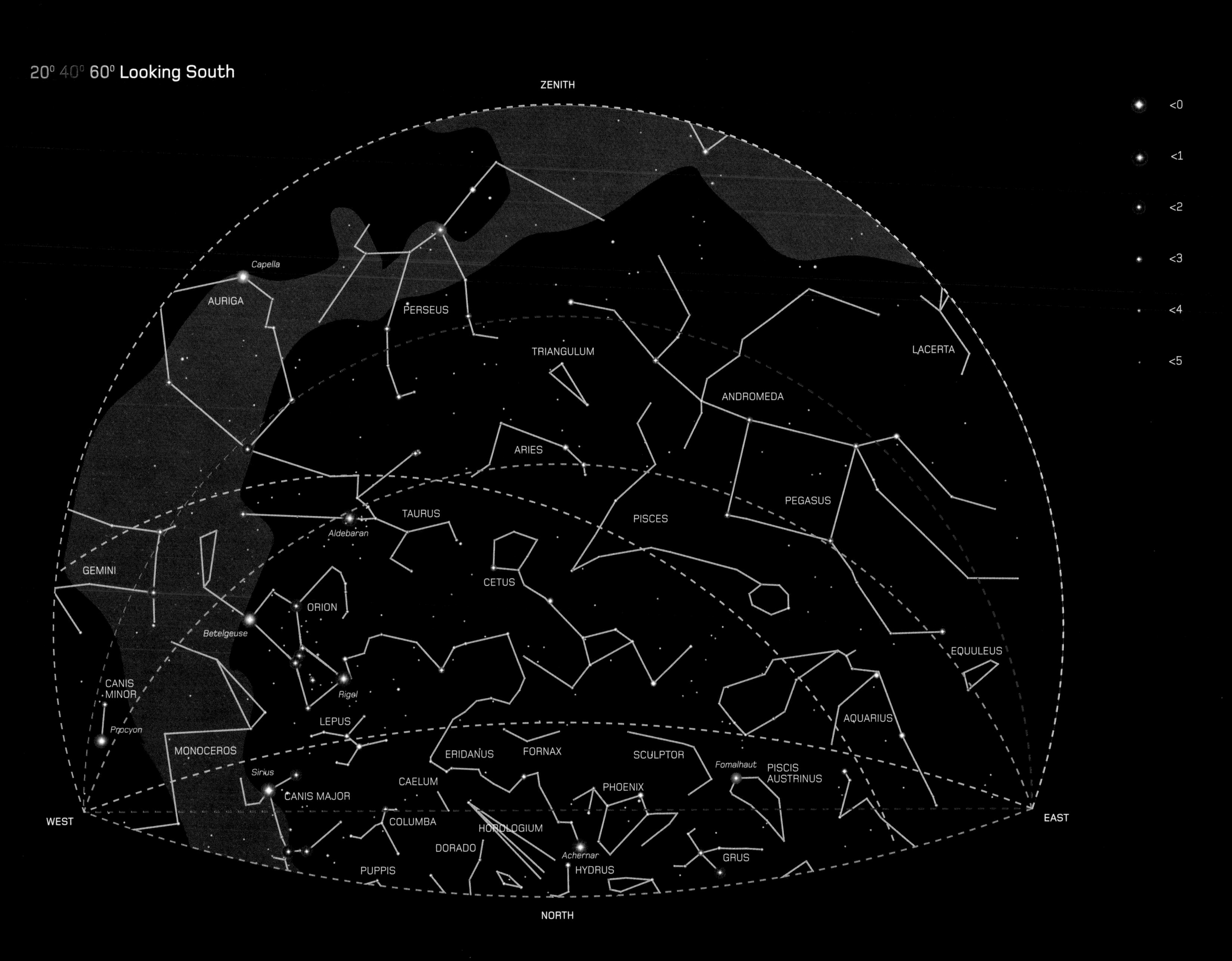
20° 40° 60° Looking South
ZENITH
<0
<1
<2
<3
<4
<5
Capella
AURIGA
PERSEUS
TRIANGULUM
LACERTA
ANDROMEDA
ARIES
TAURUS
Aldebaran
PISCES
PEGASUS
GEMINI
CETUS
ORION
Betelgeuse
EQUULEUS
CANIS MINOR
Rigel
AQUARIUS
LEPUS
Procyon
MONOCEROS
ERIDANUS
FORNAX
SCULPTOR
Fomalhaut
PISCIS AUSTRINUS
Sirius
CANIS MAJOR
CAELUM
PHOENIX
WEST
COLUMBA
HOROLOGIUM
EAST
DORADO
Achernar
GRUS
PUPPIS
HYDRUS
NORTH

November

Southern hemisphere

20° 40° 60° Looking North

The charts on these pages show the southern hemisphere skies during November. They are correct for 11pm local time (midnight Daylight Saving Time) at the start of the month, 10pm (11pm DST) in the middle of the month, and 9pm (10pm DST) at the end of the month.

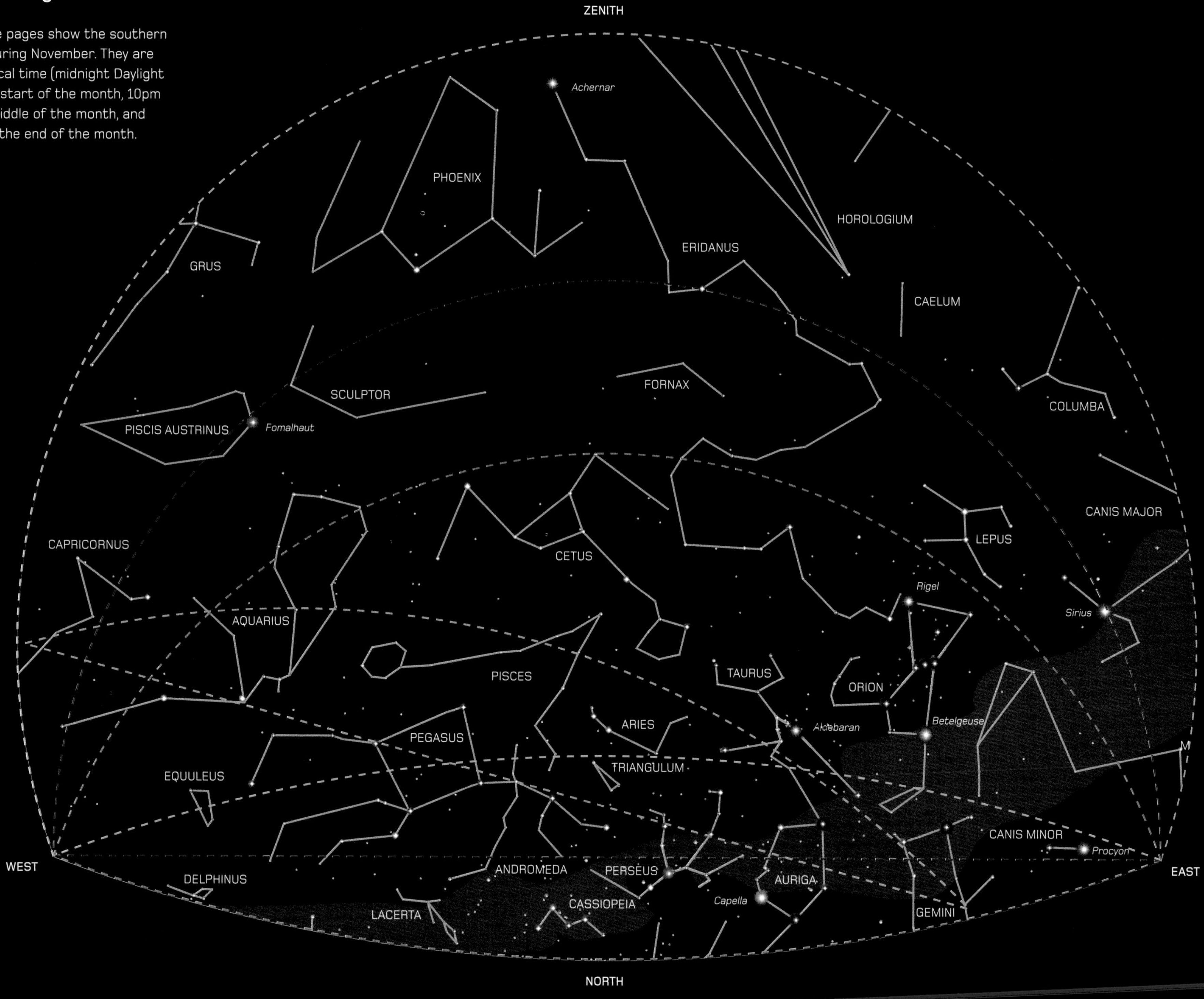

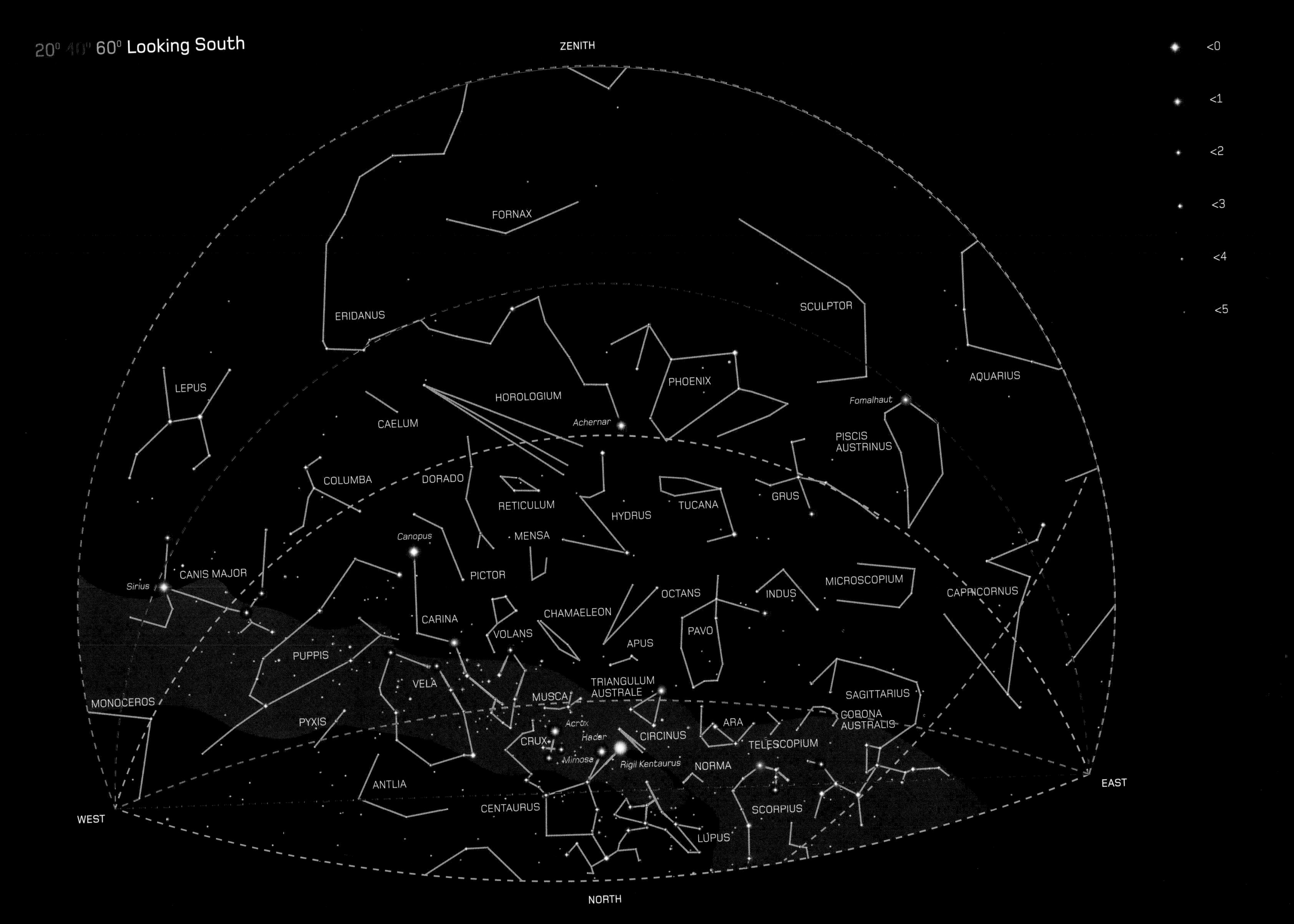
20° 40° 60° Looking South
ZENITH
<0
<1
<2
<3
<4
<5
FORNAX
ERIDANUS
SCULPTOR
AQUARIUS
LEPUS
HOROLOGIUM
PHOENIX
CAELUM
Achernar
Fomalhaut
PISCIS AUSTRINUS
COLUMBA
DORADO
RETICULUM
HYDRUS
TUCANA
GRUS
Canopus
MENSA
PICTOR
CANIS MAJOR
Sirius
OCTANS
INDUS
MICROSCOPIUM
CAPRICORNUS
CARINA
CHAMAELEON
VOLANS
APUS
PAVO
PUPPIS
VELA
TRIANGULUM AUSTRALE
MUSCA
SAGITTARIUS
MONOCEROS
PYXIS
Acrux
CRUX
Hadar
CIRCINUS
ARA
CORONA AUSTRALIS
TELESCOPIUM
Mimosa
Rigil Kentaurus
NORMA
ANTLIA
CENTAURUS
SCORPIUS
LUPUS
WEST
EAST
NORTH

December

Northern hemisphere

The charts on these pages show the northern hemisphere skies during December. They are correct for 11pm local time (midnight Daylight Saving Time) at the start of the month, 10pm (11pm DST) in the middle of the month, and 9pm (10pm DST) at the end of the month.

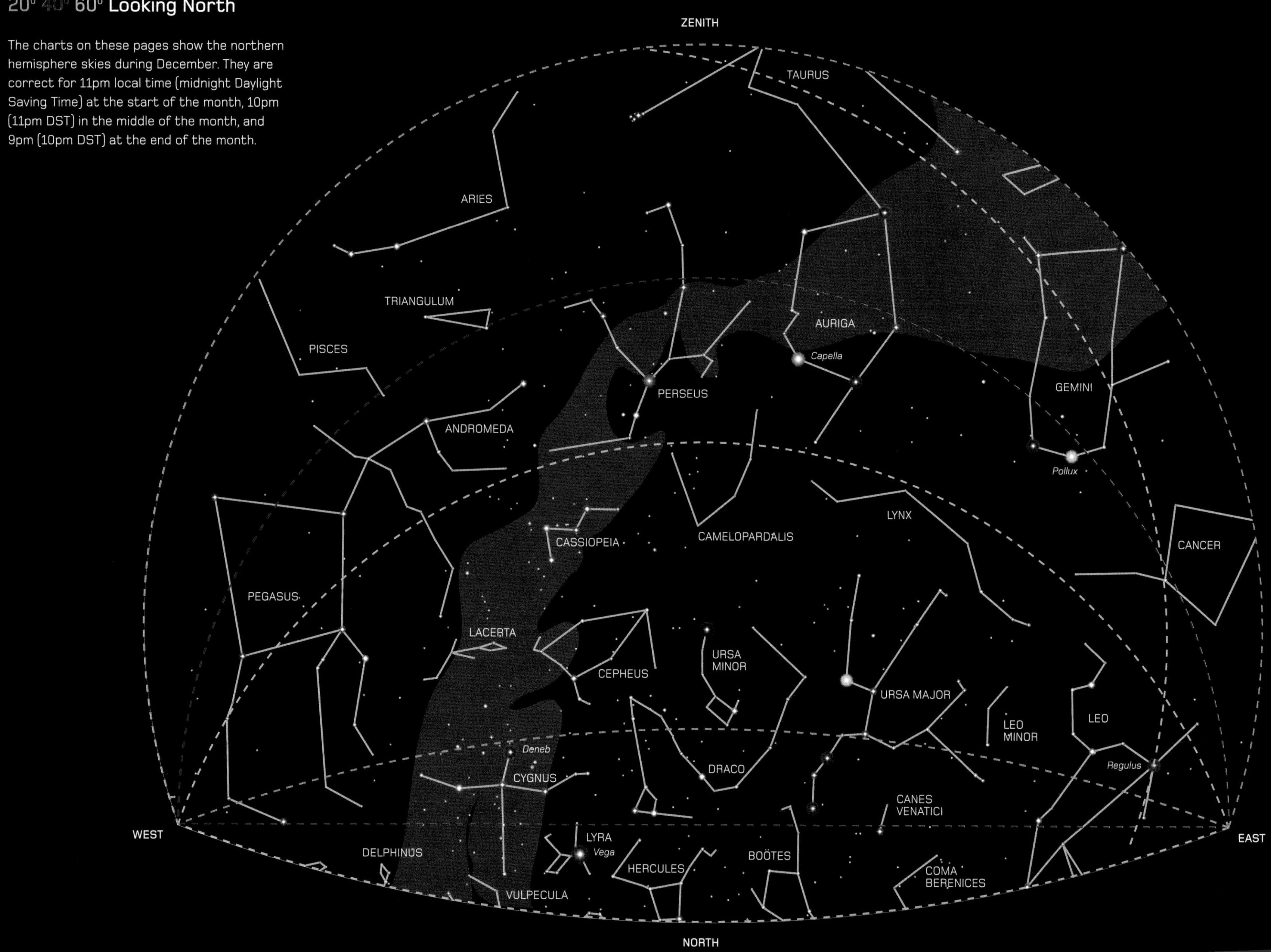

20° 40° 60° Looking South
ZENITH
<0
<1
<2
<3
<4
<5
PERSEUS
Capella
AURIGA
ANDROMEDA
TRIANGULUM
ARIES
Pollux
GEMINI
TAURUS
Aldebaran
PEGASUS
ORION
Betelgeuse
PISCES
CANCER
CANIS
MINOR
Procyon
Rigel
CETUS
MONOCEROS
ERIDANUS
Sirius
CANIS
MAJOR
LEPUS
COLUMBA
HYDRA
FORNAX
PUPPIS
CAELUM
AQUARIUS
WEST
SEXTANS
PICTOR
HOROLOGIUM
SCULPTOR
EAST
PYXIS
Canopus
DORADO
Achernar
PHOENIX
CARINA
RETICULUM
NORTH

December

Southern hemisphere

-20° -40° **-60° Looking North**

The charts on these pages show the southern hemisphere skies during December. They are correct for 11pm local time (midnight Daylight Saving Time) at the start of the month, 10pm (11pm DST) in the middle of the month, and 9pm (10pm DST) at the end of the month.

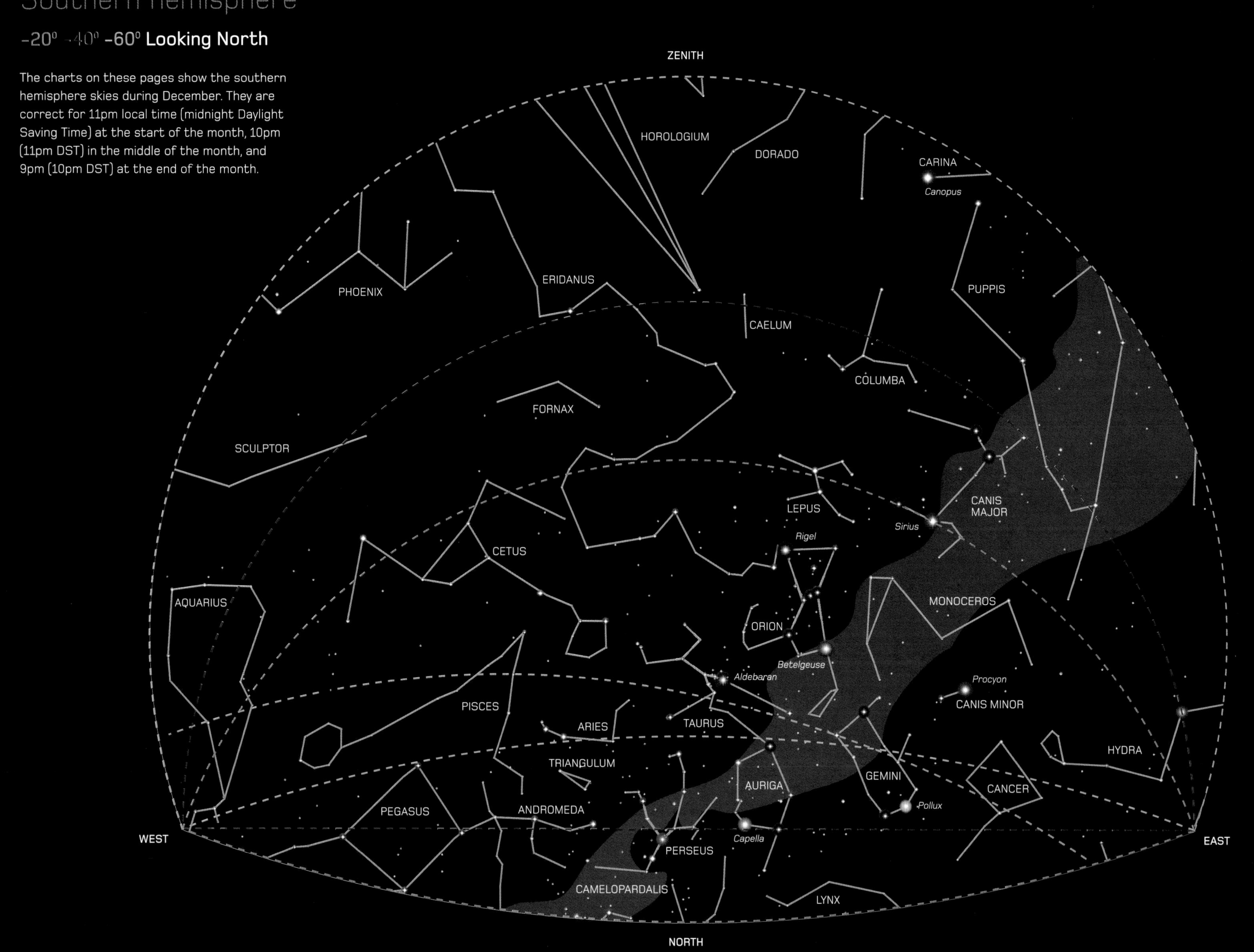

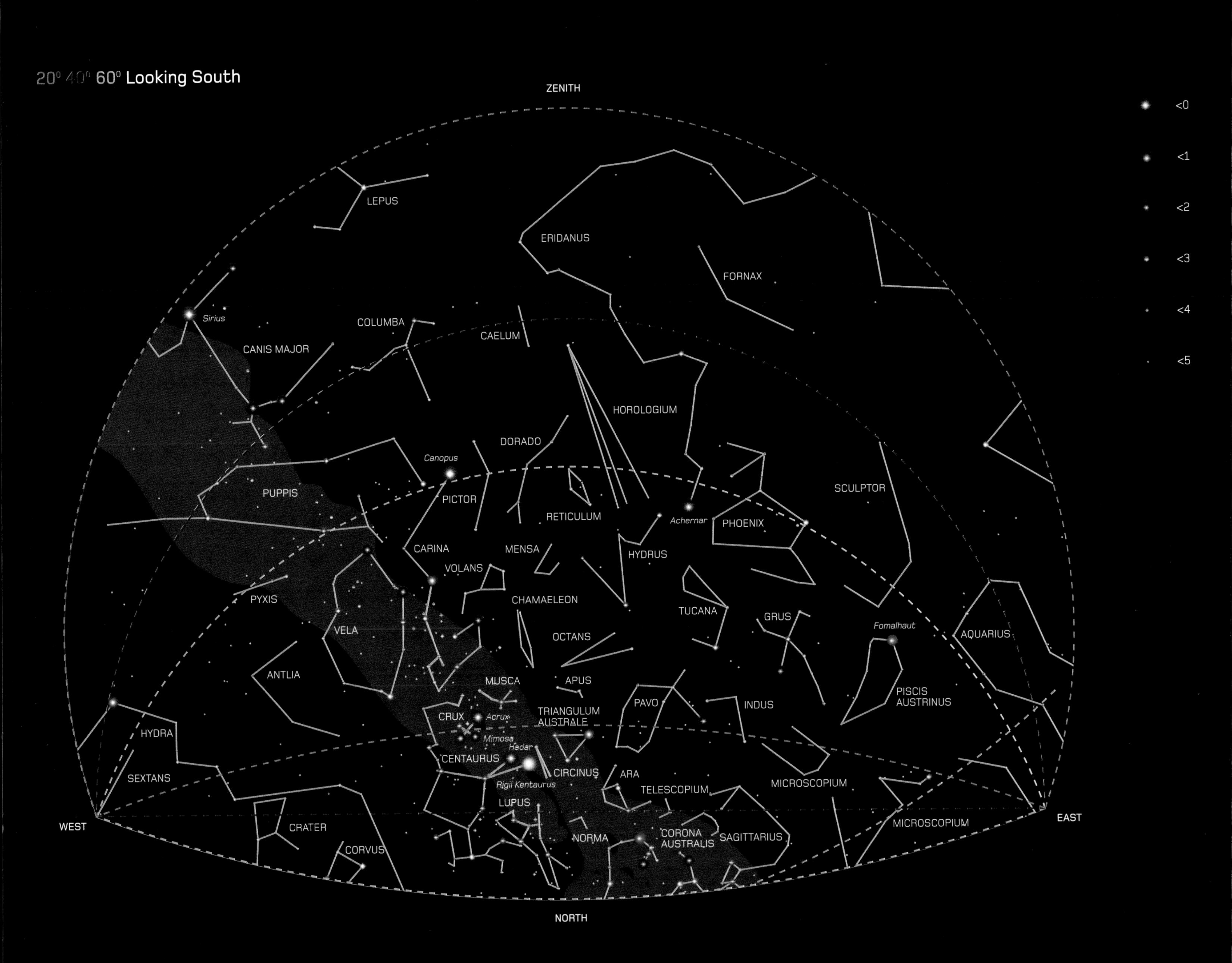

20° 40° 60° Looking South
ZENITH
<0
<1
<2
<3
<4
<5
LEPUS
ERIDANUS
FORNAX
Sirius
COLUMBA
CAELUM
CANIS MAJOR
HOROLOGIUM
DORADO
Canopus
PUPPIS
PICTOR
SCULPTOR
RETICULUM
Achernar
PHOENIX
CARINA
MENSA
HYDRUS
VOLANS
PYXIS
CHAMAELEON
TUCANA
GRUS
VELA
OCTANS
Fomalhaut
AQUARIUS
ANTLIA
MUSCA
APUS
PAVO
INDUS
PISCIS
AUSTRINUS
TRIANGULUM
AUSTRALE
CRUX
Acrux
HYDRA
Mimosa
Hadar
CENTAURUS
CIRCINUS
ARA
SEXTANS
Rigil Kentaurus
TELESCOPIUM
MICROSCOPIUM
LUPUS
WEST
EAST
CRATER
NORMA
CORONA
AUSTRALIS
SAGITTARIUS
MICROSCOPIUM
CORVUS
NORTH

Glossary

Active galaxy
A galaxy that emits large amounts of energy from its central regions, probably generated as matter falls into a supermassive black hole at the heart of the galaxy.

Alt-Azimuth
A co-ordinate system that measures the positions of celestial bodies in altitude relative to an observer's horizon, and in azimuth relative to the observer's 'due North' direction.

Asteroid
One of the countless rocky worlds of the inner solar system, largely confined in the main asteroid belt beyond the orbit of Mars.

Astronomical unit
A unit of measurement widely used in astronomy, equivalent to Earth's average distance from the Sun – roughly 150 million km or 93 million miles.

Atmosphere
A shell of gases held around a planet or star by its gravity.

Barred spiral galaxy
A spiral galaxy in which the arms are linked to the hub by a straight bar of stars and other material.

Binary star
A pair of stars in orbit around one another. Because the stars in a binary pair were usually born at the same time, they allow a direct comparison of the way that stars with different properties evolve.

Black dwarf
A former white dwarf star (the dense, exhausted core of a star like the Sun) that has cooled until it no longer emits visible light.

Blue giant
An extremely massive star with such high gravity that when it exhausts its core fuel supply and brightens to become a giant, it remains relatively compact, with a hot blue surface.

Brown dwarf
A so-called failed star that never gains enough mass to begin the fusion of hydrogen in its core, and start to shine properly. Instead, brown dwarfs radiate low-energy radiation (mostly infrared) through gravitational contraction and a more limited form of fusion.

Celestial equator
A projection of the Earth's equator onto the sky, used as a basis for measuring celestial co-ordinates.

Celestial pole
A projection of one of Earth's poles into the sky. The daily rotation of the Earth means that the celestial poles remain fixed while the rest of the sky spins around them.

Comet
A chunk of rock and ice from the outer reaches of the solar system. When comets fall into orbits that bring them close to the Sun, they heat up and their surface ices evaporate, forming a coma and a tail.

Conjunction
An alignment of two celestial bodies in Earth's skies. In terms of planetary orbits, inferior conjunction happens when an inferior planet (orbiting closer to the Sun than Earth) comes between Earth and the Sun, and superior conjunction (simply 'conjunction' for superior planets orbiting beyond Earth) occurs when a planet lies on the far side of the Sun from Earth.

Core
The central region of a star where temperatures and pressures are high enough to trigger nuclear fusion.

Dark nebula
A cloud of interstellar gas and dust that absorbs light, and only becomes visible when silhouetted against a field of stars or other nebulae.

Declination
A measure of the angle between an object in the sky and the celestial equator, as seen by an observer on Earth.

Dwarf planet
Any object that is in an independent orbit around the Sun, and has sufficient gravity to pull itself into a roughly spherical shape, but which, unlike a true planet, has not cleared the region around it of other objects. Currently there are three known dwarf planets – the asteroid Ceres, and the Kuiper Belt objects Pluto and Eris, but there are many more objects whose status is still uncertain.

Eclipse
A direct alignment between the Earth, Sun and Moon. Solar eclipses happen when the moon comes between the Earth and the Sun, and briefly blocks our view of the solar disc. Lunar eclipses happen when the Moon passes into the shadow cast by the Earth.

Eclipsing binary
A binary star in which one star regularly passes in front of another as seen from Earth, causing a drop in the overall brightness of the system.

Ecliptic
The path of the Sun through Earth's sky in the course of a year – in fact a projection of the plane of Earth's own orbit onto the heavens. Because the solar system is roughly flat, the major planets are also usually found close to the ecliptic.

Electromagnetic radiation
A form of energy consisting of combined electric and magnetic waves, able to propagate itself across a vacuum at the speed of light. The energy or temperature of an object emitting radiation affects its wavelength and other characteristics.

Elliptical galaxy
A galaxy consisting of stars in orbits that have no particular orientation, and generally lacking in star-forming gas. Ellipticals are among the smallest and largest galaxies known.

Elongation
A point in the orbit of an inferior planet (one closer to the Sun than Earth) where it reaches its maximum separation from the Sun in Earth's skies.

Emission nebula
A cloud of gas in space that glows at very specific wavelengths, producing a spectrum full of emission lines. These nebulae are usually energized by the high-energy light of nearby stars.

Equatorial co-ordinates
A co-ordinate system that measures the positions of celestial bodies in 'declination' relative to the celestial equator and poles, and in 'right ascension' relative to the First Point of Aries.

First Point of Aries
The point on the celestial equator which the Sun passes at the beginning of northern spring, when it passes from the southern to the northern hemisphere of the sky. The First Point of Aries (which today lies in Pisces thanks to precession) is used as the base point for measuring the right ascension of celestial objects.

Flare
A huge release of superheated particles above the surface of a star, caused by a short-circuit in its magnetic field.

Flare star
A dim, lightweight star (typically a red dwarf), whose surface is periodically wracked by violent flares that cause its brightness to vary.

Fusion shell
A spherical shell of nuclear fusion spreading out through a star after it has exhausted a particular fuel supply in its core.

Galaxy
An independent system of stars, gas and other material with a size measured in thousands of light years.

Gamma rays
The highest-energy forms of electromagnetic radiation, with extremely short wavelengths, generated by the hottest objects and most energetic processes in the Universe.

Giant planet
A planet comprising a huge envelope of gas, liquid, or slushy ice (various frozen chemicals), perhaps around a relatively small rocky core.

Globular cluster
A dense ball of ancient, long-lived stars, in orbit around a galaxy such as the Milky Way.

Helium fusion
Nuclear fusion of helium (formed by hydrogen fusion) into heavier elements (so-called metals). Most stars rely on helium fusion to keep on shining as they exhaust their supplies of hydrogen and near the end of their lives.

Hydrogen fusion
The nuclear fusion of hydrogen, the lightest element, into helium, the next lightest. Hydrogen fusion is the main power source for all stars for the majority of their lives, but it can proceed at different rates depending on conditions within a star.

Infrared
Electromagnetic radiation with slightly less energy than visible light. Infrared radiation is typically emitted by warm objects too cool to glow visibly.

Irregular galaxy
A galaxy with no obvious structure, generally rich in gas, dust, and star-forming regions.

Kuiper Belt
A doughnut-shaped ring of icy worlds directly beyond the orbit of Neptune. The largest known Kuiper Belt Objects are Pluto and Eris.

Light year
A common unit of astronomical measurement, equivalent to the distance travelled by light (or other electromagnetic radiation) in one year. A light year is equivalent to roughly 9.5 million million km (5.9 trillion miles).

Luminosity
A measure of the energy output of a star. Although luminosity is technically measured in watts, the stars are so luminous that it is simpler to compare them with the Sun. A star's visual luminosity (the energy it produces in visible light) is not necessarily equivalent to its overall luminosity in all radiations.

Magnitude
A measure of the brightness of objects that reflects the sensitivity of the human eye. Objects with lower magnitude are brighter than those with higher magnitudes, and a difference in magnitude of 1.0 is roughly equivalent to a factor of 2.5 difference in the real brightness of two objects.

Main sequence
A term used to describe the longest phase in a star's life, during which it is relatively stable, and shines by fusing hydrogen into helium at its core. During this period, the star obeys a general relationship that links its mass, size, luminosity and colour.

Metal
In astrophysics, a term used for any element heavier than hydrogen or helium. Metals are formed by helium fusion, and scattered across space in planetary nebulae and supernova remnants. The amount of certain metals in a new-born star can affect the entire path of its evolution, and the most massive stars can continue to shine by fusion of light metals into heavier ones even when their helium supplies have been exhausted .

Multiple star
A system of two or more stars in orbit around one another (pairs of stars are also called binaries). Most of the stars in our galaxy are members of multiple systems rather than individuals like the Sun.

Nebula
Any cloud of gas or dust floating in space. Nebulae are the material from which stars are born, and into which they are scattered again at the end of their lives. The word means cloud in Latin, and was originally applied to any fuzzy object in the sky, including some we now know to be star clusters or distant galaxies.

Neutron star
The collapsed core of a supermassive star, left behind by a supernova explosion. A neutron star consists of compressed subatomic particles, and is the densest known object – though in the most massive stars, the core can collapse past the neutron star stage to form a black hole. Many neutron stars initially behave as pulsars.

Nova
A binary star system in which a white dwarf is pulling material from a companion star, building up a layer of gas around itself that then burns away in a violent nuclear explosion.

Nuclear fusion
The joining-together of light atomic nuclei (the central cores of atoms) to make heavier ones at very high temperatures and pressures, releasing excess energy in the process. Fusion is the process by which the stars shine.

Occultation
The passage of a relatively nearby object (such as a planet) in front of a more distant one (such as a star).

Oort Cloud
A spherical shell of dormant comets, up to two light years across, surrounding the entire solar system.

Open cluster
A large group of bright young stars that have recently been born from the same star-forming nebula, and may still be embedded in its gas clouds.

Opposition
The point in the orbit of a major planet (one that orbits further from the Sun than Earth) where it lies directly opposite the Sun in Earth's skies.

Orange giant
A brilliant star that has swelled to giant size as it nears the end of its life, and whose surface has cooled and turned orange.

Planet
A world that follows its own orbit around the Sun, is massive enough to pull itself into a spherical shape, and which has cleared the space around it of other objects (apart from satellites). According to this definition, there are eight planets – Mercury, Venus, Earth, Mars, Jupiter, Saturn, Uranus and Neptune.

Planetary nebula
An expanding cloud of glowing gas sloughed off from the outer layers of a dying red giant star as it transforms into a white dwarf.

Precession
A slow wobble in the Earth's axis of rotation that causes the poles to point at different areas of the sky over a 26,000 year cycle. As a result, the north and south celestial poles shift position against the stars, and equatorial celestial co-ordinates gradually change over time.

Pulsar
A rapidly spinning neutron star with an intense magnetic field that channels its radiation out along two narrow beams that sweep across the sky.

Radio
The lowest-energy form of electromagnetic radiation, with the longest wavelengths. Radio waves are emitted by cool gas clouds in space, but also by violent active galaxies and pulsars.

Red dwarf
A star with considerably less mass than the Sun – small, faint, and with a low surface temperature. Red dwarfs fuse hydrogen into helium in their cores very slowly, and live for much longer than the Sun, despite their size.

Red giant
A star passing through a phase of its life where its luminosity has increased hugely, causing its outer layers to expand and its surface to cool. Stars usually enter red giant phases when they exhaust the fuel supplies in their core.

Reflection nebula
A cloud of interstellar gas and dust that shines as it reflects or scatters light from nearby stars.

Right Ascension
A measure of the time taken for an object to pass over or 'transit' an observer's local meridian (the north–south line across the sky) *after* the First Point of Aries. Right ascension (RA) is measured in hours, minutes and seconds.

Rocky planet
A relatively small planet composed largely of rocks and minerals, perhaps surrounded by a thin envelope of gas and liquid.

Shell star
A rapidly spinning star (usually with high mass) that is surrounded by a shell of material flung off around its equator. Also known as an emission-line star.

Spectral lines
Dark or light bands in a spectrum of light that correspond to certain wavelengths. Bright emission lines can indicate that an object is emitting certain wavelengths, while dark bands silhouetted against a broad background spectrum indicate that something is absorbing the light on its way to us. In both cases, the location of the lines offers information on which atoms or molecules are involved.

Spectroscopic binary
A binary star that can only be detected thanks to the shifting of the lines in its spectrum as its two components swing around one another.

Spectrum
The spread-out band of light created by passing light through a prism or similar device. The prism bends light by different amounts depending on its wavelength and colour, so the spectrum reveals the precise intensities of light at different wavelengths.

Spiral galaxy
A galaxy consisting of a hub of old yellow stars, surrounded by a flattened disc of younger stars, gas and dust, with spiral arms marking regions of current star formation.

Star
A dense ball of gas that has collapsed into a spherical shape and become hot and dense enough at its centre to trigger nuclear fusion reactions that make it luminous.

Stellar wind
A stream of high-energy particles blasted off the surface of a star by the pressure of its radiation, and spreading across the surrounding space.

Subgiant
A star that has just exhausted the supply of hydrogen in its core, and is now beginning to swell up, eventually to become a bloated, luminous giant.

Sun
The star at the centre of Earth's solar system. The Sun is a fairly average low-mass star, and a useful comparison for other stars. Its key properties include a diameter of 1.39 million kilometres, a mass of 2,000 trillion trillion tonnes, energy output of 380 trillion trillion watts, and a surface temperature of 5,500°C.

Sunlike star
A yellow star with roughly the same mass, luminosity and surface temperature as the Sun. Stars like this are of particular interest to astronomers because they are long-lived, stable, and any planets around them are potential havens for life.

Supergiant
A massive and extremely luminous star with between 10 and 70 times the mass of the Sun. Supergiants can have almost any colour, depending on how the balance of their energy output and their size affects their surface temperature.

Supermassive black hole
A black hole with the mass of millions of stars, believed to lie in the very centre of many galaxies. Supermassive black holes form from the collapse of huge gas clouds rather than the death of massive stars.

Supernova
A cataclysmic explosion marking the death of a star. Supernovae can be triggered when a heavyweight star exhausts the last of its fuel and its core collapses (forming either a neutron star or a black hole) or when a white dwarf in a nova system tips over its upper mass limit and collapses suddenly into a neutron star.

Supernova remnant
A cloud of superheated gas expanding from the site of a former supernova explosion.

Transit
The passage of one celestial body across the face of another – typically the movement of an inferior planet (Mercury or Venus) in front of the Sun.

Ultraviolet
Electromagnetic radiation with wavelengths slightly shorter than visible light, typically radiated by objects hotter than the Sun. The hottest stars give out much of their energy in the ultraviolet.

Variable star
A star that varies its brightness, either due to interaction with another star, or because of some feature of the star itself (most commonly a pulsation in size that may be periodic or irregular).

Visible light
Electromagnetic radiation with wavelengths between 400 and 700 nanometres (billionths of a metre), corresponding to the sensitivity of the human eye. Stars like the Sun emit most of their energy in the form of visible light.

White dwarf
A stellar remnant left behind by the death of a star with less than about eight times the Sun's mass. White dwarfs are the dense, slowly cooling cores of stars – typically very hot, but hard to see on account of their tiny size.

Wolf-Rayet star
A star with extremely high mass which develops such fierce stellar winds that it blows away most of its outer layers in a few million years, exposing the extremely hot interior.

X-rays
High-energy electromagnetic radiation emitted by extremely hot objects and violent processes in the Universe. Material heated as it is pulled towards a black hole is one of the strongest sources of astronomical X-rays.

Zodiac
The band of twelve constellations that lies along the ecliptic, and in which the Sun, Moon and planets spend most of their time.

Further reading

Books

Atlas of Stars and Planets
Ian Ridpath
Philip's Astronomy, 2004

Collins Atlas of the Night Sky
Storm Dunlop and Wil Tirion
Collins, 2005

Cosmos
Giles Sparrow
Quercus, 2006

National Geographic Encyclopedia of Space
L. Glover et al
National Geographic, 2004

Universe
R. Dinwiddie et al
Dorling Kindersley, 2005

Magazines

Astronomy
United States
www.astronomy.com

Sky and Telescope
United States
www.skyandtelescope.com

Astronomy Now
United Kingdom
www.astronomynow.com

Australian Sky and Telescope
Australia
www.austskyandtel.com.au

Sky at Night Magazine
United Kingdom
www.skyatnightmagazine.com

Software

A wide range of astronomy software is available for different applications. The following programs are all of the general 'desktop planetarium' type unless otherwise indicated

Alcyone Ephemeris
Sophisticated ephemeris software for producing data tables and charts of celestial movements.
www.alcyone-ephemeris.info

JupSatPro
Track the motions of Jupiter's satellites.
www.nightskyobserver.com/JupSatPro

RedShift
www.redshift.maris.com

Starry Night
www.starrynightstore.com

The Sky
www.bisque.com/TheSky

Websites

There are a huge variety of astonomical websites on the Internet. These are just a few highlights.

Astronomy Picture of the Day
A new picture of a stunning celestial highlight every day.
www.antwrp.gsfc.nasa.gov/apod

Hubblesite
Central repository for Space Telescope news and images.
www.hubbblesite.org

Space
Daily updated news on all aspects of astronomy and space exploration.
www.space.com

NASA
US National Aeronautics and Space Administration – includes links to websites of various missions and programs.
www.nasa.gov

Heavens Above!
Real-time charts of the sky, including tracking for satellites and spacecraft.
www.heavens-above.com

Your Sky
Produce sky maps for any date, time and location over the web.
www.fourmilab.ch/yoursky

Societies

The following organizations welcome amateurs of varying experience into their ranks:

American Association of Variable Star Observers
www.aavso.org

Association of Lunar and Planetary Observers
United States
www.lpl.arizona.edu/alpo

Astronomical Society of the Pacific
United States
www.astrosociety.org

Australian Astronomical Societies
For a list of flourishing local Australian societies, visit:
www.astronomy.org.au/ngn

British Astonomical Association
www.britastro.org/baa

International Meteor Organization
www.imo.net

Royal Astronomical Society of Canada
www.rasc.ca

Royal Astronomical Society of New Zealand
www.rasnz.org.nz

Society for Popular Astronomy
United Kingdom
www.popastro.com

Wellington Astronomical Society
New Zealand
www.was.org.nz

Index

Acknowledgements

All images by Tim Brown/Pikaia Imaging unless otherwise stated.

p10: NASA Goddard Space Flight Center Image by Reto Stöckli; p11: [top left] Image Analysis Laboratory/NASA Johnson Space Center; [top centre] Joshua Strang, USAF, Wikipedia; [top right] NASA Landsat Project Science Office and USGS National Center for EROS; p26: NSO/AURA/NSF; p28: [left] SOHO(LASCO & EIT)/NASA/ESA; [right] TRACE/NASA; p29: Göran Scharmer, Mats Löfdahl, ISP, SST, Royal Swedish Academy of Sciences; p34: [top left] NASA/JPL-Caltech; [top right] NASA; p34-35: Tom Dahl/NASA; p35: [top right] NASA/JPL-Caltech; p36: NASA/JPL-Caltech; p38: NASA/JPL-Caltech; p39: NASA/JPL-Caltech; p40: NASA/JPL-Caltech; p42: [left] NASA/JPL-Caltech; [bottom] NASA/ JPL-Caltech; p43: [bottom] NASA/JPL-Caltech; [top] NASA/JPL-Caltech; [top right] NASA/JPL-Caltech; p44: NASA/USGS; p46: [top left] ESA/DLR/FU Berlin (G. Neukum); [top right] NASA/JPL-Caltech; p46-7: NASA/JPL/Cornell; p48: NASA/JPL/Space Science Institute; p50: [clockwise, top left] NASA/ JPL-Caltech; NASA/JPL-Caltech; NASA/JPL-Caltech; NASA/JPL-Caltech; NASA/JPL-Caltech; NASA/JPL-Caltech; NASA/JPL-Caltech; NASA/JPL-Caltech; p51: NASA/JPL/University of Arizona; p52: NASA/JPL/Space Science Institute; p53: [bottom] NASA and The Hubble Heritage Team (STScI/AURA), Acknowledgment: R.G. French (Wellesley College), J. Cuzzi (NASA/Ames), L. Dones (SwRI), and J. Lissauer (NASA/Ames); p54: NASA/JPL/Space Science Institute; p55: [top] NASA/JPL/Space Science Institute; [bottom left] NASA/JPL/Space Science Institute; [bottom centre] NASA/JPL/Space Science Institute; [bottom right] NASA/JPL/Space Science Institute; p56: NASA/JPL-Caltech; p57: [centre] Erich Karkoschka (University of Arizona) and NASA; p58: NASA/ JPL-Caltech; p59: [centre] NASA/JPL-Caltech; p72: NASA, ESA, and the Hubble Heritage STScI/AURA, acknowledgement Robert A. Fesen (Dartmouth College, USA) and James Long (ESA/Hubble); p73: T.A. Rector/University of Alaska Anchorage, H. Schweiker/WIYN and NOAO/AURA/NSF; p80: NASA, R. Williams and The Hubble Deep Field Team (STScI); p81: NASA and ESA, Acknowledgment: K.D. Kuntz (GSFC), F. Bresolin (University of Hawaii), J. Trauger (JPL), J. Mould (NOAO), and Y.-H. Chu (University of Illinois, Urbana); p82: NASA, ESA, and The Hubble Heritage Team (STScI/AURA), J. Gallagher (University of Wisconsin), M. Mountain (STScI), and P. Puxley (National Science Foundation); p83: NASA, ESA, S. Beckwith (STScI), and The Hubble Heritage Team (STScI/AURA); p90: Hubble Heritage Team (AURA/STScI/NASA); p91: European Southern Observatory; p94: T. Rector/University of Alaska Anchorage and WIYN/NOAO/AURA/NSF; p95: Bruce Balick (University of Washington), Jason Alexander (University of Washington), Arsen Hajian (U.S. Naval Observatory), Yervant Terzian (Cornell University), Mario Perinotto (University of Florence, Italy), Patrizio Patriarchi (Arcetri Observatory, Italy) and NASA; p98-99: T.A.Rector and B.A.Wolpa/NOAO/AURA/NSF; p104: ASA and The Hubble Heritage Team (AURA/STScI), acknowledgment: D. Garnett (U. Arizona), J. Hester (ASU), and J. Westphal (Caltech); p105: T.A. Rector (NRAO/AUI/NSF and NOAO/AURA/NSF); p108: NASA, ESA, J. Hester and A. Loll (Arizona State University); p109: NASA, ESA and AURA/Caltech; p118: The Hubble Heritage Team (AURA/STScI/NASA); p119: NASA and The Hubble Heritage Team (AURA/STScI); p122: NASA and The Hubble Heritage Team (STScI/AURA); p123: NASA and The Hubble Heritage Team (STScI/AURA); p128: T.A. Rector and B.A. Wolpa (NRAO/AUI/NSF); [inset] NASA, Jeff Hester and Paul Scowen Arizona State University; p129: NASA and The Hubble Heritage Team (STScI/AURA); p142: NASA,ESA, M. Robberto (Space Telescope Science Institute/ESA) and the Hubble Space Telescope Orion Treasury Project Team; p143: T.A.Rector (NOAO/AURA/NSF) and Hubble Heritage Team (STScI/AURA/NASA); p148: Canada-France-Hawaii Telescope/Coelum; p149: NASA, ESA, and Hubble Heritage Team (STScI); p154: E.J. Schreier (STScI), and NASA; p155: NASA and The Hubble Heritage Team (STScI/AURA); p160: NASA, ESA and A.Zijlstra (UMIST, Manchester, UK); p161: Garrelt Mellema (Leiden University) et al., HST, ESA, NASA; p164: A. Caulet (ST-ECF, ESA) and NASA; p165: NASA/JPL-Caltech/S. Stolovy (SSC/Caltech); p166: NASA, H. Ford (JHU), G. Illingworth (UCSC/LO), M.Clampin (STScI), G. Hartig (STScI), the ACS Science Team, and ESA; p167: NASA and Jeff Hester (Arizona State University); p180: NASA and The Hubble Heritage Team (AURA/STScI); p181: ESA/NASA & Valentin Bujarrabal (Observatorio Astronomico Nacional, Spain); p184: Hubble Heritage Team (STScI/AURA/NASA); p185: Anglo-Australian Observatory/David Malin Images; p188: NASA, The Hubble Heritage Team (AURA/STScI); p189: Jon Morse (University of Colorado), and NASA; p192: Anglo-Australian Observatory/David Malin Images; p193: Raghvendra Sahai and John Trauger (JPL), the WFPC2 science team, and NASA; p204: 2MASS/T. Jarrett; p205: NASA, ESA and A. Nota (STScI/ESA); p210: ESA/NASA, ESO and Danny LaCrue; p211: Hubble Heritage Team (AURA/STScI/NASA).

Quercus Publishing plc
21 Bloomsbury Square
London
WC1 2NS

Book design: Grade Design Consultants, London

A catalogue record for this book is available from the British Library.

Printed case edition
ISBN 978-1-84724-145-0

Cloth case edition
ISBN 978-1-84724-418-5